On the Primaeval Ocean

For My Parents

CNI PUBLICATIONS 26

THE CARLSBERG PAPYRI 5

ON THE PRIMAEVAL OCEAN

BY MARK SMITH

THE CARSTEN NIEBUHR INSTITUTE OF NEAR EASTERN STUDIES
UNIVERSITY OF COPENHAGEN 2002 · MUSEUM TUSCULANUM PRESS

CNI PUBLICATIONS 26
On the Primaeval Ocean

Set in Charter by K. Ryholt
Printed in Denmark by Special-Trykkeriet Viborg a/s

ISBN 87-7289-646-9
ISSN 0902-5499 (CNI Publications)
ISSN 0907-8118 (The Carlsberg Papyri)

The publication of this book was made possible by a grant from
The Carlsberg Foundation

Published and distributed by
Museum Tusculanum Press
University of Copenhagen
Njalsgade 92
DK-2300 Copenhagen S.
Denmark

Orders and catalogue:
www.mtp.dk

CONTENTS

PART I: INTRODUCTION

The text edited here is an ancient Egyptian cosmology written in the demotic script. Although once of considerable extent, the manuscript now survives only in fragments. There are forty-eight of these in all. Some are quite large, preserving virtually complete columns of writing; others are very small, preserving no more than a line or two. Most of the fragments are now divided between two papyrus collections: that of the Carsten Niehbuhr Institute at the University of Copenhagen and that of the Istituto Papirologico 'G. Vitelli' in Florence. A few have found their way to the Beinecke Rare Book and Manuscript Library at Yale University in New Haven as well.[1]

The fragments of the MS. now in Copenhagen were first made known to scholars by the late Aksel Volten. In a paper presented during the *Journées Scientifiques d'Orientalisme* at Prague in June 1949, he surveyed the papyrus holdings of what was then known as the Egyptological Institute of Copenhagen University. This survey was published two years later.[2] In it, Volten drew attention to specimens of 'Hermetic literature' among the demotic manuscripts preserved in the Institute. One of these, he went on to say, 'treats the primitive Ocean *Nwnw* and may well be characterised as a scientific tract',[3] a clear reference to our text, which is the only one in Copenhagen that could be described in such terms.

It is not certain when Volten began the task of reconstructing this particular papyrus. Nor is it known precisely how much he had accomplished at the time of his death in 1963, although the fact that most of the larger Copenhagen fragments of our cosmology had been sorted together by him indicates that, at least in terms of identifying these, he had made considerable progress.[4] No one else appears to have worked on the text during Volten's lifetime.[5] After he died, it remained neglected until the inception of the International Committee for the Publication of the Carlsberg Papyri

[1] The Copenhagen fragments are collectively designated as P. Carlsberg 302, the Florence fragments as PSI Inv. D 7. Those at Yale are catalogued as P. CtYBR inv. 861, 4387 (10), 4403 (2), and 4921. For details of which fragments belong to which collections, see the concordance which follows this Introduction.

[2] See *Archiv Orientální* 19 (1951), pp. 70-4.

[3] Art. cit., p. 73.

[4] On Volten's sorting methods, see Tait in Frandsen (ed.), *The Carlsberg Papyri 1: Demotic Texts from the Collection*, pp. 13-17.

[5] A brief reference to it by Botti in *Anthemon: Scritti di archeologica e di antichità classiche in onore di Carlo Anti* (1955), p. 4, probably relies upon information provided by his Danish colleague.

(ICPCP) in 1989.[6] It was then that the actual cataloguing of the Copenhagen fragments belonging to the MS. began. My own involvement with this material started in 1992, when I was invited to edit it by the members of the ICPCP. During the summer of that year, I was able to visit Copenhagen and spend two months working on the fragments of the text there. In the course of this work, several further pieces of the MS. were identified, some of which joined directly to those previously catalogued. The following year, three additional fragments were identified by Dr Carla Salvaterra. In 1995, Dr Joachim Quack discovered another two, while effecting a join between one of those identified by Dr Salvaterra and a previously catalogued fragment. This brought the total of fragments constituting the Copenhagen portion of our cosmology to twenty-five.

I first learned of the existence of further fragments of the same MS. in Florence from Professor Jürgen Osing in September 1993. He had seen nine of these during a visit to the Istituto Papirologico 'G. Vitelli' and noted the resemblance which they bore to their counterparts in Copenhagen. I had the opportunity to spend ten days searching through the collection of the Istituto Papirologico for additional fragments belonging to the text in July 1994. This search yielded a further twelve, bringing the total of those in Florence to twenty-one.

The fragments of the manuscript in the Beinecke Rare Book and Manuscript Library were brought to my attention by Dr Joachim Quack in January 1997. He had identified three of them during a visit to New Haven the previous year. After obtaining photographs of the fragments in question, I was able to confirm that Dr Quack's attribution of them to our cosmology was correct and join one of them to a larger fragment in Copenhagen. A three-day visit to the Beinecke Library in May 1999 provided the opportunity to search for further pieces of the manuscript there. In the course of this visit I found one additional fragment; it too could be joined to one of those in Copenhagen.

It will be clear from what has been said above that the identification of the individual fragments which make up our cosmology has been a lengthy process in which a number of people have participated over the years. There are considerable difficulties involved in such work, and it will be appropriate to give some account of these here.

As noted at the outset, the cosmology preserved in our manuscript is written in the demotic script. The hand is characteristic of Tebtunis in the Fayyum, and may be dated on palaeographical grounds to the first half of the second century AD. It corresponds to Type 1 in the classification scheme devised by Tait.[7] At Copenhagen, as at Florence, there are literally hundreds of papyrus fragments inscribed with demotic hands of this type, all of which look more or less alike. At New Haven, the number of such

[6] For the history of this committee, see Zauzich in Frandsen (ed.), *The Carlsberg Papyri 1: Demotic Texts from the Collection*, pp. 1-11.

[7] See idem, *Papyri from Tebtunis in Egyptian and in Greek*, pp. viii-ix.

fragments, although smaller than this, is still considerable. Given any two fragments exhibiting similar hands then, in the absence of a direct join, one has to decide first whether they are the work of a single scribe or merely that of two members of the same scribal school. If the former, do the fragments originate from a single composition or from two distinct works copied by the same scribe? Such questions can be difficult to answer, particularly when the fragments involved are small, as is true of the majority of those in all three of the collections which have yielded pieces of our manuscript.

The actual writing can be misleading as well. Among previously published Copenhagen papyri, for example, P. Carlsberg 207 preserves the greater part of two columns of a literary text.[8] Although the whole was written by a single scribe, the appearance of the writing changes sufficiently in the last seventeen lines of the second column that, if the lines in question were preserved on a separate fragment from the rest, one would hardly guess that they formed part of the same manuscript. The sudden change in the appearance of the writing may be due to the pen having been sharpened or even replaced altogether.[9]

Variability is also a problem where the colour and texture of papyrus fragments are concerned. When the individual pieces of a manuscript have been separated from one another for an extended period of time, their physical appearance can differ greatly. The fragments of P. Carlsberg 230, a demotic herbal, for instance, exhibit such a marked variation in colour that Volten was led to assume that some of them belonged to a separate text or texts.[10]

Nor are the contents of a fragment always a reliable guide. When faced with two pieces of papyrus exhibiting what appears to be the same demotic hand, both inscribed with material of a cosmological nature, for example, one has to consider the possibility that these originate from two distinct works of the same type copied by a single scribe.

Fortunately, our text, like most of the demotic literary papyri in Copenhagen and, to a lesser extent in Florence and New Haven as well, was written on the back of a roll, the other side of which had previously been inscribed with a Greek document.[11] The particular document preserved on its front is a register of land compiled for purposes of taxation. The demotic and Greek texts have the same disposition relative to each other. The latter is inscribed in a distinctive hand, assignable on palaeographical

[8] See Tait in Frandsen (ed.), *op. cit.*, pp. 19-46 and plates 1-3.

[9] As suggested by Tait, ibid., p. 22.

[10] The text is published by Tait in Frandsen (ed.), *The Carlsberg Papyri 1: Demotic Texts from the Collection*, pp. 47-92 and plates 4-6. See particularly pp. 48-9 for his remarks on the colour variation. This problem affects the reconstruction of texts in other papyrus collections as well. See e.g. comments of Demarée in *Sesto Congresso Internazionale di Egittologia – Atti* 2, p. 102.

[11] By 'back' I mean the side of the roll on which the vertical fibres lie uppermost. The terms 'front' and 'back' are used here in preference to 'recto' and 'verso' following Smith and Tait, *Saqqâra Demotic Papyri* 1, p. xi.

grounds to the first half of the second century AD.[12] Not only does this help to confirm the date of the demotic on the other side, which must have been written shortly afterwards,[13] it also provides a useful check when one is considering whether a particular fragment ought to be included as part of that text or not. Logically, all fragments belonging to our cosmology should display not only the same demotic hand on the back, but the same Greek hand on the front as well, and with the same disposition. Thus a number of pieces which one might be tempted to ascribe to it on the basis of the demotic alone can be eliminated from consideration by virtue of the fact that their Greek texts are clearly unrelated to the one on the front of the fragments which do belong.[14]

There are, as it happens, exceptional cases where, for one reason or another, a demotic text is preserved on the back of a roll with two different Greek texts on the front. Thus the scribe of P. Carlsberg 600, a version of the myth of the solar eye, appears to have pasted two separate Greek manuscripts together in order to create a roll long enough for his requirements. Similarly, P. Carlsberg 465, which preserves a portion of the Inaros cycle, is inscribed on the back of a roll made up of two distinct Greek papyri that have been recut and pasted together. Here, the situation is even more complicated, since a damaged section of the roll has been patched with a piece of papyrus displaying yet another Greek hand, making a total of three that are visible on the other side of a single demotic text. However, such cases are sufficiently rare that they do not invalidate the general principle set forth in the preceding paragraph.[15]

Of course, deciding whether the hands of any two fragments are the same or not necessarily involves a certain degree of subjectivity, a fact of which I am very much aware. Moreover, as noted earlier, the potential for an error in judgement increases when one is dealing with very small fragments where the actual amount of writing available for comparison is meagre. Nevertheless, by utilising a combination of demotic and Greek evidence in the manner described above, I hope that the chances have been minimised, both of overlooking a fragment which should be assigned to our cosmology, and of inadvertently including one which has no connection with it, although for obvious reasons the possibility of making such mistakes cannot be eliminated entirely. If any pieces have been omitted through oversight, these will only be very small ones

[12] This information courtesy of Dr Carla Salvaterra. For the appearance of the Greek text, see plate 1.

[13] See above.

[14] A good example is P. Carlsberg 10881, brought to my attention by Joachim Quack, which deals with human reproduction and the role of a man's semen and a woman's blood in the reproductive process. Both the nature of its contents and the appearance of the demotic suggest a connection between this fragment and our MS., but the Greek on the front of the former rules this out.

[15] I am indebted to Kim Ryholt for bring these examples to my attention, and for his careful description of the manuscripts involved. According to him (letter of 27 October 2000), there are no other cases of this type among the papyri of the Carlsberg collection.

which preserve insufficient writing on them to make any material contribution to the understanding of the text as a whole.

Having identified the constituent fragments of a text like our cosmology, one has next to place them in their proper sequence. Although some of the individual pieces which comprise the manuscript have been made up from two or more smaller ones, regrettably none of the resulting forty-eight numbered fragments can be joined directly to any of the others. Nor has examination of the papyrus fibres of the individual fragments helped to ascertain the positions of any of them relative to the others.[16] It might have been expected that the Greek inscribed on the other side of the demotic would provide some assistance in arranging the fragments in their proper order, but I am informed by Dr Salvaterra that this has not proven to be the case. Consequently, in reconstructing the text, I have had to rely entirely upon the demotic for guidance.

Two types of evidence have been employed to order the fragments: internal and external. One important form of internal evidence involves a naming formula which recurs at intervals throughout our cosmology. It appears in a number of variants, e.g. *ḏ=w n=w* X, 'they were called X' (fragment 2, lines 9-10); *iw=[w] ḏ* X *n rn n* [...], 'they saying X as the name of [...]' (fragment 4, II, 22); and *ḏ=w n=f* X *n rn*, 'he was called X as a name' (fragment 5, II, 6).[17] The recipient of the name in such cases is normally a divinity or group of divinities. Once, it is a plant.[18] In Egyptian terms, creation is not a matter of making something out of nothing. Rather, it is a process of revealing things which already have a latent existence by summoning them forth into full existence. Thus, naming is a necessary concomitant of creation.[19]

There are important implications here for the ordering of the fragments of our cosmology. Logically, any fragment that recounts the naming of, say, a divinity should precede a fragment in which that divinity appears without any introduction as an ordinary actor in the narrative, since the giving of a name marks the recipient's initial coming into being in the particular form designated by the appellation in question. In a few instances, the recipient of such a name is then given a further name or names, to mark subsequent transitions from that initial form to other ones. By the same logic, any fragment which recounts the giving of a second name should precede a fragment in which its recipient is called by the name in question without any further comment,

[16] On the limitations of fibre analysis in this sort of reconstruction work, see Tait in Frandsen (ed.), *The Carlsberg Papyri 1: Demotic Texts from the Collection*, pp. 48-9.

[17] Comparable formulas occur in the demotic cosmogony of P. Berlin 13603, among other texts. Cf. ibid., II, 28: *i-ir=f ḏ ꞽInb-ḥḏ n rn Mn-nfr*; II, 29: *i-ir=f ḏ n=f ꜥnḫ-tꜣ.wy n rn*; IV, 6: *i-ir=w ḏ n=f ms ṯḥny n rn*; IV, 25: *ḏ=w n=f nꜣ-nfr-ḥr n rn* (Erichsen and Schott, *Fragmente memphitischer Theologie in demotischer Schrift*, plates 5-6).

[18] See fragment 1, line 12.

[19] See Finnestad, *Image of the World and Symbol of the Creator*, pp. 68-74 and 113-14; Bickel, *La Cosmogonie égyptienne avant le Nouvel Empire*, pp. 101-2.

while at the same time following any fragment which refers to that recipient by a name given to mark the commencement of an earlier phase of existence.

Thus, the fragment which I have called no. 1, since it records the giving of the name *ꜣsy*, 'reed', to a particular plant, should precede the fragment that I have called no. 2, where this reed is mentioned in the ordinary way. Similarly, the fragment which I have called no. 4, as it recounts how the creator god *pꜣ-Šy pꜣ Nwn*, 'Pshai the Primaeval Ocean' or 'Pshai (in) the Primaeval Ocean', is given a new name, simply 'Pshai', to mark his emergence from the waters of the Ocean and initial manifestation in the created cosmos, must follow fragment 1 as well, since there that divinity is still designated by his original and longer name.[20]

The latter part of fragment 1 is concerned with the god Ptah and a thicket of reeds (*bꜣ n ꜣsy*); so is the first part of fragment 2. This identity of subject matter suggests that the one should follow closely after the other. Hence, I have numbered the two fragments consecutively, placing what I have called fragment 4 last in the sequence.

As mentioned above, that fragment recounts the giving of the name Pshai to the creator god. Thus, it must precede what I have called fragment 5, since the latter speaks of Pshai in the ordinary way. Fragment 5, in turn, describes the creation and naming of the god Thoth. Therefore, it must precede those fragments to which I have assigned the numbers 6, 8, 10, 11 and 12, inasmuch as Thoth figures in all of them as an actor in the narrative without any sort of introduction. This gives us a basic sequence of four fragments (1, 2, 4 and 5) followed by a cluster of five further ones, the order of which relative to one another has still to be determined.

The basic sequence can be expanded by means of external evidence in the form of parallels and comparable material from other Egyptian cosmogonical or cosmological works. Thus, the fragment which I have designated as no. 3 extols the god Amun in the form of a black bull with eight aspects or hypostases. The demotic cosmogony preserved in P. Berlin 13603 records how the eight members of the Hermopolitan Ogdoad, after their creation, began to unite or merge together: the four male members into a black bull, and the four females into a black cow, identified as Amun and Amaunet respectively. Subsequent to this, the bull and cow merge into a single entity.[21] From other sources, too, we learn that Amun incorporates all eight members of the Ogdoad, male and female, within himself.[22] On the basis of these parallels, one can be fairly certain that fragment 3 belongs immediately after the fragment of our text which describes the creation of the Ogdoad and the first stages of their merging together (= fragment 2), even though the two do not join directly together. The eight hypostases of Amun as a black bull correspond to the eight members of the Hermopolitan corporation embodied within him.[21]

[20] See commentaries on the fragments in question.

[21] For details of this account, see commentaries on fragments 2 and 3.

[22] See commentary on fragment 3.

In the case of the fragments which I have designated as nos. 6-8, it is impossible to determine the precise nature of the contents. Nothing is preserved but the names of various divinities and one or two other words and phrases. With these and other fragments in a similar state of preservation, I have tried something of an experiment, making use of a concept introduced to Egyptology by Jan Assmann, that of the 'constellation'. A constellation is a group of divinities who always occur in a fixed relation together in a specific context.[23] Looking at the fragments of our text, it is sometimes possible to identify a particular constellation of deities even where no coherent narrative can be reconstructed, and this can at least suggest the probable nature of the contents.

What I have called fragment 6, for instance, refers to *pꜣ Nwn*, 'the Primaeval Ocean', in line 1, crocodiles (*msḥ.w*) in line 2, some unspecified gods and Thoth in line 3, and the 'great sand' (= the primaeval mound) of Hermopolis in line 6. The phrase 'his seven [...]' occurs in line 7. Other isolated words and phrases are legible as well, including some dialogue. In the commentary on the fragment in question, I have reviewed the various contexts in which this combination of elements might occur, arriving at the conclusion that the most likely setting is that of a rebellion against the sun god. Such revolts are described in a number of other ancient Egyptian sources, notably the New Kingdom Book of the Heavenly Cow and the Neith cosmogony at Esna. In some accounts of this type, the Primaeval Ocean and Thoth play important roles as advisers who assist the solar deity in suppressing those who have risen up against him. Others name the primaeval mound of Hermopolis as the place where the rebels are overcome and punished.[24] Obviously, this sort of approach does not permit one to establish the nature of a fragment's contents with certainty, but it does at least indicate in which direction the balance of probability is likely to lie.

What I have called fragment 7 must be placed somewhere after fragment 4, since it mentions the god Pshai in the ordinary way (see above). Like fragment 6, it also mentions *msḥ.w*, 'crocodiles'. On this basis, I have assumed that there is a connection between the two and ordered them consecutively, although it is not actually clear which fragment preceded the other. Perhaps no. 6 should follow no. 7, rather than vice versa.

The type of analysis employed with fragment 6 suggests that what I have called fragment 8 may be concerned with rebellion against the sun god as well.[25] There is reason to think that this uprising is mentioned for the first time in fragment 5.[25] If so, then fragments 6-8 should follow quite closely after that one. Once again it is possible that the last of these belongs before the first two rather than vice versa.

What I have called fragments 9 and 10 are concerned with the burial of the Ogdoad. The former actually records this event; the latter treats it as something which

[23] See Assmann, *Göttinger Miszellen* 25 (1977), pp. 14-15 and literature cited there.

[24] See commentary on fragment 6 for details.

[25] See commentary on that fragment.

has already occurred. This allows one to determine the order of the two fragments in relation to each other. In other Egyptian sources, the death and burial of the Ogdoad are said to take place after the rebellion against the solar deity has been suppressed.[26] Thus, what I have called fragments 9 and 10 have been placed after fragments 5-8.

According to fragment 10, the deity who performs the rite of burial for the Ogdoad is Thoth. Thereafter, that god was supposed to have assumed the form of the moon in the sky. What I have called fragment 11 appears to recount the entrance of Thoth into the heavens.[27] It seems logical, therefore, to place this fragment after nos. 9 and 10. If my suggested reconstruction is correct, then one has a series of fragments (nos. 5-11) concerned with the rebellion against the solar deity and its aftermath. What I have called fragment 12 recounts the birth of Horus in the papyrus marshes of Khemmis. Chronologically, this should follow the aforesaid sequence of events. Hence, I have placed that fragment after no. 11.

All of the fragments discussed so far have one feature in common: they are concerned with what happened in the past, forming part of a narrative recording the origin and subsequent history of the cosmos. By contrast, there are two further fragments which describe various phenomena that occur in the present rather than narrate past events. These treat, among other subjects, the nature of the relationship between life and death, the origin of particular bodies of water, how mountains and the precious stones which they contain come into being, and the influence of the full moon on the reproduction of animals and the germination of plants. The fragments illustrate the Egyptian view that creation is not a single event which took place in the distant past, but rather a cyclical, eternally recurrent one that continues into the present as well.[28] It seems logical to place these two fragments after the purely narrative ones. What I have called fragment 13 has been put first simply because it deals with broader, more fundamental matters than no. 14 appears to do. The latter is devoted largely to the elaboration of ideas which have already been adumbrated in the former.

By the process outlined in the preceding paragraphs, a sequence of fourteen fragments can be established. Apart from a few minor uncertainties, the order of these relative to one another seems fairly certain. The fragments in question provide the basic structure, the 'skeleton' so to speak, of our cosmology. For the sake of convenience, they may be said to constitute Group 1.

Fifteen further fragments of the text, which I have called nos. 15-29, comprise Group 2. None of these can be assigned a definite position within the basic sequence. Yet for each a plausible link can be suggested with one of the fragments in Group 1. Normally, it is on the basis of the contents that such links are postulated, although

[26] See commentary on fragment 9.

[27] See commentary on fragment 11, I, 2-4.

[28] On this point, see Finnestad, *Image of the World and Symbol of the Creator*, pp. 14 and 22-3.

other factors like a piece's colour and texture have been taken into account as well. There is good reason to think that each fragment in Group 2 originated from the same column of text, or an adjacent column, as the fragment in Group 1 with which it is associated. Because this cannot be demonstrated conclusively, however, I have thought it best to keep them separate.

To cite just one example, what I have called fragment 15 preserves the phrases *mtw=f ḥwy*, 'that he might deposit', in line x+1, and *n3 swḥ[y].w r-ḥwy=f*, 'the eg[g]s which he deposited', in line x+6. The divine name Ptah occurs in lines x+3 and x+8. In the former, it is followed by the verb *sty*, 'scatter'. This suggests a connection with fragment 2, where the god Ptah is said to scatter his seed and fertilise some objects, presumably eggs or egg-sacs, from which the members of the Ogdoad subsequently emerge. Thus, fragment 15 may preserve part of the same account as fragment 2, although it cannot be directly joined with that piece.

The nineteen remaining fragments of our text (nos. 30-48) I have classified as Group 3. These are pieces for which no really plausible link with any of the fragments in Group 1 can be suggested. In some cases, this is because they preserve insufficient text to permit the nature of their contents to be ascertained. In others, the contents can be identified, but are not readily associable with those of any of the Group 1 fragments. I have, nevertheless, in a few instances, ventured to suggest a link between one of these pieces and a fragment in the first group, but such proposals should not be regarded in the same light as those involving fragments in Group 2, being as they are of a more tentative nature. I have also, in certain cases, suggested the possibility of linking a fragment in Group 3 with one or more fragments in the same group.

It will be apparent from what has been said above that it is impossible, in the present circumstances, to offer a definitive text of the cosmology which is edited here. Because of the fragmentary nature of the MS. and the uncertainties affecting the ordering of its constituent pieces, no edition of this work can avoid having a somewhat tentative or provisional character. Inevitably, readers of the present book will have their own ideas about the reconstruction of the text, not all of which will coincide with mine. One might feel that a particular fragment which I have classified in Group 3 really belongs in Group 2. Another might wish to remove a fragment from Group 2 and add it to those of Group 1. It is possible, moreover, that additional fragments of the MS. will come to light in other papyrus collections. The information provided by these might necessitate modifications in the arrangement which I have proposed. In a sense, therefore, the text is destined always to be coming into being rather than having a permanently established form, exactly like the Egyptian cosmos whose creation it describes. There is an element of appropriateness here.

Despite all the uncertainties which I have enumerated, it is still possible to make some general observations about the form and characteristic features of our cosmology. A detailed consideration of the text's account of the creation of the universe and how it operates, as well as its relationship to similar accounts in both Egyptian and Greek, is reserved for the commentaries on the individual fragments and the concluding summary. Here, I shall be concerned with less specific matters.

As noted above, our cosmology falls broadly into two sections: a narrative part, concerned with the past, and a descriptive or didactic part concerned with the present. The language is straightforward, without any of the archaisms found, for instance, in mortuary works written in the demotic script. The orthography is fairly consistent. For a few words, the scribe has employed more than one demotic writing.[29] Occasionally, he varies demotic writings with hieratic ones.[30] Exceptionally, one small fragment with traces of only seven lines preserves no less than three different orthographies of the same word, two demotic and one hieratic.[31] As might be expected, given the provenience of the manuscript, Fayyumic *l* appears regularly in place of *r*.[32]

Attention has already been drawn to the importance of naming as a concomitant of creation in the text, and the regular recurrence of naming formulas in it. Another feature which our cosmology has in common with numerous other Egyptian religious texts is the use that it makes of word play.[33] In fragment 3, for instance, there is a play on words involving the divine name *Ỉmn*, 'Amun', and what may be the verb *imn*, 'to be hidden'. In fragment 30, there is a play on words between the divine name *Spt.t*, 'Sothis', and the noun *sty*, 'fire, flame'. Aetiological statements also occur frequently, brief explanations of how a particular practice arose or why some being or object came to be called by a particular name as a consequence of an event described in the text. None of these has been preserved in its entirety. They may be recognised, however, by their distinctive remains, e.g. phrases like *r-hn-r pꜣ hrw r-ḏbꜣ.ṯ=s*, '(X is done/called Y) up until today because of it'. Such statements not only provide useful information; they serve, in addition, to emphasise the continuity that links the period of the initial creation of the cosmos with the present.[34]

The constituent fragments of our cosmology are presented below in the three groups to which they have been assigned. The edition of each fragment comprises five elements: a description of the piece, including its dimensions and colour, a transliteration of the demotic, a translation, textual notes, and general commentary.

[29] Compare e.g. writings of *iny*, 'stone', in fragments 2, line 11, and 14, II, 6; writings of *iwf.w*, 'flesh', in fragments 14, II, 7, and 26, line x+5; and writings of *kꜣ*, 'bull', in fragments 3, line 9, and 40, line 2.

[30] Compare demotic writings of *Nwn*, 'Primaeval Ocean', in fragment 13, II, 4 and passim with the hieratic ones in fragments 1, line 2, and 11, I, 11. (The determinatives of a demotic writing of this word are visible at the beginning of I, 1 of the latter fragment.)

[31] See occurrences of *rꜣ*, 'mouth', in fragment 21, lines x+2 and x+4-5.

[32] See, for instance, *lhny*, 'resting place', in fragments 1, line 6, 2, line 5, and 4, II, 11; *ly.t*, 'side', in fragments 1, line 9, and 11, I, 8; *Ḫl*, 'Syria', in fragment 13, II, 7; *lym.w*, 'fish', ibid., II, 11; and *ḥlly*, 'flower', ibid., II, 20.

[33] See Guglielmi in *Lexikon der Ägyptologie* 6 (1986), columns 1287-91 (s.v. Wortspiel).

[34] On Egyptian aetiologies, see Spiegel in *Lexikon der Ägyptologie* 1 (1975), columns 80-3.

The textual notes are concerned purely with establishing readings, identifying problematic words and constructions, elucidating poorly legible traces, and so on. The commentaries analyse the contents of the fragments in wider terms. The descriptions of the fragments in Group 1 include cross-references to related fragments in Group 2, where such exist, and vice versa. More tentative suggestions concerning possible links between some fragments in Group 3 and others in the same group or in Group 1 have been relegated to the commentaries on the fragments in question.

In writing a commentary on material in such a state of preservation as this, there are two different approaches that one can adopt. The first is to consider solely what actually survives in each fragment. The second is to attempt to relate the fragments to a wider context, not only taking into account what is preserved in them, but also investigating various possible restorations of damaged passages and assessing their implications for the understanding both of the individual fragments in which they occur and of the text as a whole as well. Thus, the second approach incorporates the first but also goes beyond it. This is the approach that I have adopted in the present edition.

Obviously, any interpretation of a text that relies heavily upon restoration can never be more than a tentative one, a statement of possibilities rather than a definitive account. For this reason, readers are asked to bear in mind that the suggestions made in the commentaries here with respect to the contents of some fragments and possible parallels to these in other Egyptian sources are just that, suggestions, and nothing more. I have tried to emphasise their tentative character by frequent use of such qualifying adverbs as 'perhaps' and 'possibly'. It goes without saying that a rigid distinction should be maintained at all times between such suggestions and statements based firmly upon the evidence preserved in the text itself.

Some might think it preferable to exclude all discussion of a speculative nature from the commentaries. As the author of a recent study of hieratic ostraca has observed: 'Je kürzer und schlechter erhalten ein Text, desto langatmiger – und häufig gar überflüssiger – nicht selten der Kommentar.'[35] This is normally a view with which I would have much sympathy. In the case of our cosmology, however, it has seemed to me that its contents, where they can be established with certainty, are of sufficient interest, not only for demotists but for other Egyptologists as well, that it is a worthwhile exercise to try at least to explore the various possibilities for restoring what may have stood in the less well-preserved portions of the text and relating these to the rest of it. I hope, moreover, that the publication of the present volume will stimulate others to search for additional fragments of the manuscript in which our cosmology is preserved. For their benefit as well, I have thought it appropriate to include a certain amount of hypothetical reconstruction in order to illustrate as fully as possible the range of subject matter which any new fragments are likely to contain. Anyone who

[35] Fischer-Elfert, *Lesefunde im literarischen Steinbruch von Deir el-Medineh*, p. v.

disagrees strongly can simply avoid reading those passages of commentary where I have done this.

The editions of the individual fragments of our manuscript are followed by a chapter of summary. In this, I have attempted to present a picture of the cosmology as a whole, synthesising what is said about its contents in the various commentaries. Other questions of a broader nature are considered there as well: the work's probable date of composition, its most important themes, the function of the text, whether ritual or didactic, the extent to which, as a work of Fayyumic provenience, it reflects distinctively Fayyumic theological concerns, its relationship to other Egyptian sources of the same type, and possible affinities with contemporary Greek works of a similar nature, whether those known to have been composed or circulated in Egypt itself, or those written elsewhere and alleged to have been influenced by Egyptian sources. This chapter is followed by a bibliography of works consulted in the preparation of the present edition. The volume concludes with a glossary of the demotic words that occur in our cosmology and photographic reproductions of the text.

For abbreviations used herein to refer to other books and periodicals, see the list in Smith, *The Mortuary Texts of Papyrus BM 10507*, pp. 11-12. In citing publications which do not appear in that list, I have employed either the full title or a form which resembles it closely enough to permit ready identification. Full titles of all works cited are provided in the bibliography. Where reference has been made to a text by column and/or line number alone, with no accompanying bibliographical information, the text in question is unpublished. Some unpublished fragments in the Carlsberg collection are cited by means of a five digit registration number (e.g. P. Carlsberg 10029). Such numbers are temporary ones, and subject to alteration at a later date.

Throughout these pages, I have referred to the watery mass with which our text is concerned as the Primaeval Ocean. In doing so, I have opted to use a slightly archaic spelling of the qualifying adjective, in place of the more current 'Primeval'. Thus, the word's orthography reflects its meaning, in accordance with good Egyptian practice.

It remains to thank the numerous individuals and institutions from whose assistance I have benefited while work on this book was in progress. I am grateful to the members of the International Committee for the Publication of the Carlsberg Papyri for the invitation to edit the Copenhagen fragments belonging to the manuscript here studied. My work on these was made possible through the support of the Carlsberg Foundation. Permission to study and publish the Florence fragments of the text was kindly given by Professor Manfredo Manfredi. My visit to that city to work on these was subsidised by a grant from the British Academy. For permission to publish the fragments in New Haven, I am indebted to the authorities of Yale University Library. I was able to examine these at first hand thanks to the support of the Management Committee of the Griffith Institute Egyptological Fund.

In the course of my visits to the different collections which house the fragments that make up the cosmology edited here, I received help from several colleagues. I should like to acknowledge particularly the valuable advice and assistance given by Bente Holmen in Copenhagen, Dr Giovanna Menci in Florence, and Dr Robert Babcock in New Haven. As noted earlier, Professor Jürgen Osing and Drs Joachim Quack and Carla Salvaterra discovered several fragments belonging to the manuscript of whose existence I was unaware. I am grateful to them for bringing these to my attention. Dr Salvaterra also provided much useful information about the Greek text written on the other side of the fragments. I have benefited greatly from discussing both the Greek and the demotic with Dr Kim Ryholt and Professor W.J. Tait.

The title adopted for this book, *On the Primaeval Ocean*, was suggested to me by a colleague at Oxford University, Professor Glen Dudbridge. It can be understood in two different ways. On one level, like *De natura deorum* or περὶ Ἴσιδος καὶ Ὀσίριδος, it simply identifies the subject with which the book is concerned. At the same time, however, it is meant to evoke something of the ancient Egyptian attitude towards that subject. As will emerge from the pages that follow, the Primaeval Ocean was regarded as an active principle which played a significant part in the initial creation of the cosmos and continues to order and influence its development right up to the present. The first life emerged from it, and all subsequent existence is played out on it and in interaction with it. Thus, the title of the book not only defines its subject, it explicates the nature of our relationship with that subject, as seen from an ancient Egyptian perspective: whether living in the first half of the second century AD, or in the final days of the twentieth century, we are all fellow voyagers 'on the Primaeval Ocean'.

CONCORDANCE OF FRAGMENTS

Group 1

Frag.	*Location*	*Frag.*	*Location*
1	PSI Inv. D 7 (a)	7	PSI Inv. D 7 (b)
2	P. Carlsberg 302 (1)	8	P. Carlsberg 302 (6)
3	P. Carlsberg 302 (2) + P. CtYBR inv. 4921	9	PSI Inv. D 7 (c)
		10	PSI Inv. D 7 (d)
4	P. Carlsberg 302 (3)	11	PSI Inv. D 7 (e)
5	P. Carlsberg 302 (4) + P. CtYBR inv. 861	12	P. Carlsberg 302 (7)
		13	P. Carlsberg 302 (8)
6	P. Carlsberg 302 (5)	14	PSI Inv. D 7 (f)

Group 2

Frag.	*Location*	*Frag.*	*Location*
15	PSI Inv. D 7 (g)	23	P. Carlsberg 302 (12)
16	P. Carlsberg 302 (9)	24	PSI Inv. D 7 (k)
17	P. Carlsberg 302 (10)	25	P. Carlsberg 302 (13)
18	P. CtYBR inv. 4387 (10)	26	PSI Inv. D 7 (1)
19	P. Carlsberg 302 (11)	27	PSI Inv. D 7 (m)
20	PSI Inv. D 7 (h)	28	PSI Inv. D 7 (n)
21	PSI Inv. D 7 (i)	29	PSI Inv. D 7 (o)
22	PSI Inv. D 7 (j)		

Group 3

Frag.	*Location*	*Frag.*	*Location*
30	P. Carlsberg 302 (14)	40	PSI Inv. D 7 (r)
31	P. Carlsberg 302 (15)	41	PSI Inv. D 7 (s)
32	P. Carlsberg 302 (16)	42	PSI Inv. D 7 (t)
33	P. Carlsberg 302 (17)	43	PSI Inv. D 7 (u)
34	P. Carlsberg 302 (18)	44	P. Carlsberg 302 (22)
35	P. Carlsberg 302 (19)	45	P. Carlsberg 302 (23)
36	P. Carlsberg 302 (20)	46	P. Carlsberg 302 (24)
37	P. Carlsberg 302 (21)	47	P. Carlsberg 302 (25)
38	PSI Inv. D 7 (p)	48	P. CtYBR inv. 4403 (2)
39	PSI Inv. D 7 (q)		

PART II: TEXT

GROUP 1: FRAGMENTS 1-14

FRAGMENT 1 — Plate 2

Dimensions: 14 × 22.3 cm (main piece), 2.7 × 3.9 cm (piece from top of column), 5.2 × 9.1 cm (piece from bottom of column). Colour: darkish brown.

Transliteration

1. *] rt=f m-sꜣ pꜣ šꜥṱ=f r-hn-r pꜣ hrw ... pꜣ [...] n(?) pꜣ Nwn [*
2. *š]ll(?) m-bꜣḥ pꜣ-Šy pꜣ Nwn pꜣ [...].. pꜣ rn n [*
3. *] n-tr.t Ptḥ pꜣy=k šr ỉ-ỉr tỉ ḫpr n=n ẖlꜣ nhm [....]..[*
4. *] r-ḏbꜣ.ṱ=s ỉ-ỉr šꜥ ỉr nꜣ ỉꜣw.wt nhm ỉw=w sy [*
5. *].. qmꜣ m-šs [t]lg(?)=f s ꜥnḫ(?)=w n-ỉm=s ỉrm [*
6. *ẖ]n pꜣ mw pḥ=f r wꜥ.t lhny n šꜥꜣ ꜥḥꜥ=f [*
7. *]..[...] šꜥꜣ ḏ ỉw=f r rt ḫpr ỉw=k [...].. mꜣ ..[*
8. *]..=f [...].m rt=f n wꜥ.t bꜣ ꜥꜣ.t .....[*
9. *w]py m-šs ḫꜣy=f n tꜣ ly.t ḥry.t ..[*
10. *ỉw]=f ỉw r-ẖly [..]=f ḥr-ꜣṱ=f gm=f s ỉw=f(?) [*
11. *Ḫmn]w n rn rḫ=f s ḏ ỉ-ỉr=f pr n Yb [*
12. *] rn .. wꜣḥ(?)=f r pꜣ mḥt ḏ=w n=f ꜣsy n rn nw nꜣ ntr.[w*
13. *]..=f s sḫny=f r-r=s sty=f tꜣ[y=f] mtw[y*
14. *]...y tꜣ [...]... rn=s r p[ꜣ(?) ...]..[*
15. *]....y [...] n nꜣ ntr.w n pꜣ [*
16. *] r-r=s [...] pꜣ(?) ḥf ...[...]..[..]..[......].*
17. *]...[...]. r [..].... n sẖ ..[...]...[..]..[......]..*
18. *] qmꜣ ḏ(?) Ptḥ mt(?) r ḫpr ḥr [..] tꜣ bꜣ [......]..*
19. *]..[..] r-b-ỉr-tw pꜣ ỉs[t] wn[ḥ(?)] ..[......]=f m.. tꜣ*
20. *] ḥmsy=f ḥr tꜣ bꜣ n ꜣsy [..].[......] Nwn*
21. *]......[..] sty [........] Nwn*
22. *ẖ]n=s ....[...]......[...]. rn(?)=f ỉ-[ỉr]* =f
23. *]....[...].[..].. tꜣ/nꜣ ......[...] pꜣ ḥf n rn=f*

Translation

1.] It will grow, after being cut off, up until today ... the [...] of/in the Primaeval Ocean [

2. p]ray(?) before *pꜣ-Šy pꜣ Nwn*, the [...].. the name of [
3.] by/through Ptah your son, who created sustenance for us, so that there rejoice [....]..[
4.] because of it, who first made the cattle rejoice, they being satisfied [
5.].. rush very much. He [up]rooted(?) it so that they might live(?) on it together with [
6. i]n the water. He reached a resting place of sand and stood up [
7.]..[...] sand, saying: It will grow, because you [...].. place ..[
8.].. him/it [...].*m*. It grew into a large thicket .....[
9. o]pen greatly. He/it rose high above ..[
10. H]e came down and [..] upon it. He discovered that it was [
11. Hermopo]lis by name. He knew that it came forth from Elephantine [
12.] name .. It came to rest(?) in the north. It was called 'reed' by name. The god[s] looked [
13.] he [...] it. He settled upon it. He scattered hi[s] see[d
14.]...*y* the [...]... aforesaid to th[e(?) ...]..[
15.]....*y* [...] of the gods of the [
16.] to it [...] the(?) serpent ...[...]..[..]..[......].
17.]...[...]. to it [..].... in writing ..[...]...[..]..[......]..
18.] rush. Ptah spoke(?) a word(?) to come into being upon [..] the thicket [......]..
19.]..[..] before the grou[nd] appear[ed(?)] ..[......] him/it *m*.. the
20.] He sat down upon the thicket of reeds [..].[......] Primaeval Ocean
21.]......[..] scattered [........] Primaeval Ocean
22. i]n it ....[...]......[...]. his/its name(?). He
23.]....[...].[..].. the ......[...] the aforesaid serpent

Textual notes

Line 1

(a) A writing of *šꜥṭ*, 'cut off', identical with that in the present line occurs in P. Carlsberg 124, line 3.

(b) For better preserved instances of the phrase *r-hn-r pꜣ hrw*, 'up until today', see e.g. fragment 4, II, 15, and fragment 13, II, 5.

Line 2

(a) The partially preserved word at the beginning of the line ends in *ll* and is determined with the man with hand to mouth. This could be *šll*, 'pray' (*Glossar*, p. 519), but other alternatives are possible as well, e.g. the *ꜣll*, 'shout, cry', which occurs in P. Harkness, I, 20 (Logan in Johnson and Wente [eds.], *Studies in Honor of George R. Hughes*, p. 151), or a demotic form of Coptic ⲧⲉⲗⲏⲗ, 'rejoice' (Crum, *CD*, p. 410).

(b) The semi-hieratic form of *Nwn*, 'Primaeval Ocean', seen here recurs in fragment 11, I, 11.

Line 3

(a) With the writing of *n-tr.t*, 'by/through', in the present line, compare that of the noun *tr.ṯ*, 'hand', in fragment 6, line 4.

(b) No other example of *pꜣy=k* occurs in the text. For the writing, see *Glossar*, p. 128.

(c) The end of *ḫpr* runs into the lower stroke of *n=n*, 'for us'. For the shape of that stroke, compare the writing of the possessive article *nꜣy=n* in fragment 2, line 10.

(d) For the traces of *nhm*, 'rejoice', at the end of the line, cf. the better preserved occurrence of that verb in the line that follows.

Line 4

(a) Better preserved occurrences of *r-ḏbꜣ.ṯ=s*, 'because of it', occur in fragment 4, II, 17, and fragment 14, II, 7, q.v.

(b) For the writing of *sy*, 'be satisfied', at the end of the line, see *Glossar*, p. 407. The syntax here is slightly puzzling. One expects something like *i-ir šꜥ ti nhm nꜣ iꜣw.wt iw=w sy*. Nevertheless, the sense of the passage is clear. See commentary below.

Line 5

(a) There is a trace of writing visible before *qmꜣ*, 'rush', which I am unable to identify. It is not the definite article *pꜣ*.

(b) With the traces of the adverb *m-šs* here, cf. the better preserved example of that word in line 9 below.

(c) The verb after *m-šs* has *l* and *g* as its last two consonants and ends with the knife determinative. The initial sign must have been a low, broad one, to judge from the surviving trace of it. In my transliteration, I have suggested restoring [*t*]*lg*. This could be a demotic ancestor of Coptic **ⲧⲱⲗⲕ**, 'pluck out, uproot' (Crum, *CD*, p. 410a). Alternatively, one could restore the *plg*, 'detach, free', of *Glossar*, p. 137. Yet another possibility is that one has here a variant of *grg*, 'hunt' (ibid., p. 586). Compare the writings of *bḥs*, 'hunt', with knife determinative cited in *Glossar*, p. 121. Unfortunately, the context is insufficiently well preserved to permit one to decide among these alternatives. See further commentary below.

(d) The initial sign of the verb before *n-im=s* appears to be an *ꜥ* written over another sign, perhaps *l*. The traces that follow would suit *ꜥnḫ*, 'live'.

Line 6

(a) For the traces of *ḫn*, 'in', at the beginning of the line, cf. better preserved occurrences of that preposition in fragments 2, line 8; 4, II, 10; and 5, I, 4.

(b) A better preserved example of the verb *ꜥḥꜥ*, 'stand up', occurs in fragment 12, II, x+5.

Line 7

(a) The trace at the beginning of the line might conceivably be of *wn*, 'exist', or *wn*, 'open'. See e.g. fragment 30, line 3.

(b) A tall vertical stroke is visible before *mꜣ*, 'place'. Could this be part of the indefinite article *wꜥ*? Cf. fragment 13, II, 13.

Line 8

(a) The word before *rt=f* has *m* as its final consonant and is determined with the plant sign. To judge from the traces preceding the *m*, this cannot have been *sym*, 'plant', as in fragment 11, I, 9, or fragment 13, II, 12.

(b) For the traces of *bꜣ*, 'thicket', cf. the better preserved example of that noun in fragment 2, line 2. For those of the following *ꜥꜣ.t*, 'great', cf. the writing of *ꜥꜣ*, the masculine form of that adjective, in fragment 6, line 6.

Line 9

(a) For the initial signs of *wpy*, 'open', at the beginning of the line, cf. the writing of that verb in fragment 5, II, 18.

(b) With the traces of *n tꜣ ly.t ḥry.t*, 'above', cf. other examples of the same phrase in fragments 10, line x+10; 11, I, 8; and P. Carlsberg 10029, line x+4.

Line 10

(a) *iw* is the qualitative form of the verb 'come'. This form is used in preference to the *sḏm=f* in past narrative, hence the restoration [*iw*]=*f* immediately before it.

(b) With the traces of *ẖly* in *r-ẖly*, 'down', cf. the better preserved occurrence of *ẖly* in fragment 24, line x+7.

(c) The verb after *r-ẖly* has been lost except for a few traces. Perhaps restore something like *wꜣḥ*, 'alight'. Compare the word which I have tentatively read as *wꜣḥ* in line 12 below; also *wꜣḥ*, the marker of the perfect, in fragment 19, line x+9.

(d) For the writing of *ꜣt*, 'back', here, see note on fragment 4, II, 16.

(e) One expects something like *iw=f* after *gm=f s*. There is a trace extending downward diagonally near the very end of the line, which could be the top part of an *f*.

Line 11

The partially preserved word at the beginning of the line is determined with the toponym sign. The traces before that suit the last two signs of *Ḫmnw*, 'Hermopolis'. Cf. better preserved writings of that toponym in fragment 24, line x+10, P. Carlsberg 10032, line x+5, and P. Carlsberg 10040, lines x+5, x+8, and x+10.

Line 12

(a) I have not succeeded in reading the sign following *rn* at the beginning of the line. It does not appear to be *s* or any of the suffix pronouns which one might expect to be attached to that noun.

(b) With the signs that I have tentatively read as *wꜣḥ* here, cf. the writing of *wꜣḥ*, the marker of the perfect, in fragment 19, line x+9.

(c) For the restoration *nṯr.*[*w*], 'god[s]', at the end of the line, cf. the better preserved occurrence of that noun in fragment 5, I, 1.

Line 13

(a) *sẖny r*, followed as it is by the phrase *sty=f tꜣ*[*y=f*] *mtw*[*y*], 'he scattered hi[s] see[d]', clearly means 'have sexual intercourse with' or the like. The verb *sẖny* is written with phallus determinative. My translation 'settle upon' is intended to evoke the probable derivation of the word from the verb *sẖni*, 'alight' (*Wb.* 4, 253-4). Alternatively, one might have here a specialised usage of the verb *sẖny*, 'meet, embrace' (*Glossar*, pp. 455-6), which is sometimes construed with the preposition *r*. Cf. writings of *mr*, 'love', with the phallus determinative added to impart a specifically sexual connotation, cited ibid., p. 167.

(b) With the traces of the phrase *sty=f tꜣ*[*y=f*] *mtw*[*y*], 'he scattered hi[s] see[d]', here, cf. other occurrences of the same phrase in fragment 2, line 6, and fragment 11, I, 6.

Line 14

(a) The partially preserved verb at the beginning of the line ends in *y* and is determined with the walking legs sign. I am unable to make out the remaining traces.

(b) After the definite article *tꜣ,* one might conceivably restore the noun *lhny,* 'resting place', as in e.g. line 6 above. The traces suit and this would fill the available space well.

(c) With the traces of *rn,* 'name', cf. better preserved examples of the same noun in lines 11 and 12 above.

(d) Visible near the end of the line is a house determinative. The word to which it belonged is entirely lost.

Line 15

(a) The first sign preserved here could be either the seated child determinative or the feminine singular indefinite article *wꜥ.t.* After some illegible traces, a *y* can then be made out, followed by the same sign again.

(b) With the traces of *ntr.w,* 'gods', cf. the better preserved occurrence of that noun in fragment 5, I, 1.

Line 16

r-r=s near the beginning of the line is clear on the original. Cf. the better preserved example of that preposition and suffix pronoun in line 13 above.

Line 17

(a) After the break that follows *n sẖ,* 'in writing', there are two tall vertical signs which might conceivably be *ntr,* 'god'.

(b) The trace visible near the end of the line is the animal skin determinative. Cf. the writing of *iꜣw.wt,* 'cattle', in line 4 above.

Line 18

(a) With the traces of *qmꜣ,* 'rush', at the beginning of the line, cf. the better preserved example of the same noun in line 5 above.

(b) With the traces that I have tentatively read as *ḏ,* 'say', cf. better preserved occurrences of that verb in e.g. lines 7 and 12 above. A better preserved example of *mt,* 'word, speech', occurs in fragment 5, II, 2, q.v.

(c) For the traces of *bꜣ,* 'thicket', cf. the better preserved example of that noun in fragment 2, line 2.

Line 19

(a) A further example of the verbal prefix *r-b-ir-tw* occurs in fragment 11, I, 10, q.v.

(b) With the traces of *ist,* 'ground', cf. the better preserved occurrence of that noun in fragment 2, line 4.

(c) The verb after *is*[*t*] is perhaps *wnḥ,* 'appear, be revealed'. Cf. *Glossar*, p. 92.

Line 20

(a) For the semi-hieratic writing of the verb *ḥmsy,* 'sit', here, see fragment 8, line x+9, P. Carlsberg 10029, line x+6. The following suffix pronoun is clearly *f* on the original, and not *w* as the photograph might suggest.

(b) For the traces of *bꜣ,* 'thicket', see reference cited in note (c) on line 18 above.

Line 21

The verb *sty,* 'scatter' is clear on the original, immediately below *ꜣsy,* 'reeds', in the line above. Cf. better preserved examples of the same word in line 13 above, and fragment 15, line x+3.

Line 22

(a) With the traces of *ẖn,* 'in', at the beginning of the line, cf. better preserved examples of the same preposition in fragments 2, line 8; 4, II, 10; and 5, I, 4.

(b) The traces in between the two breaks near the middle of the line might conceivably be the top of a man with hand to mouth determinative followed by the definite article *tꜣ*.

(c) With the traces that I have tentatively read as *rn*, 'name', cf. better preserved occurrences of that noun in lines 11 and 12 above.

(d) The amount of space between the two short strokes and the *f* at the end of the line indicates that *i-*[*ir*]*=f* should be restored. Cf. the example of *i-ir=f* in line 11 above.

Line 23

For the writing of *ḥf,* 'serpent', near the end of the line, cf. the better preserved occurrence of that noun in line 16 above.

Commentary

This fragment begins with a reference to a plant, concerning which it is said: *rt=f m-sꜣ pꜣ šʿṯ=f r-hn-r pꜣ hrw,* 'It will grow, after being cut off, up until today.' One has here an aetiological statement of a type which occurs frequently in our text. On the function of such statements, see Introduction. Presumably the preceding section of the cosmology, now lost, explained how the plant in question acquired the ability to replace its lost growth. The name of this plant has not been preserved. Lines 5 and 18 below refer to a reed or rush (*qmꜣ*), lines 12 and 20 to a reed (*ꜣsy*). It could have been either of these, or a different plant entirely.

The partially preserved word [..]*ll* at the beginning of line 2 denotes some type of prayer, invocation, or expression of acclaim; see textual note (a) ad loc. According to the text, this is uttered in the presence of *pꜣ-Šy pꜣ Nwn*. The precise syntactical relationship of these two names is uncertain. In theory, they could simply be coordinates. However, both can refer to the same being. Cf. the demotic magical spell on an ostracon in Strasbourg, where Ptah is called *Ptḥ tny iṯ ntr.w pꜣ-Šy ʿꜣ pꜣ Nwn nt ḥr pꜣ ꜣsṯ,* 'Ptah-(Ta)tenen, father of the gods, Pshai the great, the Primaeval Ocean, who is upon the basis'.[36] Equally, the deity Pshai can be said to be within the ocean. The Demotic Magical Papyrus of London and Leiden invokes *pꜣ-Šʿy nt ẖn pꜣ Nwn*, 'Pshai who is in the Primaeval Ocean'.[37] Contemporary Greek and Old Coptic magical spells preserve similar invocations.[38] Thus one has three possible translations for the words *pꜣ-Šy pꜣ Nwn* in the passage under consideration: 'Pshai (and) the Primaeval Ocean', 'Pshai the Primaeval Ocean', and 'Pshai (in) the Primaeval Ocean'. The first of these can be eliminated, I think, since the words *Ptḥ pꜣy=k šr*, 'Ptah your son', in the ensuing line

[36] Spiegelberg, *ZÄS* 49 (1911), pp. 35-7.

[37] London-Leiden Magical Papyrus, II, 5 (Griffith and Thompson, *The Demotic Magical Papyrus of London and Leiden* 2, plate 2). Cf. Quaegebeur, *Le Dieu égyptien Shaï*, p. 168.

[38] See Quaegebeur, *op. cit.*, pp. 169-70; Thissen in Verhoeven and Graefe (eds.), *Religion und Philosophie im alten Ägypten*, p. 300; Fauth, *Helios Megistos*, p. 64.

make it plain that a single deity is being addressed here. Accordingly, it is either 'Pshai the Primaeval Ocean' or 'Pshai (in) the Primaeval Ocean' who is identified as the father of Ptah. In order to avoid prejudging the issue, I have referred to this deity throughout as *pꜣ-Šy pꜣ Nwn.*

Fragment 2 credits Ptah with the creation of the eight members of the Hermopolitan Ogdoad, which is the next stage in the process whereby the rest of the cosmos comes into being. According to fragment 4, a still later stage in this process is the creation of Pshai.[39] Thus that deity is at the same time both the father of Ptah and one of his descendants. Similarly, Theban cosmogonical speculation of the Graeco-Roman Period identifies Amun as both the father and a descendant of Ptah-Tatenen.[40] In fact, he can be that god's parent, son, grandson, or even identified with him. Elsewhere, Amun and other deities frequently bear epithets like *wtṯ it=f,* 'begetter of his father', and *wr r qmꜣy.w sw,* 'elder than those who created him'.[41]

The anomaly here is explicable in terms of the Egyptian conception of creation. As outlined in the Introduction, this was viewed, not as a process whereby things came into being from nothingness, but rather as a form of revelation, in which things possessing a latent existence were made manifest and thereby brought into full existence.[42] Thus, it is possible to envisage a deity like Pshai, as *pꜣ-Šy pꜣ Nwn,* initiating creative activity by producing offspring while still concealed within the waters of the Primaeval Ocean, but not actually becoming manifest himself until some generations later. An allusion to such an anterior role of that god is perhaps to be recognised in a Greek magical papyrus where he is invoked as ὁ ἐν τῷ ὠκεανῷ ὀχεύων, 'the one who accomplished the act of procreation in the ocean'.[43]

In lines 3-4, Ptah is called *i-ir ti ḫpr n=n ḫlꜣ,* '(the one) who created sustenance for us', and *i-ir šʿ ir nꜣ iꜣw.wt nhm iw=w sy,* 'who first made the cattle rejoice, they being satisfied'.[44] One has here a pair of allusions to the role of Ptah as creator of material abundance. This role is treated at length in the demotic cosmogony of P. Berlin 13603,[45] and is widely attested in other sources as well.[46] Material abundance is one

[39] See commentary ad loc.

[40] Cf. Sethe, *Amun und die acht Urgötter von Hermopolis*, pp. 56-61; Sandman Holmberg, *The God Ptah*, pp. 171-3.

[41] Cf. Zandee, *Der Amunhymnus des Papyrus Leiden I 344, verso*, pp. 24-6; Hornung, *Der ägyptische Mythos von der Himmelskuh*, pp. 4 and 38; Bergman, *Ich bin Isis*, pp. 280-2; de Buck, *The Egyptian Coffin Texts* 4, p. 76c.

[42] Cf. Finnestad, *Image of the World and Symbol of the Creator*, pp. 68-74 and 113-14.

[43] PGM IV, 1642-3 (Preisendanz, *Papyri Graecae Magicae* 1, pp. 124-5). Cf. Quaegebeur, *Le Dieu égyptien Shaï*, p. 169; Fauth, *Helios Megistos*, p. 70.

[44] For the syntax of the latter epithet, see note (b) on line 4.

[45] P. Berlin 13603, IV, 1-5 (Erichsen and Schott, *Fragmente memphitischer Theologie in demotischer Schrift*, pp. 316-17 [18-19] and plate 6).

[46] See e.g. Sandman Holmberg, *The God Ptah*, pp. 39-42; Bickel, *La Cosmogonie égyptienne avant le Nouvel Empire*, pp. 137-42.

of the four kas or boons which were supposed to have been granted to human beings, along with a long life, progeny, and a good burial.[47] Of the various divinities who are said to be responsible for creating these benefits, Ptah is the most prominent.[48]

The words *n=n*, 'for us', in line 3 must refer to those who are addressing *pꜣ-Šy pꜣ Nwn*. Their identity is uncertain. One could envisage either a group of deities or one of humans praying to or acclaiming that god. If it is the latter, then *iꜣw.wt*, 'cattle', in the following line should probably be taken in its literal sense, i.e. beasts as opposed to men. If the former, then 'cattle' could refer to human beings as distinct from divine ones; cf. *Wb.* 1, 171, 1. The demotic cosmogony of P. Berlin 13603 likewise begins with an address to the deity who is responsible for initiating the process of creation. In that case, the deity in question is Ptah.[49] Only a single speaker is involved there;[50] his or her identity is unknown. The Berlin text records at one point that the children of Ptah 'prayed before him' (*ir=w šll i-ir-ḥr=f*).[51] Unfortunately, only a few words of this prayer survive. According to Schott, the children of Ptah are the human race. However, elsewhere this is often a designation of the Ogdoad.[52] Conceivably, it may be a divine corporation of this sort that addresses *pꜣ-Šy pꜣ Nwn* at the beginning of our text. The cosmogony preserved on the inner face of the enclosure wall on the north side of the temple of Edfu is accompanied by a relief which depicts, among other beings, a group of twenty-four deities, including the Ogdoad, Thoth, and Seshat, who are present at its recitation.[53] It is equally possible that a larger group like this speaks in the passage which concerns us.

Line 5 opens with a reference to a plant called *qmꜣ*. According to Charpentier, *Recueil de matériaux épigraphiques relatifs à la botanique de l'Égypte antique*, p. 722, this is specifically the *Juncus maritimus*. In point of fact, however, the word probably has a wider range of meanings than this, just as ⲕⲁⲙ, the Coptic descendant of *qmꜣ*, can correspond to both Latin *juncus* and Arabic حلفا.[54] My translation 'rush' is purely conventional. For the possibility that *qmꜣ* here is a sort of reed, see discussion of line 12 below.

[47] See Meeks, *RdE* 15 (1963), pp. 35-47; Sauneron, ibid., pp. 60-1; idem, *BIFAO* 57 (1958), pp. 163-4; Wild, *BIFAO* 54 (1954), pp. 201-6; Gutbub, *Textes fondamentaux de la théologie de Kom Ombo*, pp. 296-7; Jasnow, *Enchoria* 12 (1984), pp. 8-9.

[48] Meeks, art. cit. For Ptah as provider of kas, see further Sandman Holmberg, *The God Ptah*, pp. 39-40; de Buck, *The Egyptian Coffin Texts* 6, p. 268i-k; Berlandini, *RdE* 46 (1995), p. 17.

[49] P. Berlin 13603, I, 1-6 (Erichsen and Schott, *Fragmente memphitischer Theologie in demotischer Schrift*, pp. 310 [12] and 347-8 [49-50] with plate 5).

[50] Cf. the vocatives *pꜣy=y ḥry*, 'my master', in P. Berlin 13603, II, 19, and *pꜣy=y nb ꜥꜣ Ptḥ*, 'my great lord Ptah', ibid., IV, 10 (Erichsen and Schott, *op. cit.*, pp. 314 [16] and 317 [19] with plates 5-6).

[51] P. Berlin 13603, II, 30 (Erichsen and Schott, *op. cit.*, p. 315 [17] and plate 5).

[52] See references cited in note 97 of the commentary on fragment 2.

[53] This relief is illustrated in Finnestad, *Image of the World and Symbol of the Creator*, plates 1-2. For the complete list of deities involved, see ibid., p. 173; Chassinat, *Edfou* 6, p. 185, lines 3-10.

[54] Cf. Keimer, *Die Gartenpflanzen im alten Ägypten* 2, p. 75; idem, *OLZ* 30 (1927), p. 145.

The text continues with the words *[t]lg(?)=f s ʿnḫ(?)=w n-im=s irm* [...], 'He [up]rooted(?) it so that they might live(?) on it together with [...]', an enigmatic statement which is not made any clearer by the uncertainty over the reading of the two verbs in it.[55] In view of what is said in the two preceding lines about Ptah causing the cattle to rejoice in being sated, it seems probable that the suffix pronoun *f* here refers to him and the *w* to them. In other words, this sentence is still part of the laudatory characterisation of that god uttered by those who are addressing his father. The dependent pronoun *s* and the suffix *s* of *n-im=s* should refer to one and the same thing. This cannot be *qmꜣ*, since that noun is masculine. Rather, they must refer to a feminine noun, now lost in the break at the beginning of the line. If this was something like *bꜣ*, 'thicket' (cf. lines 8, 18 and 20 below), and if the readings *[t]lg* and *ʿnḫ* are correct, then one may have had here a statement that Ptah plucked out or uprooted a thicket of rushes or reeds to provide sustenance for the cattle. Could this possibly refer to the fabrication of the *ḏd*-pillar, an object closely associated with Ptah and sometimes said to be made, at least in part, of cut off reed stems, which symbolises enduring life and fertility?[56]

Whatever the case, line 6 continues: *pḥ=f r wʿ.t lhny n šʿꜣ ʿḥʿ=f*, 'He reached a resting place of sand and stood up.' To whom do the third person singular masculine suffix pronouns refer here? In theory, it could be either *pꜣ-Šy pꜣ Nwn* or Ptah. However, since the subsequent fragment recounts the creative activity of the latter in some detail, one might expect the present fragment to treat that of his father. Perhaps the name of the older deity stood in the break preceding the sentence under consideration. If so, then our text moves from direct address to past narration set in the third person at this point. A further instance of direct address occurs in line 7 below, but the remainder of the text thereafter is either narrative or of a didactic nature.

The question of the relationship between the initial address and what follows is a difficult one. Presumably the former is part of a frame or setting for the ensuing narrative/didactic material. But precisely how are the two connected? Is the latter uttered by *pꜣ-Šy pꜣ Nwn* in answer to a prayer or request? Schott interpreted the cosmogonical narrative of P. Berlin 13603, which begins with an address to the creator god Ptah, in such a fashion. According to him, it was 'von Ptah selbst diktiert'.[57] The difficulty with this view, in my opinion, is that all references to that deity in the narrative portion of the Berlin papyrus are in either the second or the third person, which makes it unlikely that he is actually the speaker there. The same objection could

[55] For discussion of these, see notes (c) and (d) on line 5 above.

[56] See H. Altenmüller in *Lexikon der Ägyptologie* 1 (1975), columns 1100-1105 (s.v. Djed-Pfeiler); Sandman Holmberg, *The God Ptah*, pp. 154-66; Hermsen, *Lebensbaum Symbolik im alten Ägypten*, pp. 148-50; Berlandini, *RdE* 46 (1995), pp. 25-8 and 37-9.

[57] See Erichsen and Schott, *Fragmente memphitischer Theologie in demotischer Schrift*, pp. 347-8 (49-50).

be raised in connection with our text, since here, apart from those in the opening lines of the fragment under consideration, all references to *p3-Šy p3 Nwn* (or simply *p3-Šy* as he is called later), occur in the third person.

Another possibility is that the narrative/didactic material in our text represents a recitation uttered as a form of worship, i.e. after their initial address to the creator god, the speakers (or perhaps someone else) go on to extol that deity's greatness by recounting the story of his creation before him. This would explain the shift from second person to third person references to *p3-Šy p3 Nwn* in the text. The brief reversion to direct address in line 7 could then be seen as an aside to the creator. One could envisage such an account either as a spontaneous production or as something read out from a written document. Compare the Edfu cosmogony mentioned above, which is called a *mi.t sš ir.n Ḏḥwty ḫft ḏd.n ḏ3is.w nw Mḥ.t-wr.t*, 'copy of a writing which Thoth made in accordance with what the *ḏ3is.w* of Mehetweret said'.[58]

To return to line 6, the words 'He reached a resting place of sand and stood up' mark the first stage of the process whereby the universe is brought into existence. In order to accomplish this task, the creator god requires a platform in the midst of the waters of the Primaeval Ocean, on which he can stand. Thus, in Coffin Text Spell 80, the primordial god Atum refers to a time before the cosmos came into being: *sk wi wʿ.kwi ḥnʿ Nww m nnw.t n gm.n=i bw ʿḥʿ=i im n gm.n=i bw ḥms=i im,* 'when I was alone together with the Primaeval Ocean in a state of inertness, unable to find a place on which to stand or sit'.[59] Closer in time to our own text, in the Neith cosmogony at Esna, that goddess is made to say: *ḥ3 n=i bw pn m s3ṯw m ḥr-ib Nwn rhn=i ḥr=f,* 'Would that I had this place as a platform in the midst of the Primaeval Ocean so that I might rest upon it.' The desired platform duly emerges in the shape of a mound and Neith ascends it.[60] *s3ṯw* (*Wb.* 3, 423-4) is a technical term used for a basis or platform of this nature.[61] In demotic, the word appears as *ist* or *3sṯ*. See, for instance, line 19 below and fragment 2, line 4; also the Strasbourg ostracon cited in connection with line 2 above, where Ptah-(Ta)tenen is called *p3-Šy ʿ3 p3 Nwn nt ḥr p3 3sṯ*, 'Pshai the great, the Primaeval Ocean, who is upon the basis'. It is probable that *lhny* in the passage under consideration is a synonym for this word, and that our text is describing the initial arrival of the creator god upon the platform from which he brings the rest of the cosmos into existence.

[58] Chassinat, *Edfou* 6, p. 181, lines 10-11.

[59] De Buck, *The Egyptian Coffin Texts* 2, p. 33e-g. Cf. Bickel, *La Cosmogonie égyptienne avant le Nouvel Empire*, pp. 62-4.

[60] See Sauneron, *Esna* 3, p. 29 (206,2); idem, *Esna* 5, p. 255, note (b); Ramadan el-Sayed, *La Déesse Neith de Saïs* 1, p. 61.

[61] For the primaeval *s3ṯw* in other texts, see e.g. Drioton, *ASAE* 44 (1944), pp. 126-7, 131, and 135; Zandee, *De Hymnen aan Amon van Papyrus Leiden I 350*, pp. 25-7 and plate 2; Sethe, *Amun und die acht Urgötter von Hermopolis*, p. 118; Reymond, *The Mythical Origin of the Egyptian Temple*, p. 66; de Buck, *De Egyptische Voorstellingen betreffende den Oerheuvel*, pp. 11 and 47.

The deity's resting place is said to be made of *šʿꜣ*, 'sand'. Similarly, the Book of the Fayyum says of this substance: *šʿ šd m š sꜣṯw pw*, 'The sand which is protected in the sea (= its floor) is the basis.'[62] The *sꜣṯw*, originally a platform in the midst of the waters, subsequently forms the summit of the primaeval mound, the first land, when this rises up from the depths. Egyptian sources frequently identify sand as the constituent material of that mound.[63]

Line 7 preserves the words *iw=f r rt*, 'it will grow'. This is part of a speech, to judge from the preceding *ḏ*, 'saying'. One could imagine the words in question being uttered by *pꜣ-Šy pꜣ Nwn* in connection with the *qmꜣ*, 'rush, reed', mentioned two lines earlier. Cf. the statement *rt=f m-sꜣ pꜣ šʿṯ=f r-hn-r pꜣ hrw*, 'It will grow, after being cut off, up until today', in line 1 above, which could have reference to the same plant. The *iw=k* which occurs later on in line 7 may be part of an aside addressed to the creator by the recitants of the text, as I have suggested above.

Line 8 describes how the aforesaid plant grew into a large thicket (*rt=f n wʿ.t bꜣ ʿꜣ.t*). The partially preserved verb *wpy*, 'open', at the beginning of the next line could refer to its flowering. The subject of the ensuing sentence, *ḫꜣy=f n tꜣ ly.t ḥry.t* ..[...], 'He/it rose high above .. [...]', could be either the plant or the creator god *pꜣ-Šy pꜣ Nwn*. If the latter, then perhaps this and the sentence which follows, [*iw*]*=f iw r-ẖly* [..]*=f ḥr-ꜣṯ=f*, '[H]e came down and [..] upon it', describe how that deity first rose up and then descended and settled upon the plant in question. One is reminded of the cosmogony of the Edfu temple, according to which a drifting mass of reeds (*nbi*) appeared in the Primaeval Ocean. A part of this mass separated from the rest and came to a stop in a particular place. The immobilised thicket, which is given the name 'floater' (*ḏbꜣ*), was seen by the creator god hovering in falcon form overhead, who flew down and alighted upon it. From this perch, the rest of creation was summoned into existence.[64]

Line 11 of our fragment begins with the words [... *Ḫmn*]*w n rn rḫ=f s ḏ i-ir=f pr n Yb*, '[... Hermopo]lis by name. He knew that it came forth from Elephantine.' I assume that 'he' here is the creator *pꜣ-Šy pꜣ Nwn* and 'it' the plant mentioned in the preceding lines. The ensuing sentence, *wꜣḥ(?)=f r pꜣ mḥt*, 'It came to rest(?) in the north', should refer to the plant as well. Cf. the Edfu cosmology cited above, in which a floating mass of reeds comes to rest and is fixed or stabilised in one place as a prelude to creation. From

[62] Line 1184 (Beinlich, *Das Buch vom Fayum*, pp. 256-7 and 294).

[63] See references cited in note 61 above and in the commentary on fragment 6.

[64] See Chassinat, *Edfou* 6, pp. 181, line 14 -182, line 3 (there are parallels in other inscriptions from the same temple). On the Edfu cosmogony, see Finnestad, *Image of the World and Symbol of the Creator*, especially pp. 27-9, 34, 41, 43, 47-8, and 54-6, on the significance of the reeds and floater. Other studies of the same material include Kurth in Verhoeven and Graefe (eds.), *Religion und Philosophie im alten Ägypten*, pp. 189-202; Goyon, *Les Dieux-gardiens et la genèse des temples* 1, pp. 3-44: Barucq, *BIFAO* 64 (1966), pp. 125-67 (presenting the unfinished work of Maurice Alliot); and Reymond, *The Mythical Origin of the Egyptian Temple*, passim. Kurth, art. cit., p. 197, argues that *ḏbꜣ* does not mean 'floater' but is simply a synonym of *nbi*. Cf. Wilson, *A Ptolemaic Lexikon*, pp. 502-3 and 1230.

the Egyptian point of view, bearing in mind the direction of the Nile current, it is natural to think of any floating or drifting object as moving from south to north. Perhaps our text named Hermopolis as the place where the plant's movement ceased. In the Edfu cosmogony, the reed thicket is stabilised at the point where the first land subsequently emerges.[65] In fragment 6, line 6 of our cosmology, this land, the primaeval mound, is explicitly identified as the 'great sand of Hermopolis'.[66]

The text says of the plant which has come to rest: *ḏ=w n=f ꜣsy n rn*, 'It was called "reed" by name.' *ꜣsy*, written with tree determinative, is identical with the *isw*, 'Schilfrohr', of *Wb.* 1, 127, 21-2. In theory, it could also be a writing of *isr*, 'tamarisk' (= Coptic **ⲟⲥⲓ**), for which see ibid., pp. 114-15. However, the context here requires an aquatic or semi-aquatic plant. The writing of *ꜣsy* with tree determinative is employed consistently in our text. In much the same way, *nbi* and the other words used to describe the reed thicket in the Edfu cosmogony are regularly written with the tree determinative.[67] Presumably the purpose, in both sources, is to indicate the substantial character of the growths involved.

The question inevitably arises: how many different plants are to be distinguished in the narrative thus far? It is impossible to answer this question with certainty. For simplicity's sake, one might assume that all references to plants in the fragment under consideration have to do with the same one, i.e. only a single plant is involved. There is nothing to preclude this being the case. Two distinct botanical terms occur in fragment 1: *qmꜣ* and *ꜣsy*. Charpentier, *Recueil de matériaux épigraphiques relatifs à la botanique de l'Égypte antique*, p. 112, identifies *isw*, the hieroglyphic ancestor of *ꜣsy*, specifically as the name of the reed *Phragmites communis*, although no authority is cited for this identification. *qmꜣ*, as noted in the discussion of line 5 above, has a Coptic counterpart **ⲕⲁⲙ**, which can be used to denote more than one species of plant, including the rush and halfa grass. Occasionally it signifies 'reed' as well; cf. references cited in Crum, *CD*, p. 108a. It is possible, therefore, that *qmꜣ* and *ꜣsy* in our text are simply variant names for the same plant, a reed or reed thicket, with the latter being a new additional appellation given to the thicket after it has come to rest. Compare the use of two names (*nbi, ḏbꜣ*) to denote the reed thicket in the Edfu cosmogony.

Immediately after the naming of the newly immobilised plant, the text continues with the words *nw nꜣ nṯr.*[*w* .....], 'The god[s] looked [.....].' This statement is problematic inasmuch as, from the point of view of our narrative, the creation of the gods has yet to occur. Two possible explanations suggest themselves. Either *nw* is prospective rather than past *sḏm=f* and one should translate 'that the god[s] might (at some future time) look [.....]', or the deities in question are part of an earlier

[65] Cf. Finnestad, *op. cit.*, pp. 80-1.

[66] See commentary ad loc.

[67] See Kurth, art. cit., p. 196. In e.g. Chassinat, *Edfou* 6, p. 181, line 16, both *nbi* and *ḏbꜣ* are so determined.

generation of divine beings. Cf. the obscure pair *W3i*, 'the distant one', and *ʿ3*, 'the great one', who, according to the Edfu cosmogony, existed at the time of the appearance of the floating reed thicket and helped to stabilise it.[68]

Line 13 says: *sḫny=f r-r=s sty=f t3*[*y=f*] *mtw*[*y*], 'He settled upon it. He scattered hi[s] see[d].' The pronoun 'he' presumably refers once again to *p3-Šy p3 Nwn*. Thus far, no female deity has been mentioned in our narrative. I wonder whether the suffix pronoun *s* of *r-r=s* does not refer to the immobilised mass of reeds, here envisaged as a *b3.(t)*, 'thicket' (cf. line 8 above), and thus treated as grammatically feminine. For the scattering of a god's semen onto a living plant, compare the demotic cosmogony of P. Berlin 13603, in which the god Amun is said to ejaculate his seed into the great lake of Hermopolis, where it fertilises a lotus from which the sun emerges in the form of a child.[69]

Line 14 is very damaged, but may contain another reference to the *lhny*, 'resting place', first mentioned in line 6 above. See commentary on the latter. Line 15 speaks of 'the gods' (*n3 ntr.w*) in a broken context. On the possible identity of these deities, see discussion of line 12 above.

A serpent (*ḥf*) is mentioned in line 16. The context is too poorly preserved to permit the identity of this creature to be established with certainty. It could be *p3-Šy p3 Nwn* himself, since the serpent form is characteristic for Pshai.[70] Alternatively, it could be Ptah, who is mentioned by name two lines below. If I am correct in assuming that the sexual act described in line 13 is predicated of *p3-Šy p3 Nwn*, then one expects here some reference to the creation of his offspring Ptah, which may well have been the subject of this and the adjacent lines.[71] In some Egyptian sources, Ptah is explicitly identified with the primordial serpent called *Ỉr-t3*, 'Earth-maker', who is himself the offspring of a yet older serpent called *Km-3t=f*, 'He who has completed his moment'.[72] According to the cosmogony preserved in the temple of Khonsu at Karnak, the latter (there called *ḥf tpy*, 'the first serpent') created the sky and the sky spat forth an egg, from which *Ỉr-t3* (called *ḥf sn-nw*, 'the second serpent') emerged.[73]

A third possibility is that the serpent in our text is an inimical being of some sort. The Edfu cosmogony, in common with other Egyptian sources, refers to a combat between the primaeval creator and a monstrous serpent at the dawn of time. The latter

[68] Kurth, art. cit., p. 194; Finnestad, *Image of the World and Symbol of the Creator*, pp. 26-9 and 60-1.

[69] P. Berlin 13603, II, 6-7 (Erichsen and Schott, *Fragmente memphitischer Theologie in demotischer Schrift*, pp. 312-13 [14-15] with plate 5). For further parallels, see commentary on fragment 3.

[70] See Quaegebeur, *Le Dieu égyptien Shaï*, pp. 53, 84-5, 143-4, 163-4, and 268.

[71] For Ptah as the son of *p3-Šy p3 Nwn*, see line 3 above.

[72] See Sandman Holmberg, *The God Ptah*, pp. 185-7; Sethe, *Amun und die acht Urgötter von Hermopolis*, pp. 26-7 and 56-7.

[73] See Sethe, *op. cit.*, p. 63 and plate 2; Parker and Lesko in Baines et al (eds.), *Pyramid Studies and other Essays Presented to I.E.S. Edwards*, p. 169 and plate 34; Cruz-Uribe, *JARCE* 31 (1994), pp. 170, 173, 177; Zivie-Coche in *Hommages à Jean Leclant* 4, pp. 426-7.

is a symbol of chaos, who must be overcome so that the subsequent stages of creation can proceed in an orderly fashion.[74]

The botanical term *qmꜣ* is visible at the beginning of line 18. For the possibility that this should be translated 'reed' rather than 'rush' here, see discussion of line 12 above. The text continues with the words *ḏ(?) Ptḥ mt(?) r ḫpr ḥr* [..] *tꜣ bꜣ* [......], 'Ptah spoke(?) a word(?) to come into being upon [..] the thicket [......].' Two lines later, in line 20, it says: *ḥmsy=f ḥr tꜣ bꜣ n ꜣsy*, 'He sat down upon the thicket of reeds.' I assume that Ptah is the subject of this sentence as well. If so, and if the interpretation of lines 9-10 proposed above is correct, then that god is said to settle down upon the reed thicket alongside, or in succession to, *pꜣ-Šy pꜣ Nwn*. I am not aware of any other ancient Egyptian source in which Ptah is described as having done anything of this sort.[75] However, in his form as Ptah-Tatenen, known from the Ramesside Period onwards, he is identified with the risen land that emerges from the Primaeval Ocean.[76] In the Edfu cosmogony Tatenen is said to emerge from the water and 'arrive at' (*sꜣḥ*) the floating reed thicket.[77] As Finnestad aptly remarks, 'This means that the water has subsided so as to lay bare the earth at the place where the reed is, so that it is no longer in the water.' The earth rising out of the water, according to her, 'is the main theophany' of the god.[78]

Obviously, the act of emerging from the water beneath a reed thicket, as described in the Edfu cosmogony, and that of settling upon it, as narrated in our text, are two quite different things. Nevertheless, the result is the same in both cases: Ptah/Tatenen reaches the thicket of reeds. Accordingly, I wonder whether lines 18-20 of the fragment under consideration do not parallel the Edfu text in somehow recounting the theophany of Ptah as the newly risen land, or at least his initial appearance in association with it. This could help to explain the partially preserved sentence in line 19: [....] *r-b-ir-tw pꜣ is*[*t*] *wn*[*ḥ*(?)] ..[......], '[....] before the grou[nd] appear[ed(?)] ..[......].' The *ist* (< earlier *sꜣṯw*) is the platform or basis in the depths of the Primaeval Ocean on which the creator god stands in order to initiate the process of creation.

[74] See Goyon, *Les Dieux-gardiens et la genèse des temples* 1, pp. 112 and 121-4; Sternberg-El Hotabi, *Der Propylon des Month-Tempels in Karnak-Nord*, pp. 38-9, note 9; idem, *Ein Hymnus an die Göttin Hathor und das Ritual 'Hathor das Trankopfer Darbringen'*, pp. 43-4. On the connection between serpents and evil in general, see Goyon, *op. cit.*, p. 182.

[75] For a survey of Egyptian deities who sit or rest upon plants, see Schlögl, *Der Sonnengott auf der Blüte*, pp. 11-16.

[76] The syncretism of Ptah and Tatenen is first attested in the reign of Ramesses II; see Schlögl, *Der Gott Tatenen*, p. 62, note 47. On the equation of that deity with the risen land, see ibid., pp. 71-2; Sandman Holmberg, *The God Ptah*, pp. 56-63; de Buck, *De Egyptische Voorstellingen betreffende den Oerheuvel*, pp. 49-62.

[77] Chassinat, *Edfou* 6, p. 183, line 3; idem, *Edfou* 3, p. 7, lines 10-11.

[78] *Image of the World and Symbol of the Creator*, pp. 80-1. For translations of the passage in question, see ibid., pp. 34 and 54-5; Barucq, *BIFAO* 64 (1966), pp. 145-6.

Subsequently, this forms the summit of the primaeval mound, the first land to emerge from the waters of the Ocean.[79] Perhaps the sequence of events narrated in our text was this: (1) Ptah spoke a word in order to come into being upon the reed thicket 'before the grou[nd] appear[ed]'; (2) he sat down upon the thicket of reeds (and the primaeval mound arose).

The verb *sty*, 'scatter', occurs in a broken context in line 21. Presumably this referred to the scattering of the male deity's seed, as in line 13 above. Reference is made to *pꜣ ḥf n rn=f*, 'the aforesaid serpent', at the end of line 23. Again, the context is damaged. This serpent must be the same one as that mentioned in line 16. For the identity of the latter, see commentary ad loc.

FRAGMENT 2 — Plate 3

Dimensions: 15.5 × 13.9 cm. Colour: varies from dark brown (left) to medium brown (right). Related fragments: No. 15.

Transliteration

1. [....]=*f/pꜣy* (vac) [...........] *n tmy pꜣy i-ir=f ḏ.ṱ=s ḥr*
2. [....].. *nt ḥr* [...........] *ḫpr tꜣ bꜣ n ꜣsy ẖn*
3. [...] *ꜣsy ḫpr* [...........].. *n tmy nb iw=f mtw=f*
4. [....]. *r-ḏb*[*ꜣ* ..] *m-sꜣ* [.......... *r*]*ṱ=w ḥr pꜣ ist*
5. [....].*w n lhny* [....... *m-sꜣ*] *nꜣy ir Ptḥ mꜣwy*
6. [....] *ir=f hyn*[*y.w* .... *st*]*y=f tꜣy=f mtwy*
7. [.. *t*]*y*(?) *ḫpr=f n pꜣy=w qty* ..[........ *ḫ*]*pr pꜣy=w ṱꜣ n ḫꜥ*
8. [....] *ntr 8 r-bnr ẖn=w ḥw*[*t*] *4* [*n ḥr n qll*] *s-ḥm.t 4.t n ḥr*
9. [*n ḥf* .. *i*]*r*(?) *Ptḥ pꜣy=w iṱ twt*(?) *ḏ nꜣy=*[*n*] *ḫ*[*m mn*]*y ḏ=w n=w nꜣ*
10. [*Ḫmnw r-ḏbꜣ pꜣ*] *rn nꜣy=n ḫm mny r-ḏ=f* [*r-r=w*(?)] ⸢*ḫ*⸣*m-ẖl.w nꜣ.w i-ir=f*
11. [... *mn*]*y rn n twt n in*[*y*] *pꜣy r-ḏbꜣ ti ꜥḥꜥ*(?)*=f n rmt mtw=f* ⌞*ḫpr*⌟(?) [*n*(?)]*=f*
12. [....] *tꜣ pꜣ nt-iw=f ḏ n-im=f ḏ* ... *pꜣy=w twt nꜣ Ḫmnw n rn=w*
13. [... *hr*]*w*(?) *nꜣ.w tw=s nꜣy=w rn ꞽImn ꞽImn.t Ḥḥ Ḥḥ.t Kk Kky.t Nwn*
14. [*Nwn.t pꜣ rn*] *n pꜣ 8 ntr ꜥn nꜣ.w pꜣ rn n pꜣ 4 ḥwt* [*mt*]*re=w r pꜣ 4 rꜣ*
15. [... *pꜣ*] *rn n tꜣ 4.t s-ḥm.t mtre=w r pꜣ 4 r*[*ꜣ*] *tꜣ p.t i-ir Ptḥ pꜣy=w*
16. [....] ⌞*ḏ/rn*⌟(?) *4 ḏ=w pꜣ rn n pꜣ Nw*[*n* ....].. *pꜣ t*[*ꜣ*(?) *šꜥ*(?)] *ḏ*[*t*] *m-qt*[*y*]
17. [........] ⌞*iṱ*⌟ *šꜥ-h*[*n*]*-r pꜣ hrw* [...........]
18. [...............] *s*[*ẖ* ....]

Translation

1. [....] him/it. (vac) It is [...........] of a town. He said it concerning
2. [....].. which is on [...........]. The thicket of reeds came to be in

[79] See discussion of line 6 above.

3. [...] reeds. There occurred [...........].. in/of every town which is his.
4. [....]. becau[se ..] after [..........] their [fe]et on the ground
5. [....].*w* as a resting place [........ After] this, Ptah took thought
6. [....] He fashioned som[e ....]. He [scatter]ed his seed.
7. [.. he]re(?). He came to be around them ..[........]. Their time of appearance [occ]urred.
8. [....] eight divinities out of them, four male[s with frog heads], and four females with heads
9. [of serpents ..]. Ptah their father [ma]de(?) an assemblage/union(?), saying: Ou[r] lit[tle imag]es. They were called the
10. [Ogdoad in respect of the] name 'Our little images' which he uttered [concerning them(?)]. They are ⌜chi⌝ldren. He
11. [... ima]ge. It is a name for a statue of ston[e] because of erecting(?) it for a man that it might ⌞exist⌟(?) [for(?)] him.
12. [....] land, what he says, namely ... their union, the Ogdoad according to their names.
13. They are [... da]y(?). Here are their names: Amun, Amaunet, Heh, Hauhet, Kek, Kauket, Nun,
14. [and Naunet]. These again are [the names] of the eight divinities. The name(s) of the four males [me]t at the four doors
15. [... The] name(s) of the four females met at the four do[ors] of the sky. Ptah their
16. [....] ⌞say/name⌟(?) four. They said the name of the Primaeval O[cean ....] .. the la[nd(?) for(?)] ev[er] lik[e]
17. [........] ⌞father⌟ up to today [...........]
18. [...............] wri[te ....]

Textual notes

Line 1

Apparently nothing was written originally in the space immediately after *=f/pꜣy* near the beginning of the line.

Line 2

(a) The first sign visible here is the plant determinative. Cf. e.g. the writing of *bꜣ*, 'bush, thicket', near the end of the line, which is perhaps to be restored here.

(b) The combination *bꜣ n ꜣsy*, 'reed thicket', also occurs in fragment 1, line 20, q.v. For the writing of *ꜣsy*, see commentary ad loc.

Line 3

I owe the reading *ꜣsy* to Kim Ryholt. Cf. the better preserved instance of that word in the line immediately above.

Line 4

(a) For undamaged writings of *r-ḏbꜣ* elsewhere in the text, cf. e.g. line 11 below, fragment 4, II, 15, fragment 12, I, 15, fragment 13, II, 15. Probably restore *r-ḏb*[*ꜣ.ṱ=s*] or similar.

(b) I have restored [*r*]*ṱ*=*w* before *ḥr pꜣ ist* following a suggestion of Kim Ryholt. According to him what looks like a vertical stroke in front of the flesh determinative is an illusion created by damage to the papyrus surface. For better preserved writings of *rṱ*, 'feet', elsewhere in the text, see fragments 3, line 20, and 36, line 1.

Line 5

(a) The *w* visible before *lhny* could be either the plural marker or a suffix pronoun. For the noun *lhny*, 'resting place', see also fragments 1, line 6, and 4, II, 11.

(b) For the suggested restoration [*m-sꜣ*] *nꜣy*, see e.g. P. Carlsberg 462, II, 16 (Smith in Frandsen and Ryholt [eds.], *The Carlsberg Papyri 3: A Miscellany of Demotic Texts and Studies*, plate 15), P. Carlsberg 10040, line x+9, P. Carlsberg 257, line 1.

Line 6

The damaged verb before *tꜣy=f mtwy* is *sty*, 'scatter'. Cf. better preserved occurrences of the phrase *sty=f tꜣy=f mtwy* in fragment 1, line 13, and fragment 11, I, 6.

Line 7

The trace visible at the beginning of the line is perhaps part of *ty*, 'here'. Cf. writings of that adverb cited in *Glossar*, p. 604.

Line 8

For the restoration of *ḥw*[*t*] after *ẖn=w*, see the better preserved writing of the same word in line 14 below. The third consonant there is clearly *t* rather than *ṱ* . Only faint traces are visible after the numeral 4 that follows. Context requires the restoration [*n ḥr n qll*]; see commentary below. Reference is to the four male members of the Ogdoad, who are represented with the heads of frogs.

Line 9

(a) For the restoration [*n ḥf*] at the beginning of the line, see commentary below. Reference is made to the four female members of the Ogdoad, who are represented with the heads of serpents. It is not clear to me how the rest of the break should be restored, however. The trace immediately before *Ptḥ*, 'Ptah', could be part of *ir*, 'make, do', or even *ḏ*, 'say'. There appears to be insufficient space before this word to supply a phrase like *m-sꜣ nꜣy*.

(b) The initial stroke of *iṱ*, 'father', has been omitted through haplography, owing to its resemblance to the final stroke of the preceding *pꜣy=w*. Cf. fuller writings of that noun in line 17 below, fragment 5, I, 9, and fragment 11, I, 1. The poorly preserved word after *iṱ* is perhaps *twt*, 'assemblage, union'. Cf. the writing of that noun in line 12 below.

(c) For the restoration *nꜣy=*[*n*] *ẖ*[*m mn*]*y*, see the better preserved occurrence of this phrase in the line immediately following. One has here a pun based upon the standard demotic form of the name of the Ogdoad: *nꜣy.w-Ḫmnw* (see *Glossar*, p. 361), which appears in our text as *nꜣ Ḫmnw*. The final element in the phrase, *mny*, is identical with the *mn*, 'Götterbild', of *Glossar*, p. 158.

Line 10

(a) Context indicates that *Ḫmnw* should be restored in the break at the beginning of the line. For *nꜣ Ḫmnw* = 'Ogdoad', see two lines below. However, this restoration would not fill up the entire gap. My suggestion to supply *r-ḏbꜣ pꜣ* before *rn* is entirely guesswork; nevertheless, the words in question would yield good sense in the present passage and fit the available space nicely. Despite any uncertainties owing to the damaged state of the text, the sense is still clear: the Ogdoad receive their name *n3 Ḫmnw* as a result of what they have been called by Ptah. See commentary below.

(b) Immediately after *r-ḏ=f*, perhaps restore *r-r=w*. A trace of the initial *ḫ* of the *ḫm-ḫl.w* that follows is visible when the text resumes.

Line 11

(a) For the restoration *[mn]y* at the beginning of the line, see the better preserved writing of that noun in the line above. *mny* alone is insufficient to fill up the entire gap, but I am unable to suggest what may have stood before it.

(b) I owe the reading *in[y]* after *twt n* to John Tait. The text identifies *mny* as a name for a stone statue.

(c) The clause introduced by *r-ḏb3* presumably gives the explanation for why the term *mny* is used to designate statues made of stone. The signs between *r-ḏb3 ti* and *n rmt* are difficult to make out, but the traces would suit *ꜥḥꜥ=f*. For *ti ꜥḥꜥ* used of setting up statues, see e.g. P. Carlsberg 462, II, 6 (Smith in Frandsen and Ryholt [eds.], *The Carlsberg Papyri 3: A Miscellany of Demotic Texts and Studies*, plate 15); line 22 of the demotic version of the Rosetta Decree (Spiegelberg, *Der demotische Text der Priesterdekrete von Kanopus und Memphis*, p. 56).

(d) My restoration at the end of the line is only a guess. *ḫpr n=f*, assuming that the reading is correct, could also mean 'belong to him'.

Line 12

(a) I am unable to suggest a restoration for the break at the beginning of the line. Presumably one had here the continuation of the explanation of the term *mny* that begins in the line above.

(b) The damaged word between *ḏ* and *p3y=w twt*, now illegible, is determined with the seated man sign. Cf. the writing of the same determinative after *rmt*, 'man' in the line preceding.

(c) For the union (*twt*) of the Ogdoad 'according to their names', see commentary below.

Line 13

(a) The traces at the beginning of the line represent the signs used to write *hrw*, 'day' (cf. line 17 below). The same signs are also employed as the determinative of various other words, e.g. *ṯ3*, 'time', in line 7 above.

(b) The listing of the members of the Ogdoad here follows tradition; see commentary below.

Line 14

(a) Context indicates that *Nwn.t p3 rn* should be restored at the beginning of the line. On the basis of this restoration, it can be ascertained that the extent of the missing text here and elsewhere in the column is roughly 3.2 cm. For use of *p3 rn* with plural reference, see instances later on in this line and in the one that follows where it is resumed with the suffix pronoun *w*.

(b) For the restoration *[mt]re=w* after *p3 rn n p3 4 ḥwt*, cf. the following sentence, which is very similar to this one.

Line 15

(a) Lost in the break at the beginning of the line is the name of the place at the four doors of which the names of the four male members of the Ogdoad are said to have met. For the possibility that this could be *p3 Nwn*, 'the Primaeval Ocean', see commentary below. According to the sentence that follows, the names of the four female members met at the four doors of the sky. On these doors, see ibid.

(b) For the restoration *r p3 4 r[3]* immediately after *mtre=w*, cf. the undamaged writing of the same phrase at the end of the preceding line.

Line 16

(a) The traces in front of the numeral four at the beginning of the line could represent either *ḏ*, 'say', or *rn*, 'name'. I am unable to suggest what might have stood before this damaged word.

(b) Perhaps restore [*t*]*ȝ*, '[la]nd', after the third occurrence of *pȝ*. Cf. undamaged writings of that word in e.g. fragment 5, I, 10 and II, 6, fragment 13, II, 5 and 13.

(c) For the suggested restoration [*šʿ*] *ḏ*[*t*] before *m-qt*[*y*], cf. the undamaged writing of that phrase in fragment 13, II, 1.

Line 17

(a) The partially preserved word at the beginning of the line is *iṯ*, 'father'. Cf. the writing of this noun in line 9 above.

(b) For the restoration *šʿ-h*[*n*]*-r* before *pȝ hrw*, cf. writings of the similar expression *r-hn-r* in e.g. fragments 4, II, 15; 5, II, 16; and 13, II, 5.

Line 18

The only traces visible in this line are the top of the initial sign and the looped cord determinative of *sẖ*, 'write'. For the latter, cf. the writing of [*pr*]-*ʿnḫ*, '[house of] life', in fragment 34, line x+3.

Commentary

Line 1 begins with a reference to a town, in a damaged context. This is part of a speech uttered by a male being, most probably the god Ptah, who figures prominently in the closing lines of the previous fragment and throughout the present one. The town in question could be one of those (Hermopolis or Elephantine) mentioned in line 11 of fragment 1. In line 2, the words *ḫpr tȝ bȝ n ȝsy ẖn* [....], 'The thicket of reeds came to be in [....]', occur. The thicket in question is the one on which Ptah was described as resting in the previous fragment. See commentary ad loc.

Line 4 concludes with the phrase [*r*]*ṯ=w ḥr pȝ ist*, 'their [fe]et on the ground'. It is not clear to whom the suffix pronoun *w* refers. As discussed in the commentary on fragment 1, the term *ist* denotes the platform or basis in the midst of the Primaeval Ocean on which the creator god stands to initiate the process of creation, and which subsequently becomes the summit of the first land to emerge from the depths. The noun *lhny*, 'resting place', which occurs in a damaged context near the beginning of line 5, is probably a synonym for *ist*.[80]

At the end of that line and the beginning of the next one is a description of how Ptah takes thought (*ir mȝwy*) and fashions some objects. The noun identifying these has been lost in a break, but was probably *swḥy.w*, 'eggs' (see below), or a synonym of that word. The text goes on to narrate how the god scatters his seed and comes to be around the objects in question.[81] At the appropriate time (lines 7-8), the divinities of

[80] See commentary on fragment 1, line 6.

[81] *ḫpr=f n pȝy=w qty*. The suffix pronoun *f* clearly refers to Ptah and not his seed, since the noun *mtwy* is feminine in gender.

the Ogdoad emerge from these mysterious objects, four of them male with the heads of frogs, four of them female with the heads of serpents.

The god Ptah is well attested in the role of creator. His creative activity is said to take various forms in Egyptian sources. Some texts, notably the so-called Memphite Theology, characterise it as an intellectual process, what exists coming into being as the result of a thought conceived by the deity's heart and expressed by his tongue.[82] Others describe the god as a master craftsman, forming or fashioning all things.[83] Yet other texts stress the role of Ptah as primordial father and progenitor.[84] His sexual potency is emphasised in sources that speak of him as a 'lusty bull' (*kȝ nḏmnḏm*),[85] 'begetter of the gods' (*wtṯ nṯr.w*),[86] or 'begetter of the Ogdoad' (*wtṯ Ḫmny.w*).[87]

In our text, all three modes of creative activity are combined: Ptah is said to take thought, fashion, and scatter his seed. It might be considered that this represents a late attempt to harmonise divergent accounts of creation. However, such combinations are found in earlier sources as well, e.g. P. Berlin 3048, II, 20-22, where Ptah is called *ḥm tȝ m sḫrw n ib=f ms nty nb wtṯ qmȝ wnn.t*, 'he who fabricated the earth in accordance with the plan of his heart, who made everything, the begetter who created what exists'.[88] It is possible, moreover, that this sort of multi-faceted view of the nature of the creative process actually predates the development of separate doctrines or systems which concentrate upon a single aspect of it.[89]

As regards the nature of the objects which Ptah is said to fashion, given the context it is difficult to avoid thinking of eggs or something similar. Support for this interpretation is provided by fragment 15, which may well refer to the same act as that described here. Line x+1 of this fragment speaks of a male being depositing (*ḥwy*) something. In line x+3, the divine name Ptah is followed by the verb *sty*, 'scatter' (cf. line 6 of the present fragment). Line x+6 refers to *nȝ swḥ*[*y*].*w r-ḥwy=f*, 'the eg[g]s which he (scil. Ptah) deposited'.

Similarly, the cosmogony in the temple of Khonsu at Karnak calls Ptah *qmȝ swḥ.t pr m Nwn*, 'the creator of the egg that appeared in the Primaeval Ocean', and says of him:

[82] See e.g. Sandman Holmberg, *The God Ptah*, pp. 42-5; Sauneron and Yoyotte in *La Naissance du Monde*, p. 40; Allen, *Genesis in Egypt*, pp. 42-7. In a Ramesside stela, Ptah is called *mȝwy m nty*, 'who thought of what exists' (Koefoed-Petersen, *Recueil des inscriptions hiéroglyphiques de la Glyptothèque Ny Carlsberg*, p. 57).

[83] Sandman Holmberg, *op. cit.*, pp. 45-6.

[84] *Op. cit.*, pp. 31-7.

[85] Chassinat, *Edfou* 6, p. 175, line 7.

[86] Kitchen, *Ramesside Inscriptions* 2, p. 260, line 2.

[87] Sethe, *Amun und die acht Urgötter von Hermopolis*, p. 58.

[88] Wolf, *ZÄS* 64 (1929), p. 18. The papyrus is of 22nd Dynasty date.

[89] Cf. remarks of Pépin in Schoske (ed.), *Akten des Vierten Internationalen Ägyptologen Kongresses, München 1985*, vol. 3, pp. 339-40; also Bickel, *La Cosmogonie égyptienne avant le Nouvel Empire*, pp. 285-98; Willems, *The Coffin of Heqata*, pp. 305-7.

sty=f r swḥ.t ḫp(r)  *m-ẖnw=s m-swꜣ Ḫmny.w*, 'He scattered seed over the egg and eight came into existence within it around the Ogdoad.' The eight objects in question are subsequently transported to Thebes where the actual birth of the Hermopolitan corporation takes place.[90] The inscription does not explain what the 'eight' that come into existence within the egg are, but presumably one has here a reference to some sort of sacs or compartments inside it occupied by the individual members of the Ogdoad prior to their emergence.[91] Perhaps a similar group of modules was mentioned in our text.

The statement 'he (scil. Ptah) came to be around them' is somewhat cryptic. Two possible interpretations suggest themselves. In the Khonsu cosmogony discussed in the preceding paragraph, the god Ptah is said to absorb the eight sacs containing the Ogdoad within himself by swallowing them after they have been created.[92] A similar sequence of events could have been described in our text. Another possibility, more likely in my view, is that Ptah is said to transform himself into a sort of liquid solution or medium in which the Hermopolitan deities have their latent existence. From the reign of Ramesses II onwards, that god is identified not only with the primaeval mound, as Ptah-Tatenen,[93] but with the Primaeval Ocean from which that mound arose as well, in the form of Ptah-Nun.[94] Compare the demotic magical spell on an ostracon in Strasbourg, previously cited in the commentary on fragment 1, where the deity is invoked as *Ptḥ tny iṯ nṯr.w pꜣ-Šy ꜥꜣ pꜣ Nwn nt ḥr pꜣ ꜣsṯ*, 'Ptah-(Ta)tenen, father of the gods, Pshai the great, the Primaeval Ocean, who is upon the basis'.[95] Similarly, in P. Harris 1 (temp. Ramesses III), Tatenen is called *snṯ tꜣ m ir.n=f ḏs=f pẖr sw m Nwn wꜣḏ-wr*, 'he who founded the earth through what he did himself, and encircled it as the Primaeval Ocean and the sea'.[96] Thus, two different aspects of that god as primordial creator can be combined within a single text. Our papyrus may have incorporated a further development of the same idea.

[90] See Sethe, *Amun und die acht Urgötter von Hermopolis,* plate 2, lines 10-18; Parker and Lesko in Baines et al (eds.), *Pyramid Studies and Other Essays Presented to I.E.S. Edwards,* plates 34-5. For Ptah as creator of the egg, see also Graefe in Clarysse, Schoors and Willems (eds.), *Egyptian Religion: the Last Thousand Years* 1, pp. 118-23.

[91] For an egg said to be *m-swꜣ* the divinity within it, see e.g. Sauneron, *Esna* 3, p. 31 (206, 8).

[92] *ꜥmꜣ.n=f st iwr st bꜥn.t=f*, 'He swallowed them, his throat becoming pregnant with them' (Sethe, *op. cit.*, plate 2, line 13; Parker and Lesko, art. cit., plate 35).

[93] See references cited in note 76 on fragment 1.

[94] Cf. Schlögl, *Der Gott Tatenen*, p. 62, note 47. On this equation, see further Zandee, *Der Amunhymnus des Papyrus Leiden I 344, verso*, pp. 722 and 757; Sandman Holmberg, *The God Ptah*, pp. 115-20; Fischer-Elfert, *Literarische Ostraka der Ramessidenzeit in Übersetzung*, pp. 29-30; Baines, *Fecundity Figures*, p. 41; Berlandini, *RdE* 46 (1995), pp. 28-9.

[95] Spiegelberg, *ZÄS* 49 (1911), pp. 35-7. *ꜣsṯ* there is the same word as the *ist*, 'ground, basis', that occurs in line 4 of the fragment under consideration.

[96] Erichsen, *Papyrus Harris 1*, p. 49, lines 11-12.

Numerous Egyptian sources refer to Ptah in his form of Tatenen as the father or creator of the Ogdoad.[97] In reliefs illustrating the Edfu cosmogony, the Hermopolitan deities are shown in association with the reed thicket that marks the place where the land first emerged.[98] As might be expected from what has been said above, not only Tatenen, but the Primaeval Ocean as well, can be described as the begetter or maker of the Ogdoad.[99]

When the Hermopolitan corporation emerge from the latent stage of their existence, the four male deities are said to have the heads of frogs and the four female deities the heads of serpents, or so our text should be restored. This is how they are frequently depicted in temple reliefs of the Graeco-Roman Period.[100] In the demotic cosmogony of P. Berlin 13603, a reference to the four serpent-headed members of the Ogdoad occurs in a broken context.[101] An unpublished fragment in the Carlsberg Collection (no. 10961) preserves the remains of a vignette depicting three of the members of this corporation, two females with serpent heads and a lone male with a frog's head.

In lines 9-10 of the present fragment, Ptah refers to his offspring as *nꜣy=n ḫm mny*, 'our little images'.[102] Through a play on words, this gives rise to their actual name: *nꜣ Ḫmnw*.[103] A similar pun is found in the Neith cosmogony at Esna, where a group of deities acquire the name *Ḫmny.w* as a consequence of their saying *ḫm=n*, 'we are ignorant'.[104] In a number of inscriptions in temples of the Graeco-Roman Period, the name of the Ogdoad is written within a cartouche as if it were to be read *ḫm nn*, 'the child who has grown weary'.[105] Compare line 10 of the present fragment, where the eight deities are characterised as *ḫm-ẖl.w*, 'children'.

[97] For a selection of examples, apart from those already cited, see Sethe, *Amun und die acht Urgötter von Hermopolis*, pp. 41, 43, 51, 56, 58, and texts on plates 3-5; Rochemonteix and Chassinat, *Edfou* 1, p. 289, line 1; Chassinat, *Edfou* 3, p. 312, line 6; *Edfou* 4, p. 140, line 13; *Edfou* 5, p. 85, line 8; idem, *Le Mammisi d'Edfou*, p. 81, line 8.

[98] See Reymond, *JEA* 48 (1962), pp. 81-2.

[99] See e.g. Sethe, *Urkunden* 8, p. 80 (95c) (Ogdoad *nbi.n Nwn*, 'whom the Primaeval Ocean fashioned'); ibid., p. 111 (139b) (Amun as *Nwn wr* called *wtṯ Ḫmny.w*, 'begetter of the Ogdoad'); Chassinat, *Edfou* 5, p. 85, line 10 (Ogdoad *nbi.n Nwn*). Ryhiner, *L'Offrande du lotus,* p. 48, note 26, emends *n* to *m* unnecessarily in the last instance.

[100] See Sethe, *Amun und die acht Urgötter von Hermopolis*, pp. 61-2; Lepsius, *Über die Götter der vier Elemente bei den Ägyptern*, plates 1-5.

[101] P. Berlin 13603, I, 18 (Erichsen and Schott, *Fragmente memphitischer Theologie in demotischer Schrift*, p. 311 [13] and plate 5).

[102] With the use of the possessive pronoun 'our' here, compare Genesis 1:26, ποιήσωμεν ἄνθρωπον κατ' εἰκόνα ἡμετέραν καὶ καθ' ὁμοίωσιν.

[103] On the significance of word play in assigning names, see Guglielmi in *Lexikon der Ägyptologie* 6 (1986), columns 1287-91 (s.v. Wortspiel) and literature cited there.

[104] Sauneron, *Esna* 3, p. 31 (206, 7). Cf. idem, *Esna* 5, p. 262.

[105] See Sethe, *op. cit.*, pp. 44-6.

As remarked in the textual notes above, writings of the name of the Hermopolitan corporation with a preceding element *nꜣ* or (more often) *nꜣy.w* are the norm in demotic texts.[106] Such forms do occur in earlier sources as well, however. Cf. the Magical Papyrus Harris of New Kingdom date, where  (III, 11) varies with 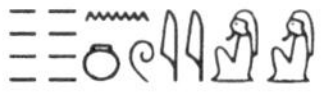 (IV, 8) as a writing of the word for Ogdoad.[107] Presumably, the passage under consideration could not have been composed before such forms came into general use, inasmuch as the play on words between *nꜣy=n ḫm mny* and *nꜣ Ḫmnw* found in it would lose much of its force in the absence of the element *nꜣ* from the latter.

In line 11, the text goes on to discuss the significance of the term *mny* employed by Ptah to describe the Ogdoad, explaining it as a name for a stone statue (*twt n iny*). The latter expression, in turn, evokes the *twt*, 'assemblage, union', which occurs in the following line and possibly in line 9 of the present fragment as well. This can be seen as an extended play on words. Whereas *mny* is said to be a synonym of *twt*, 'statue, image', at the same time, the 'assemblage' (*twt*) of the eight Hermopolitan deities is characterised as a group of images (*mny*). The end of line 11 and the beginning of line 12 apparently explain how the term *mny* acquired its meaning, but too little is preserved of this explanation for it to be understood.

The ensuing text has as its subject the union of the members of the Ogdoad *n rn=w*, 'according to their names' (line 12). A similar union is described in the demotic cosmogony of P. Berlin 13603. There, the four male Hermopolitan deities, at a certain point after their creation, are said to merge into a single black bull, and the four female deities into a black cow.[108] Subsequently, the narrative continues, *ḫpr pꜣ rn n pꜣ kꜣ irm tꜣ iḥ.t twt*, 'It happened that the name of the bull and (that of) the cow were united.'[109] Thus, they become a single entity. As will be seen below, our papyrus describes a similar process of merging among the members of the Ogdoad. It should be remembered that, in the Late Period, the term *rn*, 'name', was also used to signify a being's *kꜣ*, or vital force.[110] Thus, what is implied by the concept of union according to name is more than a mere juxtaposition or sharing of appellations.[111]

[106] Cf. note (c) on line 9. For exceptions, see Thissen, *Die demotischen Graffiti von Medinet Habu*, p. 193.

[107] Lange, *Der magische Papyrus Harris*, pp. 32 and 38. Other examples are given in Sethe, *op. cit.*, p. 42.

[108] P. Berlin 13603, II, 5 (Erichsen and Schott, *Fragmente memphitischer Theologie in demotischer Schrift*, p. 312 [14] and plate 5): *ir nꜣ ḥwṯ.w wꜥ kꜣ km ir nꜣ s-ḥm.wt wꜥ.t iḥ.t km.t.*

[109] Ibid., II, 15.

[110] See Quaegebeur in Boswinkel and Pestman (eds.), *Textes grecs, démotiques et bilingues*, p. 252; Zandee, *Death as an Enemy*, p. 180.

[111] In the case of P. Berlin 13603, II, 15, there may even be an implicit pun between *kꜣ*, 'bull', and *kꜣ*, 'vital force' (= *rn*). Cf. Chassinat, *Edfou* 5, p. 85, line 12, where the Ogdoad are called *kꜣ.w kꜣkꜣ ḥm.wt šsp kꜣ.w wr.w ḫp*(*r*) *m ḥꜣ.t*, 'bulls who ejaculate, cows who receive (it), the great kas who came into existence at the beginning'.

The individual members of the Ogdoad are listed in lines 13-14, in the following sequence: Amun, Amaunet, Heh, Hauhet, Kek, Kauket, Nun, and Naunet.[112] In temple inscriptions of the Graeco-Roman Period, this particular grouping is found only at Edfu and in the Theban region, e.g. at Medinet Habu and Karnak.[113] However, there is no precise parallel for the sequence of the present fragment, with Amun and Amaunet at the beginning and Nun and Naunet at the end. In the Book of the Fayyum, once again the names of the members of the Ogdoad are the same as in our text, but they appear either in a different sequence, or else divided up into two groups of four.[114] There was originally a list of the Hermopolitan deities given in the cosmogony of P. Berlin 13603. Unfortunately, all that is preserved of this now is a pair of names, that of Hauhet followed by that of Kek.[115] Presumably the complete list did not include the names of Amun and Amaunet, since these are not considered as integral members of the Ogdoad by the author of the Berlin papyrus.[116] In the remains of the vignette on the unpublished demotic Carlsberg fragment discussed above, there are depicted the deities Hauhet, Nun, and Naunet, in that order. It would seem, therefore, that the sequence of divinities in our papyrus is without a precise parallel.

Lines 14-15 describe how the names of the four male members of the Ogdoad meet at the four doors of a place or region, the name of which is lost in a break. The names of the four female members, it is said, meet at the four doors of the sky. I take this to refer to the process of merging or assimilation of the males and females, each group into a single being, which has already been discussed two paragraphs above. As noted there, the first allusion to this union 'according to name' occurs in line 12 of the fragment under consideration. The culmination of the process, in which all eight members of the Ogdoad are subsumed into one entity, is the subject of the fragment that follows this one.

Perhaps the four doors at which the names of the male Hermopolitan deities are said to meet are those of the Primaeval Ocean (*pꜣ Nwn*). According to a funerary papyrus of Middle Kingdom date, this watery mass possessed the requisite number of doors, which gave entry to the heavenly vault.[117] The idea that the sky itself had four doors or gates, aligned with the four cardinal points, can be traced back to the Old Kingdom. Thus, in Pyramid Text Spell 529, reference is made to the western, southern,

[112] The last is restored.

[113] See Sethe, *Amun und die acht Urgötter von Hermopolis*, p. 66 and table on plate 1.

[114] See Beinlich, *Das Buch vom Fayum*, pp. 152-3, 190-1, and 256-9. For late Fayyumic traditions concerning these divinities, see ibid., pp. 313-14.

[115] P. Berlin 13603, I, 19 (Erichsen and Schott, *op. cit.*, p. 311 [13] and plate 5).

[116] See particularly ibid., II, 1.

[117] See Borghouts in Schoske (ed.), *Akten des Vierten Internationalen Ägyptologen Kongresses, München 1985*, vol. 3, p. 135. On these portals, see further Pépin, ibid., pp. 342-3.

eastern, and northern doors (*sbꜣ*) of heaven.[118] In later sources, all four of these heavenly portals are located in the eastern horizon of the sky.[119] In one sense, both the celestial doors and those of the Primaeval Ocean interface, being identical with, or mirror images of, each other. The sun, during its daily course, in emerging from the latter openings, simultaneously enters through the former to shine forth in the sky.[120] Further, Assmann has commented upon the mysterious, unfathomable nature of the doors of the sky. As he notes, all such portals mentioned by Egyptian sources are in reality a single portal, through which the sun both leaves and enters the celestial expanse, one which it is impossible to locate in any concrete spatial sense.[121] Similarly, the four doors of the sky in the present fragment can be seen as four united in one, like the four female members of the Ogdoad whose names are said to meet at them.

Subsequently (lines 15-16), our text goes on to describe further activities of the creator Ptah. Unfortunately, the passage in question is too poorly preserved to allow the nature of these to be ascertained. Visible at the beginning of line 16 is the phrase *ḏ=w pꜣ rn n pꜣ Nw*[*n* ....], 'They said the name of the Primaeval O[cean ....].' Thus the text records the giving of the name of the Ocean to some being or thing. It is not clear whether this is a newly created entity, or one previously mentioned who is now assigned an additional name. If the latter, then perhaps restore *n Ptḥ*, 'to Ptah', after *Nw*[*n*]. One could have had here an explanation of how that god came to be identified with the primordial flood. As remarked above, there may be a further reference to Ptah's transformation into this form in line 7 of the present fragment. It is noteworthy that the writer of our cosmology makes a clear distinction between *Nwn*, the member of the Ogdoad and consort of Naunet, and *pꜣ Nwn*, the primaeval watery mass from which the Ogdoad emerge.[122]

Line 17 preserves the remains of another broken passage: ⌞*it*⌟ *šꜥ-h*[*n*]*-r pꜣ hrw*, '⌞father⌟ up to today', probably part of an aetiological statement.[123] One may have had here a remark to the effect that Ptah, in view of his creative activity as described in the preceding lines of the fragment, is called father up to the present day. Alternatively, if the suggestion made above is correct that line 16 records the giving of the name *pꜣ*

[118] Sethe, *Die altägyptischen Pyramidentexte* 2, pp. 207-8 (1252c-f). Cf. Thomas, *JEA* 42 (1956), p. 78, note 2; Černý, *JEA* 41 (1955), p. 79, note 2.

[119] See Piankoff, *Le Livre du Jour et de la Nuit*, p. 4; Assmann, *Der König als Sonnenpriester*, p. 30; Roulin, *Le Livre de la Nuit*, p. 44.

[120] Cf. discussion of the words *pr/wbn m Nwn r nn.t*, 'go forth/rise from the Primaeval Ocean to the sky', in Ryhiner, *L'Offrande du lotus*, pp. 81-2.

[121] See Assmann, *op. cit.*, pp. 33-4. Compare the shaman's gate discussed in West, *The Orphic Poems*, p. 147, which gives simultaneous access to sky, earth, and underworld.

[122] The same distinction is made in e.g. the Book of the Fayyum; see Beinlich, *Das Buch vom Fayum*, pp. 312-13.

[123] On statements of this type, see Introduction.

Nwn to Ptah, then perhaps the present line went on to say that the Primaeval Ocean is called father as well in consequence of its identification with that god.[124]

FRAGMENT 3 Plate 4

Dimensions: 8.7 × 22.2 cm. Colour: varies from medium brown (left) to dark brown (right). Related fragments: Nos. 16, 17 and 18.

Transliteration

1. *k*]*ꜣ km nꜣ.w-sbk* [
2.]*=f iwt nꜣy=f ꜥꜥe.w* [
3. *sḥ*]*t*(?) *pḥ.ṯ=f sẖf* ..[
4.] *mḏ sḥt šy.*[*ṯ=f*
5.] *nꜣ.w-qꜣ=f n pḥ.ṯ=f nꜣ.w-*[
6. *k*]*ꜣ km iw=f*.[..] *imn* [
7.].. *ntr mḥ-4 ỉmn tꜣ wnw.t* [
8.]. *ẖꜣy ꜥb.w nꜣ.w-sbk* [
9. *nt*]*r mḥ-5 ỉmn ḥr-ib Šw kꜣ km* [
10.].. *ẖꜣy ꜥb.w pꜣ ḏf* ..[
11. *ntr*] *mḥ-6 ỉmn twn=f ḥbꜥ ꜣḥ* [
12.] *nꜣ.w-sbk nꜣy=f ꜥb.w ir.t n* ..[
13.] *8 ntr mḥ-7 ỉmn n*[*ꜣ.w-ꜥ*]*n intm* [
14.].. *ẖꜣy st* .[..] *m-ḥ.ṯ=f sḥt* [
15.] *ntr mḥ-8 ỉmn* .[... *m*]*w*(?) *Nw*[*n.t*
16. *ẖ*]*ꜣy* .[..]..[..] *i-ir* [
17. *kꜣ.w iḥ*].*w*(?) *mꜣꜣ ẖrt nt-iw nꜣ* [
18.] *tꜣy=f ir.t n* [
19.]..
20.].[...] *r rt* [
21.]*=f nꜣ.w-sḥt* [
22. *ꜥ*]*b.w*

Translation

1.] black [bu]ll. Short are [
2.] him/it among his limbs [

[124] For examples of the Ocean designated as *it nṯr.w*, 'father of the gods', see el-Banna, *BIFAO* 86 (1986), pp. 156-8. Cf. also Thissen in Verhoeven and Graefe (eds.), *Religion und Philosophie im alten Ägypten*, pp. 297-8. For Ptah-Nun so designated, cf. el-Banna, art. cit., p. 164, and text cited in note 95 above.

3. shin]ing(?) are his hindquarters *sẖf* ..[
4.] phallus. Radiant are [his] two feat[hers
5.] He is exalted in his hindquarters. *nꜣ.w*-[
6.] black [bu]ll who is .[..] hidden(?) [
7.].. god. Fourth: Amun, the hour [
8.]. with long horns. Short are [
9. go]d. Fifth: Amun dwelling within Shu, black bull [
10.].. with long horns, the pupil ..[
11. god.] Sixth: Amun, he will arise. Play *ꜣḥ* [
12.] Short are his horns. Eye of ..[
13.] eight divinities. Seventh: Amun, f[ai]r is the *intm* [
14.].. with long tail .[..] before him. Radiant is [
15.] god. Eighth: Amun .[... li]ke(?) Nau[net
16. l]ong .[..]..[..] *i-ir* [
17. bulls and cow]s(?) see a child, which the [
18.] his eye of [
19.]..
20.].[...] to the feet [
21.] him/it. Radiant is [
22. hor]ns.

Textual notes

Line 1

For the writing of *kꜣ*, 'bull', here, see the undamaged example of this word in line 9 below. For that of *sbk*, 'short', see lines 8 and 12. In the latter, the subject of the adjective verb *nꜣ.w-sbk* is *nꜣy=f ꜥb.w*, 'his (scil. the god's) horns'.

Line 3

(a) Could the partially preserved word at the beginning of the line be *sḥt*, 'shine, illuminate'? Cf. the undamaged writing of this verb in the line immediately below. The reading is offered with some hesitation, since there is a vertical stroke (deliberate?) which gives the damaged sign before the sun determinative the appearance of a *b* rather than a *t*.

(b) The word after *pḥ.ṱ=f* is almost certainly *sẖf*. The reading of the final consonant as *f* was suggested to me by Kim Ryholt. I am unable to make out the damaged traces following this. Immediately before the break at the end of the line is a tall upright sign which could be either an *s* or the man with hand to mouth determinative. Cf. words written *sẖf* with that determinative cited in *Glossar*, p. 454.

Line 4

šy.[*ṱ*] at the end of the line is written like the noun meaning 'nose' (cf. *Glossar*, p. 484). Elsewhere in demotic, however, *šy.ṱ* is sometimes used as an unetymological writing of *šw.ty*, 'two feathers'. See, for example, O. BM 50601, a demotic version of the hymn to the ten bas of Amun best known from the hieroglyphic copy preserved on the walls of that god's temple at Hibis, where *sḥḏ.n=f šn n p.t m šy.ṯ=f*, 'he has illuminated the circuit of heaven with his feathers', in line 9 corresponds to *sḥḏ.n=f šnw n p.t m*  in the hieroglyphic parallel (Smith, *Enchoria* 7 [1977], p. 123 and plate 18). In view of the passage just cited, 'two feathers' seems preferable to 'nose' as the subject of the verb *sḥt*, 'shine', in our text.

Line 5

(a) For the writing of *qꜣ*, 'be high, exalted', here, see e.g. P. Tebt. Tait 9, line x+5 (Tait, *Papyri from Tebtunis in Egyptian and in Greek*, plate 3).

(b) *nꜣ.w* at the end of the line must represent the beginning of an adjective verb like *nꜣ.w-sbk* or *nꜣ.w-qꜣ*.

Line 6

(a) For the traces of *kꜣ km* here, see line 1 above and line 9 below.

(b) The word at the end of the line appears to be *imn*; the determinative is lost. The preserved traces of the preceding word suggest the preposition *n*. Puns involving the name of the god Amun and homophonous words, in particular those derived from the root *imn*, 'be hidden', are a common feature of Egyptian religious texts concerned with that god. See Assmann, *Re und Amun*, pp. 195-203; Zandee, *Der Amunhymnus des Papyrus Leiden I 344, verso*, pp. 123-4, 131-3, 197, 737, 1052-5; Sethe, *Amun und die acht Urgötter von Hermopolis*, pp. 87-90, for examples of these and their significance. The present line may contain a further instance of such wordplay.

Line 8

For the traces of *ḫꜣy* here, see lines 10 and 14 of this fragment. Should *nꜣy=f ʿb.w* be restored after *nꜣ.w-sbk*, as in line 12 below?

Line 9

(a) Traces of the determinative of *ntr*, 'god', are visible immediately before *mḥ-5*. The same noun stands before the ordinal number *mḥ-4* in line 7 above, before *mḥ-7* in line 13, and *mḥ-8* in line 15 below.

(b) , sometimes written with a sun determinative as well, occurs frequently in unpublished Carlsberg papyri, where it is clearly a divine name. Among published texts, another example of the same name occurs in P. Vindob. 3873, recto VIb, 17 (Vos, *The Apis Embalming Ritual*, pp. 291-2). That this is *Šw* is evident from a pair of unpublished demotic fragments in the collection of the Istituto Papirologico 'G. Vitelli' in Florence. The first, which has the number PSI Inv. D 171, calls the god in question *sꜣ Rʿ*, 'son of Re', a characteristic epithet of Shu. The second (PSI Inv. D 146) refers to a goddess who is said to be the offspring of this divinity and Shu's wife and sister Tefnut.

Line 10

(a) The damaged word before *ḫꜣy* at the beginning of the line is determined with the brazier sign. Cf. writings of the same determinative in e.g. *ʿšm*, 'quench', in fragment 13, II, 8; *sty*, 'flame', in fragment 30, line 5; and *mḫ*, 'burn', ibid., line 6.

(b) For *ḏf*, 'pupil' (of the eye), generally used to denote the surrounding iris as well, see Hoffmann, *GM* 132 (1993), pp. 37-8. On this epithet of Amun, see commentary below. I am unable to read the traces that follow.

Line 11

(a) The restoration of *ntr* before *mḥ-6* is based upon the occurrence of the same noun before the ordinal numbers *mḥ-4*, *mḥ-5*, *mḥ-7*, and *mḥ-8* elsewhere in the fragment.

(b) *ḥbꜥ*, with stone determinative, is identical with the *ḥbꜥ*, 'spielen', of *Glossar*, p. 300. The following word *ꜣḥ*, written with spine and flesh determinatives, should denote a part of the body. I am unable to suggest a translation. Cf. demotic writings of *iḥꜥ.w*, 'limbs', as *ꜣḥe* discussed in Smith, *Enchoria* 8 (1978), part 2, p. 25, note 23.

Line 12

For the curious statement *nꜣ.w-sbk nꜣy=f ꜥb.w*, see commentary below. Cf. also notes on lines 1 and 8 above.

Line 13

(a) With the traces of *nꜣ.w-ꜥn* here, cf. the better preserved writings of *ꜥn* in P. Carlsberg 57, I, 9, P. Tebt. Tait 1, line 21 (Tait, *Papyri from Tebtunis in Egyptian and in Greek*, plate 1).

(b) *intm* written with flesh determinative occurs only here. Could this be a noun, e.g. 'phallus', derived from the root *ndm*? Cf. words from the same root like the noun *ndmndm*, 'Lust', and the verb *ndmndm*, 'eine Frau beschlafen', cited in *Wb.* 2, 381.

Line 14

With the traces of *sḥt* at the end of the line, cf. the better preserved occurrence of the same verb in line 4 above.

Line 15

(a) The partially preserved word after the break in the middle of the line ends with *w*. Perhaps restore *mw*, 'like' (< earlier *mi*). For this writing elsewhere in demotic, see Smith, *Enchoria* 8 (1978), part 2, pp. 21-2.

(b) Visible before the end of the line are three *nw*-pots. Clearly one has here the beginning of a hieratic or semi-hieratic writing of *Nwn* or *Nwn.t*, the seventh and eighth members of the Ogdoad as listed in fragment 2, lines 13-14. Compare the semi-hieratic writings of *pꜣ Nwn*, 'the Primaeval Ocean', in fragment 1, line 2, and fragment 11, I, 11. The same of the Ocean cannot have stood here, however, since it invariably includes the definite article (see commentary on line 16 of fragment 2).

In the present line, I have restored *Nwn.t* rather than *Nwn*, in view of the preceding ordinal number *mḥ-8*. As explained in the commentary below, the fragment under consideration extols the god Amun in the form of a black bull with eight hypostases, one representing each member of the Ogdoad. Each hypostasis is described in a separate section introduced with the appropriate ordinal number. In our cosmology, Naunet figures as the eighth member of the Ogdoad. Logically, therefore, one would expect her name to appear in the section of the text devoted to the eighth hypostasis of Amun, rather than that of her male counterpart.

Line 16

With the traces of *ẖꜣy* here, cf. the undamaged writings of the same word in lines 10 and 14 above. *i-ir* at the end of the line could be either the past participle or the marker of the second tense.

Line 17

(a) The partially preserved sign at the beginning of the line is the cattle determinative. For the sign in question, cf. the writings of *kꜣ*, 'bull', in lines 1 and 9 above, also that of *iḥ.(t)*, 'cow', in fragment 8, line x+7. The copula *pꜣy* looks rather different; see e.g. fragment 2, lines 1 and 11, fragment 13, II, 3, 4 and 14. For the suggested restoration [*kꜣ.w iḥ*].*w* here, see commentary below.

(b) For this writing of *ẖrt*, 'child', see e.g. P. Carlsberg 462, II, 4 and 15 (Smith in Frandsen and Ryholt [eds.], *The Carlsberg Papyri 3: A Miscellany of Demotic Texts and Studies*, plate 15). In the latter line, *ẖrt* is preceded by the article *pꜣ*.

(c) *nt-iw*, rather than *ḥr*, has to be read at the end of the line. The second vertical stroke of the preposition would extend much further below the line than the stroke here does. Cf. writings of *ḥr* in e.g. fragment 2, lines 1 and 4, fragment 4, II, 16, and fragment 5, I, 1 and 21.

Line 18

Although I have translated the *n* after *tꜣy=f ir.t* as 'of', it could also be the attributive *n* introducing an adjective like *iꜣb*, 'left', or *wnm*, 'right'.

Line 21

For the traces of *sḥt*, 'radiant', here, cf. better preserved examples of that word in lines 4 and 14 above.

Line 22

All that is visible in this line is a trace of the *w* of *ꜥb.w* followed by its two horn determinatives and a vertical plural stroke. For better preserved writings of the word for 'horns', see lines 8, 10, and 12 above. In the last two instances, and in fragment 18, line x+5 as well, the plural marker follows immediately after the horn signs, as here, omitting the flesh determinative found in line 8. Only in the present instance, however, is a vertical plural stroke used in preference to horizontal ones.

Commentary

Fragment 3 preserves a panegyric to the god Amun, who has the form of a black bull with eight hypostases. The description of each of these is given its own section, roughly two lines in length, and introduced with an ordinal number. (Only the last five of these numbers have been preserved.) Each description of which the opening has been preserved begins with the divine name *Ỉmn*. Where the ending has been preserved, this is invariably *nṯr*, 'god'.[125] Some attempt has been made to have the words that follow *Ỉmn* in each description sound alike, e.g. *tꜣ wnw.t*, 'the hour', in number four, and *twn*, 'arise', in number six, but this is not carried through consistently.[126] One may compare the earlier hymn to the god in P. Leiden I 350, where the initial and final words of each chapter are homophones of the number assigned to it.[127]

The image of Amun as a bull occurs frequently in Egyptian sources, in contexts where emphasis is placed upon the god's procreative powers.[128] His colour here, black, is a further indicator of this aspect of fecundity.[129] As will be seen below, most of the descriptions of individual features of Amun's anatomy in our text can be paralleled elsewhere.

[125] So I have divided the text in my translation. It would also be possible, however, to take *nṯr* in each case with the following ordinal number as the marker of a new section, e.g. *nṯr mḥ-4*, 'fourth deity', and so on.

[126] E.g. *ḥr-ib* follows *Ỉmn* in number five, *n*[*ꜣ.w-ꜥ*]*n* in number seven.

[127] Cf. Gardiner, *ZÄS* 42 (1905), p. 12; Zandee, *De Hymnen aan Amon van Papyrus Leiden I 350*, p. 2.

[128] For a selection of examples, see Zandee, *op. cit.*, pp. 96-7; idem, *Der Amunhymnus des Papyrus Leiden I 344, verso*, pp. 88-90 and 421-2.

[129] See Vernus, *Athribis*, p. 429, and references cited there.

Possibly in line 3, and certainly in lines 4, 14, and 21, the term *sḥt*, 'radiant, shining', is used of various parts of the god's body. This reflects his celestial/solar nature. In the hymn to Amun-Re in the temple of Ramesses III at Karnak, for instance, that monarch says to the deity: *inw=k šw*, 'Your skin/colour is the light.'[130] Similarly, in the 18th Dynasty hymn to Amun on the stela of Suti and Hor, the two brothers say: *ṯḥn iwn=k r inm=s*, ' Your colour is more radiant than its (scil. heaven's) colour.'[131] In a hymn from the temple of Hibis (temp. Darius I), the Ogdoad describe Amun's skin as shining like gold,[132] and his visages (*ḥr.w*) as gleaming like fayence.[133]

In line 4 of our fragment, it is specifically the *šy.[ṯ]* of the god that is said to shine. As discussed in the textual note on that line, one probably has here an unetymological writing of *šw.ty*, '(pair of) feathers'. A number of Egyptian sources explicitly identify the twin feathers of Amun and other celestial deities as the sun and moon. This explains their power to illuminate.[134]

Near the beginning of the same line, the noun *mḏ*, 'phallus', occurs in a broken context. One might expect a reference to this particular attribute in a description of Amun in taurine form. Compare the Hibis hymn to that deity cited two paragraphs above, where the Ogdoad address him as *nb mṯ*, 'lord of the phallus'.[135] In the hymn to Amun in P. Leiden I 350, the god is called *sḫpr mṯꜣ*, 'creator of the phallus'.[136] Either of these epithets, or another one similar to them,[137] would be suitable for restoration at the beginning of the present line.

In line 5, it is said of Amun: *nꜣ.w-qꜣ=f n pḥ.ṯ=f*, 'He is exalted in his hindquarters', probably a further reference to the god's sexual potency. This feature of the divine bull's anatomy is mentioned in line 3 above as well, where perhaps one should restore *[sḥ]t pḥ.ṯ=f*, '[Radi]ant are his hindquarters.' At the end of line 6, one may have had an allusion to the 'hidden' character of Amun. For this, see textual note (b) on the line in question and the references cited there.

Line 8 calls the god *ḫꜣy ꜥb.w*, '(the one) with long horns'. The same epithet occurs in line 10 below. Frequent reference is made in Egyptian sources both to the length and the sharpness of the horns of Amun. In the Hibis hymn to that deity cited twice above, he is called *qꜣ ꜥb.w spd ḥn.ty*, 'the one with long horns, with sharp horns'.[138] In the hymn

[130] Epigraphic Survey, *Ramses III's Temple within the Great Enclosure of Amon* 1, plate 23, A, line 4.

[131] Edwards, *Hieroglyphic Texts from Egyptian Stelae, etc.* 8, p. 24 and plate 21.

[132] Davies, *The Temple of Hibis in El Khārgeh Oasis* 3, plate 33, middle, lines 1 and 22.

[133] Ibid., line 38. For an explanation of the plural *ḥr.w* in this passage, see Barucq and Daumas, *Hymnes et prières de l'Égypte ancienne*, p. 340, note (ca).

[134] See Zandee, *Der Amunhymnus des Papyrus Leiden I 344, verso*, pp. 104, 284, 365-9, 538, and 990-2.

[135] Davies, *op. cit.*, plate 33, middle, line 38.

[136] Ibid., V, 1. Cf. Zandee, *De Hymnen aan Amon van Papyrus Leiden I 350*, p. 92 and plate 5.

[137] E.g. *nḫt mṯꜣ*, 'mighty of phallus', predicated of Amun in Sethe, *Urkunden* 8, p. 106 (134b).

[138] Davies, *op. cit.*, plate 33, middle, line 21.

of P. Leiden I 344 verso, Amun is called *spd ʿb.wy*, 'the one with sharp horns'.[139] Similarly, in an inscription on the second pylon at Medinet Habu, Ramesses III addresses the god as *kꜣ spd ḥn.ty*, 'the bull with sharp horns'.[140] Such epithets evoke the might and ferocity embodied in the animal in question. At the same time, particularly when they occur in conjunction with references to the two feathers (cf. line 4 of the present fragment), they proclaim the kingly status and power of the being to whom they are attributed, inasmuch as the horns and feathers constitute important elements of the royal regalia.[141]

In line 9, the fifth hypostasis of Amun is described as dwelling *ḥr-ib*, 'within', another deity, the god Shu.[142] The latter here personifies the celestial disk in which the solar deity Amun is manifested.[143] Compare an inscription from one of the Osiris chapels on the roof of the temple of Dendera, where Atum is called *ḥr-ib n Rʿ*, 'he who is within Re'.[144] There Re fulfils the same function as Shu does in our text. Lunar as well as solar divinities can be described as occupying a place *ḥr-ib* the disk which constitutes their visible form. In the twelfth hour of the Book of Night, for instance, Osiris is called *ḥr-ib iʿḥ=f*, 'he who is within his lunar disk'.[145] In an inscription on a Late Period coffin lid from Thebes, now in the Museo Gregoriano Egizio (Inv. 38073.2.1), the god Thoth identifies himself as *ḥr-ib ir.t=f*, 'the one who is within his eye'.[146] Just as the moon is frequently identified in Egyptian sources as the left eye of celestial deities, so too the sun disk is equated with the right eye, and Shu is well attested in the role of the sun god's right eye.[147]

The reference to Amun dwelling within the solar disk is apposite at this point, inasmuch as the next line of the present fragment (line 10) calls the deity *pꜣ ḏf*, 'the pupil' (scil. of the eye), another epithet with explicit solar associations. Compare a passage in the Magical Papyrus Harris, where the god is addressed as *Ỉmn imn sw m dfḏ=f bꜣ psḏ m wḏꜣ.t=f*, 'Amun who conceals himself in his pupil, ba who shines forth in

[139] Ibid., VII, 1-2. Cf. Zandee, *Der Amunhymnus des Papyrus Leiden I 344, verso*, plate 7 and p. 603; de Meulenaere in Clarysse, Schoors and Willems (eds.), *Egyptian Religion: the Last Thousand Years* 2, p. 1130, note (e).

[140] Epigraphic Survey, *Medinet Habu* 1, plate 44, line 16.

[141] For this aspect, see Zandee, *op. cit.*, pp. 537-8, 610-12, and 622-7.

[142] For the writing of Shu's name, see note (b) on the line in question.

[143] For Shu as the sun disk, see Smith, *The Mortuary Texts of Papyrus BM 10507*, pp. 121-2 [note (c) on XI, 19] and references cited there.

[144] Cauville, *Dendara* 10, vol. 1, p. 112, line 9, and vol. 2, plates 46 and 72.

[145] Roulin, *Le Livre de la Nuit* 1, pp. 327-8; 2, p. 151.

[146] Gasse, *Les Sarcophages de la Troisième Période Intermédiaire du Museo Gregoriano Egizio*, p. 210 and plate 58; Haikal, *RSO* 63 (1989), p. 17, note 31.

[147] See Smith, *The Mortuary Texts of Papyrus BM 10507*, p. 122.

his eye'.[148] A parallel for this address occurs in the great Hibis hymn to the deity.[149] As Zandee has noted, both pupil and eye in this context refer to the visible disk of the sun.[150] Other solar deities share the epithet *ḏf(ḏ)*, 'pupil', for instance, Harsomtus at Dendera.[151] It is an apt designation for Amun in the form of a black bull, inasmuch as the pupil of the eye itself was sometimes called by the Egyptians *km*, 'the black one'.[152]

In line 12, it is said of the deity: *n3.w-sbk n3y=f ꜥb.w*, 'His horns are short.' Perhaps the same statement is to be restored in lines 1 and 8 as well. This is puzzling, inasmuch as one does not expect attention to be drawn to such a seemingly negative characteristic. The statement also contradicts one made in lines 8 and 10 of the present fragment to the effect that Amun's horns are long (*ẖ3y*). Nevertheless, the reading and meaning of the words here are unambiguous. Perhaps one should emend *sbk* to something like *spd*, 'sharp', a term frequently used in descriptions of the horns of a bull.[153]

At the end of line 12, *ir.t*, 'eye', occurs in a broken context. Compare e.g. the Hibis hymn to Amun in which the Ogdoad say that the deity's eyes are made of electrum.[154] Perhaps a similar statement is to be restored in our text. For the identification of Amun's eye with the solar disk, see discussion of the epithet *p3 ḏf*, 'the pupil', two paragraphs above.

In line 13, the *intm* of the god is said to be fair or beautiful. If my suggestion that this might be a term for the phallus is correct,[155] then one has here a further reference to Amun's sexual potency and attractiveness. See the discussion concerning the term *mḏ*, 'phallus', in line 4 above. Compare further the hymn to Amun in P. Leiden I 344 verso, where the god in his aspect of *k3 mw.t=f*, 'bull of his mother', is called *ꜥbꜥw m nfr.w=f*, 'the one whose beauty (= phallus) is vaunted'.[156]

In line 14, Amun is called *ẖ3y st*, 'the one with a long tail'.[157] The bull's tail, like the feathers and horns discussed above, is an attribute of royalty. The god Amun is

[148] Ibid., IV, 10 (Lange, *Der magische Papyrus Harris*, p. 38).

[149] See Davies, *The Temple of Hibis in El Khārgeh Oasis* 3, plate 33, middle, line 7.

[150] *Der Amunhymnus des Papyrus Leiden I 344, verso*, pp. 123, 195-6, and 737-8.

[151] See e.g. Chassinat, *Dendara* 1, pp. 63, lines 6-7, and 139, line 17; *Dendara* 2, p. 178, line 2; *Dendara* 5, p. 75, line 4; idem and Daumas, *Dendara* 6, p. 122, lines 9-10. For examples with other deities, see *Wb.* 5, 573, 6-7.

[152] See *Wb.* 5, 124,12; Jackson, *CdE* 61 (1986), pp. 126-7. Compare Chassinat, *Edfou* 4, p. 29, line 11, where Horus is called *km wr m-ẖnw ir.t=f*, 'black bull within his eye'.

[153] See *Wb.* 4, 108, 12; also references cited in note 141 above.

[154] Davies, *op. cit.*, plate 33, middle, line 21.

[155] See textual note (b) on the line in question.

[156] P. Leiden I 344 verso, IV, 8 (Zandee, *op. cit.*, p. 333 and plate 4). Numerous parallels are cited ibid., pp. 333-40.

[157] Compare the similar expression *q3 sd* discussed by Caminos in *Late-Egyptian Miscellanies*, p. 196.

sometimes depicted wearing it even when he is shown in anthropomorphic form.[158] The line goes on to refer, in a damaged phrase, to the front or forepart of the deity.

A further reference to the eye of Amun occurs in line 18. As discussed in the textual note ad loc., perhaps an adjective like *i3b*, 'left', or *wnm*, 'right', should be restored after the words *t3y=f ir.t n*. For the identification of the right and left eyes of the god as the sun and moon respectively, see commentary on line 9 above. The deity's feet are mentioned in line 20, and there is a final reference to his horns in line 22, both unfortunately in a damaged context.

As a black bull (*k3 km*), Amun plays a significant part in the cosmogony of P. Berlin 13603. According to that text, the four male members of the Ogdoad have the forms of bulls and the four female members those of cows. At the behest of their creator (possibly Ptah, but the deity's name is not preserved), the four bulls merge into a single black bull and the four cows into a single black cow, identified as Amun and Amaunet respectively.[159] The text goes on to say: *p3y p3 k3 r t3 iḥ.t rn=s iw=f bš t3 mṯwy r p3 mw ẖn p3 šy ꜥ3 n Ḫmnw*, 'The bull hastened to the aforesaid cow, ejaculating semen into the water in the great lake of Hermopolis.' The semen fertilises or creates a lotus, from which the sun emerges in the form of a child.[160] Subsequently, the Ogdoad revert to their original forms,[161] but apparently only temporarily, for at a later point the Berlin papyrus goes on to say: 'It happened that the name of the bull and (that of) the cow were united', i.e. they were merged into a single entity.[162]

Allusions to this account, or one very similar, occur in hieroglyphic inscriptions of the Graeco-Roman Period. The Ogdoad are described as bulls and cows in two scenes in the temple of Edfu showing the rite of offering the lotus.[163] One of these refers to the deposit of their seed in the water as well,[164] while scenes at Edfu and Esna link this seed with the formation of the lotus from which the youthful solar deity was thought to have emerged.[165]

The incorporation of the four male members of the Ogdoad in the form of a single bull has a parallel in an inscription on the propylon of the temple of Montu at Karnak,

[158] Zandee, *op. cit.*, pp. 668-71.

[159] P. Berlin 13603, II, 4-6 (Erichsen and Schott, *Fragmente memphitischer Theologie in demotischer Schrift*, p. 312 [14] and plate 5). Cf. ibid., III, 3-4, where this account may have been recapitulated.

[160] Ibid., II, 6-7.

[161] Ibid., II, 9.

[162] Ibid., II, 15. For this union 'according to name', see commentary on fragment 2 of our text.

[163] Chassinat, *Edfou* 4, p. 140, lines 11-2; idem, *Edfou* 5, p. 85, line 12. On these scenes, see Ryhiner, *L'Offrande du lotus*, pp. 36-41 and 45-9, with bibliography. In the former, the male Hermopolitan deities are called specifically *k3.w p3y*, 'ejaculating bulls'. Cf. the homophonous verb *p3y*, 'hasten', predicated of the black bull in P. Berlin 13603.

[164] Chassinat, *Edfou* 4, p. 139, lines 13-14.

[165] Cf. Chassinat, *Edfou* 6, p. 16, lines 2-9 (Ryhiner, *op. cit.*, pp. 146-7); Sauneron, *Esna* 3, pp. 173-5 (no. 272) (idem., *Esna* 5, pp. 142-4; Ryhiner, *op. cit.*, pp. 130-2).

where that god is said to be *ṯ3 fdw nt p3wty.w tpy.w dmḏ ḏt=sn m rnp*, 'the four males of the first primordial ones, their bodies united as a bull'.[166] In another inscription from the same monument, this deity is called *k3 nḫt spd ḥn.ty ṯ3.w nw p3wty.w dmḏ m Mnṯ r wtṯ šw*, 'mighty bull with sharp horns, the males of the primordial ones being united as Montu in order to engender the sunlight'.[167] In some contemporary Theban sources, Montu thus constituted is explicitly identified with Amun or Amun-Re.[168]

Ultimately, according to the account of P. Berlin 13603, all eight members of the Ogdoad are assimilated within a single being who incorporates both the male and the female members. The being in question is Amun. One finds this identification already in the hymn to that god preserved in P. Leiden I 350 (temp. Ramesses II), where he is told *Ḫmny.w ḫprw=k tpy r km=k nn iw=k wʿty*, 'The Ogdoad were your initial form, until you completed these, being one.'[169] Elsewhere in the same papyrus, it is said of the god: *ky ḫprw=f m Ḫmny.w*, 'Another of his forms is the Ogdoad.'[170] In a further passage, Amun is called *wʿ wʿw nṯri imn rn=f imy nṯr 8*, 'the sole unique one, divine, who concealed his name among the eight gods'.[171]

One section of the Leiden papyrus, unfortunately damaged, begins with a reference to the four goddesses of the first occasion (*sp tpy*), then goes on to describe Amun as *k3 wr km* [*wr*(?)] *nfr.wt* [....] *r km psḏ.t ir ḥm.t sḫpr mṯ3 š3ʿ=f nḏmy.t r nfr.wt hi=f 3mm=f nn wn ḥm.t ḫʿ m Nwn ms ntt iwtt it it.w mw.t mw.wt p3 k3 n nfr.wt tf3 4.t*, 'great bull, black bull of the maidens [....] to complete the Ennead, who made the vagina, who created the phallus, initiating sexual desire for maidens, masturbating when there was no vagina, one who arose from the Primaeval Ocean, creator of what exists and what does not exist, father of fathers, mother of mothers, the bull of those four maidens'.[172] The four maidens mentioned at the end of this passage are identical with the four goddesses of the first occasion, i.e. the female members of the Ogdoad. The words *p3 k3 n nfr.wt tf3 4.t* should signify that Amun is the possessor or lover of the maidens in question.[173] However, in view of the emphasis laid upon the androgynous nature of the deity here,

[166] Sethe, *Urkunden* 8, p. 26 (30b). For the use of the term *dmḏ* in this context, see Gutbub, *Textes fondamentaux de la théologie de Kom Ombo*, pp. 457-60.

[167] Sethe, *op. cit.*, p. 5 (6b). Cf. Sternberg-El Hotabi, *Der Propylon des Month-Tempels in Karnak-Nord*, p. 42, note 1. For further parallels, see Sethe, *Amun und die acht Urgötter von Hermopolis*, p. 85; Drioton, *CdE* 6 (1931), pp. 259-70; Herbin, *Le Livre de parcourir l'éternité*, p. 156.

[168] See Sethe, *Amun und die acht Urgötter von Hermopolis*, p. 60; Quaegebeur in Verhoeven and Graefe (eds.), *Religion und Philosophie im alten Ägypten*, p. 264.

[169] P. Leiden I 350, III, 22-3 (Zandee, *De Hymnen aan Amon van Papyrus Leiden I 350*, plate 3). Cf. discussion ibid., p. 63; Assmann, *Re und Amun*, pp. 222-3.

[170] P. Leiden I 350, IV, 14 (Zandee, *op. cit.*, plate 4).

[171] Ibid., II, 2 (Zandee, *op. cit.*, plate 2).

[172] P. Leiden I 350, IV, 26 – V, 4 (Zandee, *op. cit.*, plates 4-5, with some readings adopted from Gardiner, *ZÄS* 42 [1905], p. 36).

[173] See *Wb.* 2, 258, 9; Sethe, *Amun und die acht Urgötter von Hermopolis*, pp. 84-6.

it is worth considering whether they could not refer instead or as well to Amun's incorporation of the creative power of the female members of the Ogdoad within his own being.

The god Montu who, as we have seen, can be identified with Amun,[174] is also said to incorporate all eight members of the Ogdoad within himself, and not just the four males. Thus, one of the inscriptions previously quoted from the propylon of his temple at Karnak states: *wnn bꜣ.w nt pꜣwty.w ꜥb sm=sn m Wr.ty sw m rnp*, 'The bas of the primordial ones unite their forms in Werti (= Montu), he being as a bull.'[175] In another inscription from this propylon, a form of Montu called *pꜣ kꜣ ꜥꜣ wr šps*, 'the very great and noble bull', is said to be *Nwn wr m-ꜥb ms.w=f m ti.t=sn nt rnp*, 'the great Primaeval Ocean together with his children (= the Ogdoad) in their manifestation as a bull'.[176]

A further allusion to the union of the eight Hermopolitan divinities, omitting the name of the god in whom they are incorporated, occurs in one of the scenes from the temple of Edfu showing the rite of offering the lotus to which attention has already been drawn above.[177] There, the members of the Ogdoad, described at one point as bulls and cows, are said to have deposited their seed in the water *dmḏ m ḳd wꜥ*, 'united in a single form'.[178]

This idea finds expression in Egyptian iconography as well. Thus, Amun or Amun-Re is sometimes represented with a human body surmounted by four rams' heads. Equally, in the Graeco-Roman Period, the name *Ḫmny.w*, 'Ogdoad', can be written with a hieroglyph depicting four rams' heads linked together. This writing evokes the concept of the Hermopolitan deities as a form or manifestation of Amun.[179]

A similar idea may underlie Graeco-Roman Period writings of *Ḫmny.w*, 'Ogdoad', within a cartouche, i.e. united as a single (royal) name.[180] Cf. discussion of the concept of union 'according to name' in the commentary on fragment 2; also kingly attributes ascribed to Amun in the fragment under consideration. More equivocal are references in demotic texts from the Theban area to the deity *Ỉmn-nꜣy.w-Ḫmniw*.[181] This name is conventionally rendered as 'Amun of the Ogdoad'. It could, however, be interpreted as 'Amun (who is) the Ogdoad' as well.

[174] See references cited in note 168 above; Sternberg-El Hotabi, *Der Propylon des Month-Tempels in Karnak-Nord*, p. 145.

[175] Sethe, *Urkunden*, 8, p. 6 (6g). For Werti as a name of Montu, see *Wb.* 1, 336, 16.

[176] Sethe, *op. cit.*, p. 24 (27b). For Montu *pꜣ kꜣ ꜥꜣ wr šps*, see Sternberg-El Hotabi, *op. cit.*, pp. 102-3, note 1, and 139-41. For *Nwn wr*, cf. Leitz, *Tagewählerei*, p. 70.

[177] See first reference cited in note 163 above, together with note 164.

[178] Chassinat, *Edfou* 4, p. 139, lines 14-15.

[179] See Quaegebeur in Verhoeven and Graefe (eds.), *Religion und Philosophie im alten Ägypten*, pp. 261-4.

[180] Sethe, *Amun und die acht Urgötter von Hermopolis*, pp. 45-6 and 86 (with a different interpretation).

[181] Cf. Thissen, *Die demotischen Graffiti von Medinet Habu*, pp. 35-6, 193, and references cited there.

It is within this context that one should understand the panegyric to Amun as a black bull with eight hypostases in the present fragment of our cosmology. The eight hypostases, in my view, are best interpreted as the members of the Ogdoad united in that god's being. This accounts for the presence of the words [*m*]*w*(*?*) *Nw*[*n.t*], '[li]ke(?) Nau[net]', in line 15 of our fragment.[182] It is in that line that the section of the panegyric devoted to the eighth hypostasis of Amun begins, and in our cosmology Naunet figures as the eighth member of the Ogdoad.[183] Thus, it is appropriate to have a reference to her at this point.

Fragment 3 may preserve other allusions to these gods. Note the possible reference to *8 nṯr*, 'eight divinities', in line 13 of our fragment.[184] This is well attested as a variant designation of the Hermopolitan deities.[185] It recurs in fragment 2, lines 8 and 14, of our text, in the former passage with the order of the two elements inverted. There may be a further allusion in line 17 of the present fragment, which mentions some beings in bovine form who see (*mꜣꜣ*) a child.[186] This recalls a passage in the cosmogony of P. Berlin 13603 where the members of the Ogdoad, in the form of bulls and cows, are said to look upon the infant sun god after his birth.[187] Perhaps one should restore [*kꜣ.w iḥ*].*w* as the subject of the verb *mꜣꜣ* in this line.

As we have seen, fragment 2 describes the creation of the Ogdoad by Ptah and the subsequent unions of its four male members and its four female members into two distinct entities. The present fragment describes the culmination of this process, in which all eight beings have merged into one. Its panegyric enumerates in detail the various attributes and powers which Amun was believed to possess as the embodiment of the Ogdoad in taurine form. Most prominent among these are sexual potency and fecundity, might and ferocity, the dignity of divine kingship, and celestial glory as the manifestation of solar light.

FRAGMENT 4 — Plate 4

Dimensions: 8.9 × 21.6 cm. Colour: medium brown. Related fragments: No. 19.

Transliteration

Column I

x+1.]..
x+2.]..

[182] For the reading, see textual notes ad loc.
[183] See commentary on fragment 2, lines 13-14.
[184] It is not absolutely certain that the numeral and noun go together. See note 125 above.
[185] Sethe, *op. cit.*, p. 81.
[186] The broken noun denoting them is determined with the cattle sign; see textual note (a) ad loc.
[187] P. Berlin 13603, II, 13. See commentary above, also that on fragment 4, II, 22.

x+3.]..
x+4.] *šm(?)*
x+5.]..
x+6.] *rt*
x+7. *t*]*ȝ*
x+8.] *ṯȝw(?)*

Column II

1. *n fy* [
2. *n nȝ nt nb n*[*t/ḥr*(?)
3. *i-ir=f ḏ n=f* [
4. *tȝ p.t pȝ s.*[
5. *pȝ 4 ṯȝw n tȝ p.*[*t*
6. *n rn=w i-ir* [
7. *pȝ t*[*ȝ*(?)] *m-sȝ* .[
8. .[..] *ḫpr ṯȝw ḫ*[*ȝy*(?)
9. [*p*]*ȝ Nwn* [..].[
10. *ḫn pȝ kky* [
11. *lhny r-iw=f r* [..]..[
12. *nȝy=f ʿʿe.w tr=w my/šy*[..].[
13. *šm iy n-im=f pȝy tw=s*(?) ..[
14. *bš=f tȝy=f ḫny m-qty* [............ *r*]-
15. *hn-r pȝ hrw r-ḏbȝ.ṯ=s ir=f* [
16. *ḥmsy=f ḥr-ȝ*[*ṯ*]*=s ir=f* [
17. ... *ṯsy=f* ...[
18. *ir tȝ p.t* [*pȝ tȝ*] *irm tȝ tw*[*ȝ.t*
19. *m*[..]*n/y=w*(?) *swḥy.t wn-mtw=w* [
20. *mtw* [*pȝ*]*-Šy pȝ Nwn* [
21. *pr.w n-im=f pḥ nȝy=f n.*[
22. *iw=*[*w*] *ḏ Šy n rn n* [
23. *pȝ* [...]. *nt*(?) *iwt=w ḏ=f n=w šp*[*s*(?)

Translation

Column I

x+1.]..
x+2.]..
x+3.]..
x+4.] go(?)
x+5.]..
x+6.] grow

x+7. la]nd
x+8.] wind(?)

Column II

1. of/for carrying [
2. of everything whi[ch/on(?)
3. He said to him [
4. the sky, the *s*.[
5. the four winds of the s[ky
6. as their names. *i-ir* [
7. the ea[rth(?)] after .[
8. .[..] A h[igh(?)] wind arose [
9. [th]e Primaeval Ocean [..].[
10. in the darkness [
11. resting place while he will [..]..[
12. all of his limbs *my/šy*[..].[
13. It means going and coming in it. Behold(?) ..[
14. He shed his skin like [............ up]
15. until today because of it. He made [
16. He sat upon it and made [
17. ... He elevated ...[
18. make the sky, [the earth], and the under[world
19. They(?) *m*[..]*n/y* an egg. They have [
20. and [*pꜣ*]-*Šy pꜣ Nwn* [
21. came forth from it. His *n*.[..] reached [
22. [they] saying 'Shai' as the name of [
23. the [...]. which(?) is between them. He said to them: Nob[le(?)

Textual notes

Column I

Line x+4

The upper portion of what I have read as *šm* with a query might conceivably be part of a flesh determinative instead.

Line x+8

The signs read here as *ṯꜣw* could also be the sail determinative of a longer word now lost.

Column II

Line 2

The partially preserved sign after *nꜣ nt nb* could be either *nt* or *ḥr*.

Line 7

The restoration *t*[*ꜣ*] after *pꜣ* is not absolutely certain, but the traces suit and the noun in question would fit the available space nicely. Cf. undamaged writings of the same word in e.g. fragment 5, I, 10 and II, 6, fragment 13, II, 5 and 13.

Line 8

After *ṯꜣw*, perhaps restore *ẖ*[*ꜣy*], 'h[igh]'. Cf. undamaged writings of that word in fragment 3, lines 10 and 14. On the possible significance of a reference to a high wind here, see commentary below.

Line 11

For the noun *lhny*, 'resting place', see also fragments 1, line 6; and 2, line 5. Perhaps the *lhny* mentioned in the present line is identical with the one mentioned there.

Line 12

The word after *tr=w* appears to have begun with either *my* or *šy*.

Line 13

For the suggested reading *tw=s*, 'behold', after *pꜣy*, cf. the examples of that word in fragment 2, line 13, and fragment 23, II, 2.

Line 14

ẖny written with flesh and 'leather' determinatives (not in *Glossar*) is identical with the *ẖn.t*, 'Tierfell, Schlauch', of *Wb.* 3, 367, 12. Further demotic instances of the word occur in fragment 10, line x+10, fragment 14, II, 6, and P. BM 10508, XXI, 5. The last passage speaks of the shedding (*bš*) of the skin of a human being (Thissen, *Die Lehre des Anchscheschonqi*, p. 107).

Line 16

(a) With the traces of *ḥmsy*, 'sit', at the beginning of the line, cf. the better preserved example of that verb in line 20 of fragment 1. Further instances are cited in note (a) on the latter.

(b) For other writings of *ꜣt*, 'back', with the spine determinative written below the *t* as here, see fragment 14, II, 18, *Glossar*, p. 12.

Line 18

Only a trace of *tꜣ*, 'earth', is visible after the break in the middle of the line. The restoration is conditioned by the spacing and the presence of *tꜣ p.t* and *tꜣ tw*[*ꜣ.t*] immediately before and after the damaged area.

Line 19

(a) The word at the beginning of the line starts with *m* and appears to end with *n* or *y* and a man with hand to mouth determinative. Probably no more than a single sign has been lost in the middle. The word in question could be a verb or, alternatively, a preposition like *m-qty*.

(b) With the traces of *swḥy.t*, 'egg', here, cf. the better preserved example of that noun in fragment 15, line x+6. What I have read as the suffix pronoun *w* before *swḥy.t* could also be the indefinite article *wꜥ.t*, agreeing with the latter. On the significance of the egg mentioned in this passage, see commentary below.

Line 21

The *w* after *pr* at the beginning of the line could be the plural marker, the stative ending, or the third person plural suffix pronoun. See commentary ad loc. for further discussion.

Line 22

For the naming formula *iw=* [*w*] *ḏ Šy n rn*, see discussion on the ordering of the fragments in the Introduction.

Line 23

(a) What I have tentatively read as *nt* before *iwt=w* could also be a *ṯ* attached to the end of the preceding word.

(b) For the suggested restoration *šp*[*s*], 'nob[le]', see *Glossar*, p. 503. This adjective is written *šbs* in fragment 19, line x+19; cf. note (b) ad loc.

Commentary

Only isolated words can be made out in the initial column of this fragment. In line 3 of the second column one male being addresses another, but too little is preserved to allow the identity of either to be ascertained. The words *tꜣ p.t*, 'the sky', stand at the beginning of line 4. The following line contains a reference to the four winds of the sky.

The winds in question are those associated with the four points of the compass. In later sources, each of these was attributed with special powers. Among other things, the north wind was supposed to bring the inundation to fields and fertilise them, the south wind strengthened the eggs of the unborn, the east wind opened the doors of the sky, preparing the way for the sun, and the west wind received the bas of the gods when they would set in the West.[188] Each of the four winds could be identified with a particular divinity. For example, in Spell 161 of the Book of the Dead, the north wind is equated with Osiris, the south wind with Re, the west wind with Isis, and the east wind with Nephthys.[189]

Additionally, all four winds could be embodied in a single deity. At Kom Ombo, for instance, it is said of Haroeris: *ii=f m 4 ṯꜣw*, 'He comes as the four winds.'[190] Of the divinities mentioned thus far in our cosmology, Amun is the one most closely associated with these winds. In the hymn addressed to the ten bas of that god in his temple at Hibis, it is said: *twt ḏsr ḫprw m 4 ṯꜣw n p.t*, 'You are one with incomprehensible form as the four winds of the sky.'[191] In the demotic story of the Contest for the Benefice of Amun, the four winds are described as the god's diadem (*sḥn*).[192]

[188] See Kurth in *Lexikon der Ägyptologie* 6 (1986), columns 1266-72, with references to earlier literature; Willems, *The Coffin of Heqata*, pp. 463 and 472; Leitz, *Tagewählerei*, pp. 17, 95, and 132-3; Kákosy in van Dijk (ed.), *Essays on Ancient Egypt in Honour of Herman te Velde*, pp. 219-30.

[189] See Budge, *The Book of the Dead* 3, p. 21, lines 14-15.

[190] De Morgan et al, *Kom Ombos* 1, p. 262, no. 332. Cf. Gutbub, *Textes fondamentaux de la théologie de Kom Ombo*, pp. 100 and 279.

[191] Davies, *The Temple of Hibis in El Khārgeh Oasis* 3, plate 31, middle, seventh vertical line.

[192] P. Spiegelberg, I, 7 (Spiegelberg, *Der Sagenkreis des Königs Petubastis*, plate 1). Cf. idem, *ZÄS* 49 (1911), p. 128.

Line 6 refers to the names of the winds in question. The context is damaged, but perhaps one had here a reference to the occasion when their names were given. Line 8, if my suggestion to restore *ẖ[ꜣy]* after *ṯꜣw* there is correct,[193] says: 'A h[igh] wind arose.' Perhaps this means that the four winds merged into a single one. The assimilation of four entities into one has already been met with in fragment 2 of our text, where it involved the names of the male and female members of the Ogdoad. One may have had here a reprise of this theme.

Line 9 preserves the name of the Primaeval Ocean; the following line refers to someone or something in darkness. In line 11, a *lhny*, 'resting place', is mentioned in association with a male being or deity whose name is not preserved. It is not clear whether this is the same resting place as the one which appears in fragments 1, line 6, and 2, line 5. As noted in the commentary on the former, the Primaeval Ocean and a resting place are mentioned together in the much earlier Coffin Text Spell 80 as well. There, the primordial god Atum describes the creation of his child Shu (= the wind) at a time *sk wi wꜥ.kwi ḥnꜥ Nww m nnw.t n gm.n=i bw ꜥḥꜥ=i im n gm.n=i bw ḥms=i im*, 'when I was alone together with the Primaeval Ocean in a state of inertness, unable to find a place in which to stand or sit'.[194] In this spell, the creation of the wind is linked with a dialogue between Atum and the Ocean, which other, related spells say took place *m kkw*, 'in darkness'.[195] The reference in these texts is to the time before the sky and earth were separated, an act accomplished subsequently, according to some sources, through the agency of the wind.[196] Perhaps lines 9-11 of the present fragment refer to the same stage in the process of creation.

The limbs of the anonymous male being or deity are mentioned in line 12. In line 14, it is said of him: *bš=f tꜣy=f ẖny*, 'he shed his skin'. This is a rather cryptic statement. On the one hand, the shedding of skin suggests a serpent. On the other, a serpent could not be said to possess limbs, as the being with which we are concerned clearly does. Perhaps the term 'skin' here is used in a metaphorical sense. Compare the already-cited Coffin Text Spell 80, in which Shu says of himself *ḥbsw=i pw ṯꜣw n ꜥnḫ*, 'the breath of life is my clothing',[197] and *inm=i pw mḏdw ṯꜣw*, 'the force of the wind is my skin'.[198] Shu goes on to say, in the same spell: *rḏw=i pw qrr n p.t iꜣd.t=i pw nšn iḫḫw*, 'The storm of the sky is my efflux, and the tempest at dusk is my outpouring.'[199] In the light of these passages, if the male being mentioned in lines 12-14 of this column is a god of the wind, then perhaps what is meant when he is said to shed his skin is that he sends forth

[193] See textual note on the line in question above.

[194] De Buck, *The Egyptian Coffin Texts* 2, p. 33e-g.

[195] Cf. ibid., pp. 7f-8a (Spell 76); Bickel, *La Cosmogonie égyptienne avant le Nouvel Empire*, pp. 46-9.

[196] See below.

[197] De Buck, *The Egyptian Coffin Texts* 2, p. 29f. Cf. ibid. 1, pp. 372d-375a.

[198] De Buck, *The Egyptian Coffin Texts* 2, p. 30c. Cf. Zandee, *ZÄS* 101 (1974), p. 67.

[199] De Buck, *op. cit.*, p. 30e-f.

a powerful blast of air. Cf. the 'h[igh(?)] wind' mentioned in line 8 above. The limbs to which reference is made in line 12 could also be linked with this element. Compare the hymn to Amun at Hibis temple where the Ogdoad say of that deity: *ḥꜥ.w=f m swḥ*, 'His limbs are the wind.'[200] Similarly, in the hymn to Amun-Re in the temple of Ramesses III at Karnak, that king says to the deity: *ḥꜥ.w=k ṯꜣw r fnḏ nb*, 'Your limbs are the breath for every nose.'[201]

The words [*r*]-*hn-r pꜣ hrw r-ḏbꜣ.ṯ=s*, '[up] until today because of it', in lines 14-15 are part of an aetiological statement. On the function of such statements, see Introduction. The text must have described here how something was still done, or some being or object called by a particular name, as of the time of writing, in consequence of the events narrated in the previous lines. What precisely this was is impossible to say, owing to their poor state of preservation.

Line 17 contains the words *ṯsy=f*, 'he elevated'. The identity of the actor here cannot be determined with certainty. The following line speaks of the creation (*ir*) of the sky, earth and underworld, i.e. the cosmos as known to the Egyptians.[202] This cosmos comes into existence when the sky and earth are separated.[203] A number of Egyptian sources explicitly ascribe this separation to the agency of the wind or a god associated with that element, most prominently Shu.[204] In one text, from the temple of Esna, for example, Khnum-Shu is called *ḏsr ḫprw m 4 ṯꜣw iwn wr iwd ḥw.t r tꜣ*, 'incomprehensible of form as the four winds, the great pillar who separated the sky from the earth'.[205] In another, inscribed on the sanctuary of the temple of Edfu, Shu is called *twꜣ p.t m hh n rꜣ=f*, 'the one who elevated the sky with a blast of air from his mouth'.[206] There is even an Egyptian word for wind, *twꜣ*, the literal meaning of which is 'that which raises' (scil. the sky).[207] Some texts use the verb *ṯsy*, 'elevate', to describe this act of raising the heavens above the earth. An inscription on the propylon of the temple of Khonsu at Karnak, for instance, calls Khonsu-Shu *ṯs p.t m hh n rꜣ=f*, 'the one who elevated the sky with a blast of air from his mouth'.[208] Perhaps lines 17-18 of the column under consideration allude to this act as well.

[200] Davies, *The Temple of Hibis in El Khārgeh Oasis* 3, plate 33, middle, line 5.

[201] Epigraphic Survey, *Ramses III's Temple within the Great Enclosure of Amon* 1, plate 23, A, line 4.

[202] Cf. Assmann, *Re und Amun*, pp. 242-5; Kurth, *Den Himmel Stützen*, p. 70. In a text from the temple of Esna cited by the latter (unnumbered pages between pp. 24 and 25), the work of the creator of the universe is said to consist of *ꜥḫy p.t*, 'raising up the sky', *smn tꜣ*, 'establishing the earth', and *smḏ dwꜣ.t*, 'making the underworld deep'.

[203] See Allen, *Genesis in Egypt*, pp. 14-25; te Velde, *Studia Aegyptiaca* 3 (1977), pp. 161-70; Bickel, *La Cosmogonie égyptienne avant le Nouvel Empire*, pp. 182-98.

[204] Allen, loc. cit.; Kurth, *Den Himmel Stützen*, pp. 70-6 and 78-80.

[205] Sauneron, *Esna* 2, p. 52 (no. 19). Cf. Kurth, *op. cit.*, pp. 33-5.

[206] Rochemonteix and Chassinat, *Edfou* 1, p. 59, line 18; Kurth, *op. cit.*, pp. 36-7.

[207] *Wb*. 5, 250, 18.

[208] See Sethe, *Urkunden* 8, p. 51 (63b). Further examples of *ṯs*(*y*) so used are cited in *Wb*. 5, 405, 17.

Line 19 mentions an egg (*swḥy.t*), again in a damaged context. The egg figures prominently in a number of ancient Egyptian cosmogonies, most notably that associated with Hermopolis, in which the sun is said to have emerged from an egg.[209] Different sources give varying accounts of the origin of this object. Some say that it was laid by a bird;[210] others that it was fashioned by the Ogdoad.[211] Yet others say that it emerged from the primordial lotus which is attested elsewhere as the matrix of the solar deity.[212] A few maintain that the sun god fashioned or produced his own egg.[213]

Among other creation accounts, the cosmogony in the temple of Khonsu at Karnak refers to two eggs, one spat forth by the sky, the other created by Ptah. The second of these is said to have appeared in the Primaeval Ocean with the Ogdoad inside it.[214] In the Neith cosmogony at Esna, the sun god is said to emerge from an egg fertilised in the Primaeval Ocean, one which contains the exudations of the body of Neith.[215]

The egg in our text is mentioned in conjunction with the wind, which is well attested in Egyptian sources as a prerequisite of life.[216] A number of texts refer to the generative or life-quickening powers believed to be inherent in that element. See, for example, the account of the birth of the divine child as a result of the union of wind and fire preserved in the mammisi of Kom Ombo.[217] In the mammisi at Edfu, Khnum is said to be *sipy sꜣ m ẖ.t nf nfr m rꜣ=f*, 'the one who fashions the son in the womb (with) the fair breath from his mouth'.[218] The creative power of the wind that stirs the reed thicket which plays such an important role in the Edfu cosmogony is a further illustration of this phenomenon.[219]

209 For this doctrine, see Roeder, *Hermopolis 1929-1939*, p. 186; Sauneron and Yoyotte in *La Naissance du Monde*, pp. 59-62; Bickel, *La Cosmogonie égyptienne avant le Nouvel Empire*, pp. 233-9, all with references to earlier literature.

210 Roeder, loc. cit. Cf. Bergman, *Isis-Seele und Osiris-Ei*, pp. 77-81.

211 Sethe, *Amun und die acht Urgötter von Hermopolis*, p. 80.

212 See e.g. Chassinat, *Edfou* 6, p. 16, lines 6-7; Ryhiner, *L'Offrande du lotus*, pp. 146-7. The theme of the sun's emergence from a lotus is treated exhaustively in the latter work.

213 See Assmann, *Liturgische Lieder an den Sonnengott*, pp. 125-6; Fischer-Elfert, *Lesefunde im literarischen Steinbruch von Deir el-Medineh*, pp. 151-2.

214 Sethe, *Amun und die acht Urgötter von Hermopolis*, pp. 62-3 and plate 2; Parker and Lesko in Baines et al (eds.), *Pyramid Studies and other Essays Presented to I.E.S. Edwards*, plate 34. For the egg created by Ptah, see commentary on fragment 2 above.

215 Sauneron, *Esna* 3, p. 31 (206, 7-8). Cf. idem, *Esna* 5, pp. 262-4.

216 See Morenz, *Religion und Geschichte des alten Ägypten*, pp. 469-76; also commentary on fragment 13 below.

217 Gutbub, *Textes fondamentaux de la théologie de Kom Ombo*, pp. 321-63.

218 Chassinat, *Le Mammisi d'Edfou*, p. 17, lines 16-17.

219 See Alliot apud Barucq, *BIFAO* 64 (1966), pp. 135-6; Finnestad, *Image of the World and Symbol of the Creator*, p. 41, note 131.

A few sources make explicit reference to the wind's ability to create or fertilise eggs. Thus, in an inscription on the propylon of the temple of Montu at Karnak, the Theban nome is called *swḥ.t pr m swḥ*, 'the egg which came forth from the wind'.[220] In Coffin Text Spell 80, Atum says of his offspring Shu, whose clothing is described as the breath of life, and skin as the force of the wind,[221] *rḫ.n=f sꜥnḫ imy swḥ.t m ẖ.t*, 'He has learned how to bring to life the one who is in the egg in the womb.'[222] Morenz has devoted a lengthy study to this particular aspect of the wind.[223] It may be that the egg in our text was said to have been made or fertilised by that element as well.

Line 20 mentions *pꜣ- Šy*, 'Pshai', and *pꜣ Nwn*, 'the Primaeval Ocean', in conjunction. The same pairing occurs in fragment 1, line 2. As noted in the commentary on that line, the precise syntactical relationship of the two names is uncertain. One could translate either 'Pshai the Primaeval Ocean' or 'Pshai (in) the Primaeval Ocean'. If the latter is the correct translation, and *pr.w* in line 21 of the present column is a stative (see textual note ad loc. for the various possibilities), then perhaps the words *pr.w n-im=f*, 'came forth from it', there refer to the emergence of the deity from the flood.

Line 22 contains the phrase *iw=*[*w*] *ḏ Šy n rn n* [...], '[They] saying "Shai" as the name of [...].' In other words, one has here a description of how (P)shai was given his name. Since the naming of a thing or being is normally a concomitant of its creation, this line should mark the conclusion of a passage recording how that deity came into being.[224] As *pꜣ-Šy pꜣ Nwn*, however, he already figures in fragment 1 as the eldest of the gods and initiator of all creation. Therefore, it is not a question of his coming into existence for the first time. As explained in the commentary on fragment 1, one has in the present passage rather an account of the initial manifestation in the created cosmos of a pre-existent deity; in other words, creation viewed as an act of revelation as opposed to a process of coming into being from nothingness. The transition of this deity from hidden to revealed existence is marked by the change in his name from *pꜣ-Šy pꜣ Nwn*, 'Pshai the Primaeval Ocean' or 'Pshai (in) the Primaeval Ocean' to the simpler form (P)shai which is used hereafter in the text. As the change of name indicates, the key element in this transition is the deity's emergence from the waters of the Primaeval Ocean, to which allusion may have been made in the line immediately preceding.[225] In view of the reference to an egg in line 19, it is probable, although not absolutely certain, that Pshai was said to have been revealed from within such an object.

[220] Sethe, *Urkunden* 8, p. 36 (49k). Cf. idem, *Amun und die acht Urgötter von Hermopolis*, p. 118, note 1.

[221] See above.

[222] De Buck, *The Egyptian Coffin Texts* 2, p. 33c. For an explanation of the expression 'egg in the womb', see Sauneron in *Mélanges Maspero* 1, fascicule 4, pp. 115-16.

[223] See first reference cited in note 216 above.

[224] See discussion of naming formulas in our cosmology in the Introduction.

[225] See discussion of the words *pr.w n-im=f* above.

As noted above, in most Egyptian creation accounts which mention an egg, the being who emerges from it is the sun god. In Quaegebeur's monograph on (P)shai, the evidence pertaining to the solar character of that deity has been assembled. For instance, in an inscription on a sarcophagus of the Ptolemaic Period, now in the Louvre, it is said to Re: *twt is ẖꜣ.t Šꜣy*, 'You are the body of Shai.'[226] In a scene depicting the rite of offering the lotus at Dendera, the newly emergent solar deity Harsomtus is called *Šꜣy šꜣʿ ii m š ʿꜣ*, 'Shai, the first to come (into existence) in the great lake'.[227] In a similar scene from the Roman Period mammisi at that site, the god is called *Šꜣy šꜣʿ ḫpr m-ḥꜣ.t*, 'Shai, the first to come into existence at the beginning'.[228] Likewise, the Greek magical texts cited in the commentary on fragment 1 equate Pshai with Helios, the illuminator of the entire cosmos.[229] Moreover, in the demotic cosmogony of P. Berlin 13603, at the conclusion of an account of the emergence of the sun god from a lotus in the form of a child, it is said that the deities who created him beheld (*nw r*) Pshai.[230] It seems likely, therefore, that the revelation or manifestation of Pshai in our passage is to be understood as the initial appearance of the solar deity.

If the reconstructions that I have proposed above are correct, then one has in the present fragment of our cosmology an account of how the four winds of the sky merged into a single high wind which brought the universe into existence by separating the sky from the earth and forming or fertilising an egg from which the sun (= Pshai) emerged.[231] The wind is personified by a being whose name is nowhere preserved. The question arises, therefore, who might this be? As I have already noted, of the deities mentioned thus far in our cosmology, Amun is the one most closely associated with the wind. He can be identified with the four winds, or they can be said to constitute his diadem.[232] Numerous Egyptian sources attest to his role as the incarnation of this element or the god who dispenses its life-giving power.[233] In the mammisi of Nekhtnebef at Dendera, for example, Amun is addressed thusly: *ntk di ṯꜣw r fnḏ nb* ....

[226] Quaegebeur, *Le Dieu égyptien Shaï*, pp. 100-101.

[227] Ryhiner, *L'Offrande du lotus*, pp. 110-11 and 113, note 12. The *š ʿꜣ* mentioned here is that of Hermopolis; cf. ibid., pp. 196-7.

[228] Daumas, *Les Mammisis de Dendara*, p. 88, line 11. Cf. Ryhiner, *op. cit.*, p. 115, note 4; Quaegebeur, *op. cit.*, p. 84.

[229] Quaegebeur, *op. cit.*, pp. 169-70.

[230] P. Berlin 13603, II, 13 (Erichsen and Schott, *Fragmente memphitischer Theologie in demotischer Schrift*, p. 313 [15] and plate 5). Erichsen finds references to Pshai in two other passages in the Berlin papyrus as well (I, 4 and II, 29), but the reading is far from certain in these cases.

[231] For links between the emergence of the sun god from his egg and the splitting asunder of the firmament in other Egyptian sources, see Bickel, *La Cosmogonie égyptienne avant le Nouvel Empire*, p. 186; Graefe, *Untersuchungen zur Wortfamilie bj3-*, pp. 60-66.

[232] See references cited in notes 191 and 192 above.

[233] For references, see Smith, *The Mortuary Texts of Papyrus BM 10507*, pp. 73 [note (a) on III, 5] and 104-5 [note (a) on VIII, 8].

ntk sʿnḫ ṯꜣ m-ḫn štꜣ.t=f, 'It is you who gives breath to every nose ...., it is you who causes the fledgeling to live within its egg.'[234] Similarly, in the New Kingdom hymn to that god in P. Boulaq 17, he is called *rdi ṯꜣw.w n nty m swḥ.t*, 'giver of breath to the one who is in the egg'.[235] Among other epithets, Amun is addressed as *swḥ mn m ḫ.t nb*, 'the wind that endures in all things'.[236] As Morenz has pointed out, *swḥ* functions as the masculine counterpart to *swḥ.t*, 'egg'. Thus Amun as the wind is, in effect, the egg's consort.[237] The deity is also attested as the separator of sky and earth.[238] Accordingly, it is Amun who is most likely to personify the wind in our fragment.

This identification is supported by the fact that the fragment of our text of which the chief subject is the creative activity of the wind follows immediately after one which delineates the nature of Amun as the embodiment of the Ogdoad. A further link can be suggested as well. In lines 14-15 of the second fragment of our cosmology, the names of the four female members of the Ogdoad are said to meet at the four doors of the sky, and the names of the four male members at the four doors of another region, the name of which is not preserved. The two sets of doors may have formed an interface with each other, at which all eight members of the Ogdoad ultimately unite in Amun's being.[239] The four winds, as an alternate form of that deity, share a common place of origin with him in this respect, since in Spell 161 of the Book of the Dead, the winds are said to come from the four openings (*wbꜣ.t*) of the sky.[240]

If one accepts the identification proposed above, then Amun is the being who makes or fertilises the egg from which Pshai (= the sun) emerges. The anomaly whereby Pshai is both the offspring and the ancestor of Amun has been discussed in the commentary on fragment 1, line 2. He is literally 'elder than the one who created him'. In certain other Egyptian sources, however, it is Amun himself, in his solar aspect, who figures as the occupant of the egg. For instance, in the Hibis hymn of the Ogdoad to that god, they say to him: *ḫʿ=k m n.t m swḥ.t imn.t*, 'You appeared in the flood within the hidden egg.'[241] In another passage from the same hymn, Amun is called *imn sw m ḏf*,

[234] Daumas, *Les Mammisis de Dendara*, p. 31, lines 4-6.

[235] P. Boulaq 17, VI, 5 (Mariette, *Les Papyrus égyptiens du Musée de Boulaq* 2, plate 12).

[236] Sethe, *Urkunden* 8, p. 17 (18b). Cf. idem, *Amun und die acht Urgötter von Hermopolis*, p. 102.

[237] Morenz, *Religion und Geschichte des alten Ägypten*, pp. 484-5. For a further instance of *swḥ* and *swḥ.t* used together, see Sauneron, *Esna* 3, p. 218 (302, 13).

[238] See Kurth, *Den Himmel Stützen*, p. 83, and references cited there.

[239] See commentary on the lines in question.

[240] Budge, *The Book of the Dead* 3, pp. 21-2. Cf. Gutbub in Vercoutter (ed.), *Hommages à la mémoire de Serge Sauneron* 1, pp. 409-10. Yet another connection may lie in the fact that Amun as the embodiment of the Ogdoad has the form of a bull, since the winds are sometimes envisaged as possessing taurine form as well. Cf. Coffin Text Spell 162, where they are invoked as *ṯꜣw.w 4 ipw nw p.t kꜣ.w nw p.t*, 'these four winds of the sky, bulls of heaven' (de Buck, *The Egyptian Coffin Texts* 2, p. 399a).

[241] Davies, *The Temple of Hibis in El Khārgeh Oasis* 3, plate 33, middle, line 23.

'the one who concealed himself in the pupil'.[242] The pupil in question, that of the solar eye (= the sun disk), can be identified with the egg that contains the sun god.[243] Amun and Pshai can be equated with each other as well.[244] Thus, as Morenz has noted, the former is attested both as the creator of the solar egg and as the being who emerged from it.[245] The paradox is well expressed in the hymn to Amun in P. Leiden I 350, where the deity is called *nbw swḥ.t=f ḏs=f*, 'the one who formed his own egg'.[246]

FRAGMENT 5 — Plate 5

Dimensions: 11.6 × 21.9 cm. Colour: medium brown. Related fragments: Nos. 19, 20 and 21.

Transliteration

Column I

1.] *nꜣ ntr.w ḥr pꜣ tꜣ*
2.] *tꜣwy ḥtp*
3.]*š n tꜣ p.t*
4.].*=w ẖn*
5.] *syw.w*
6. *gꜣ*]*y n pr r*
7.] *n pꜣ Nwn*
8.] *ḏ/ir Ḏḥwty ḫpr=f*
9.] *iṱ r-r=w*
10.]*=f* ⸢*sṱ*⸣*=f pꜣ tꜣ*
11.]...*=f iwt*
12.] *r pꜣ ḥr n*
13.] ⸢*r-bnr*⸣*-n rꜣ=f*
14.]. *m-qty pꜣ mtw*
15.]... *mtw pꜣ mw*
16.]..[..]
17.]...*=f ꜥn*
18.]....*y*

[242] Davies, *op. cit.*, plate 33, middle, line 7.

[243] See Drioton, *RdE* 1 (1933), pp. 81-5; Bickel, *La Cosmogonie égyptienne avant le Nouvel Empire*, p. 238. Cf. fragment 3, line 10, of our text, where Amun is called *pꜣ ḏf*, 'the pupil'.

[244] See Quaegebeur, *Le Dieu égyptien Shaï*, pp. 79-80.

[245] Morenz, *Religion und Geschichte des alten Ägypten*, p. 481.

[246] P. Leiden I 350, IV, 10-11 (Zandee, *De Hymnen aan Amon van Papyrus Leiden I 350,* p. 74 and plate 4).

19.]..... *nꜣ.w*
20. *f]y pꜣ hb*
21.] *ḥr pꜣ-Šy*
22.] *nw r pꜣ mw*

Column II

1. *iw=f in-[sn]y.w r-bnr* ...[
2. *rꜣ=f n mt ḏ mtw=k(?)* ..[
3. *ḏ=f iw=y hb=w r tꜣ p.t* [
4. *tꜣ sty nꜣ nt nb i-ir* ⸤*pr*⸥ [
5. *n-im=k ḏ pꜣ-Šy mtw=k* [
6. *iw r tꜣ ḏ=w n=f Ḏḥwty n rn* [
7. *tyly n špš.w=f* .[
8. *pꜣ Nwn ir pꜣ-Šy* [
9. *n-im=w ḏ Ḏḥwty i-ir=w ḏ ntr.w* .[
10. *ḏ pꜣ-Šy ḏ* ...[
11. *n=f Ḏḥwty n rn mtw=*⸢*f*⸣ *ir* ..[
12. *ntr nb* ..[..]..*=f* ...[
13. *pꜣ hb* [..] *wn-mtw* ...[
14. *bnr b-ir=f w*⸢*ꜣ*⸣*y r pꜣ* [..]..[
15. *n pꜣ N*[*w*]*n irm pꜣ* [........... *r*]-
16. *hn-*[*r pꜣ hr*]*w ḏ pꜣ nt* [
17. *r-bnr-n* ⸢*rꜣ*⸣ *pꜣ-Šy* [
18. *wp*[*y*] *rꜣ rn=f* [
19. *r-b*[*nr-n rꜣ*]*=f r wp*[*y(?)*
20. *pꜣ* [*Nw*]*n i*[*r.t(?)*
21. *n* [....]..[

Translation

Column I

1.] the gods upon the earth
2.] dawn, rest/peace
3.]*š* in/of the sky
4.]. they/them in
5.] stars
6. mann]er of going forth to
7.] of/in the Primaeval Ocean
8.] say/do, Thoth. It happened/he became
9.] father to/concerning them
10.] him/it. He ⸢conducted⸣ the earth

11.] he ... between
12.] to the face of
13.] ⸢out⸣ of his mouth
14.]. like that which
15.]... and the water
16.]..[..]
17.]... him again
18.].....*y*
19.] They are .....
20. There rose/will rise] up the ibis
21.] on/because of Pshai
22.] look at the water

Column II

1. he pa[ss]ing out ...[
2. his mouth to speak, saying: You(?) ..[
3. He said: I will send them to the sky [
4. the flame, everything which came forth [
5. in you. Pshai said: You [
6. came to earth. He was called 'Thoth' by name [
7. rise/raising of his arms .[
8. the Primaeval Ocean. Pshai made [
9. in them. Thoth said: They say 'gods' .[
10. Pshai spoke, saying ...[
11. to him Thoth by name, and ⸢he⸣ makes ..[
12. every god ..[..].. him/it ...[
13. the ibis [..] ... has [
14. outside. He is not/lest he be di⸢st⸣ant from the [..]..[
15. of/in the Prim[aeval] Ocean and the [............ up]
16. until [tod]ay, saying: The one which [
17. out from ⸢the mouth⸣ of Pshai [
18. open[ing] the mouth, his name [
19. o[ut of] his [mouth] to op[en(?)
20. the [Primaeval Oce]an, e[ye(?)
21. of/to [....]..[

Textual notes

Column I

Line 3

The partially preserved word at the beginning of this line ends with *š* and the man with hand to mouth determinative.

Line 4

The trace at the beginning of the line is of a walking legs determinative, presumably belonging to a verb of motion. Cf. better preserved instances of the same determinative in writings of *pr* in line 6 below, *in*-[*sn*]*y* in II, 1, and *hb* in II, 3. The following *w* could be either a suffix pronoun, a plural stroke, or a stative ending.

Line 6

The traces at the beginning of the line are those of the determinative and the last stroke of the *y* of *g3y*, 'manner'. Cf. the undamaged writing of that word in fragment 44, line x+9.

Line 8

The traces before *Ḏḥwty* could represent either *ḏ*, 'say', or *ir*, 'make, do'.

Line 10

For a better preserved example of *sṯ*, 'draw, conduct', see fragment 13, II, 14.

Line 13

For the restoration ⸢*r-bnr*⸣ at the beginning of the line, cf. the undamaged writings of the same adverb in Column II, lines 1 and 17 of the present fragment.

Line 14

mtw after *p3* here must represent the relative converter *nt-iw*. Or should one read *nt ṯ*, 'which concerns'?

Line 17

With the writing of *ʿn*, 'again', at the end of the line, cf. better preserved examples of that adverb in fragments 2, line 14; 14, II, 7 and 8; and 24, line x+2. The preceding *f* must be a suffix pronoun, perhaps the object of a preposition. The sign immediately before *f* might be a flesh determinative.

Line 18

The damaged word at the end of the line ends in *y*. Only a trace of its determinative is still preserved. In view of the available space, this must have been quite small.

Line 20

(a) For the restoration of *fy*, 'rise up', before *p3 hb*, cf. the identical writing of that verb, with strong arm determinative as here, in P. Tebt. Tait 1, lines 3 and 24 (Tait, *Papyri from Tebtunis in Egyptian and in Greek*, plate 1). For *fy* used to describe the movements of ibises elsewhere, see Smith in Clarysse, Schoors and Willems (eds.), *Egyptian Religion: the Last Thousand Years* 1, p. 431; idem, *The Liturgy of Opening the Mouth for Breathing*, pp. 63-4 [note (c) on P. Berlin 8351, V, 8]; Chauveau, *CdE* 73 (1998), p. 77.

(b) *hb*, 'ibis', recurs in II, 13 of the present fragment. The verb *hb*, 'send', intervenes in II, 3. The use of these two homophones in such close proximity to each other suggests that one had here a play on words between them. Paronomasia involving this pair of words occurs in fragment 20, lines 2 and 4 as well. For examples elsewhere, see commentary below.

Line 22

With the traces of *nw*, 'look at', here, cf. better preserved examples of that verb in fragments 1, line 12; 7, line x+3; and 22, line 7.

FRAGMENT 5

Column II

Line 1

For the qualitative form *in-sny.w*, see *Glossar*, p. 437 (s.v. *snj*); Johnson, *The Demotic Verbal System*, pp. 26-7.

Line 2

With the traces that I have read tentatively as *mtw=k*, cf. the better preserved example of that independent pronoun in line 5 below.

Line 3

For the use of the verb *hb*, 'send', here in conjunction with the noun *hb*, 'ibis', which occurs in I, 20 and line 13 below, see textual note (b) on the former.

Line 4

The reading *pr* at the end of the line is certain. Compare the well-preserved example of this verb in I, 6 of the present fragment.

Line 6

For the naming formula *d̲=w n=f D̲ḥwty n rn*, see discussion on the ordering of the fragments in the Introduction.

Line 7

(a) *tyly* with sail determinative (not in *Glossar*) I take to be the demotic counterpart of Coptic ⲧⲟⲓⲗⲉ, 'rise up' (Crum, *CD*, p. 410a). Cf. use of the sail determinative elsewhere in demotic in words like *pꜣy*, 'rise up, fly' (*Glossar*, p. 130); *nhs*, 'arise' (ibid., p. 222); and *ḥl*, 'fly' (ibid., p. 327).

(b) The sign above the suffix pronoun *f* after *špš* is the plural marker. For identical writings of *špš.w=f*, see fragment 16, line x+6, fragment 19, line x+20.

Line 11

The end of the tail of the suffix pronoun *f* is visible after *mtw*.

Line 13

For the use of *hb*, 'ibis', here in conjunction with the verb *hb*, 'send', which occurs in line 3 above, see textual note (b) on I, 20 of the present fragment.

Line 16

For the restoration [*r*]-*hn*-[*r pꜣ hr*]*w* here, cf. better preserved occurrences of the same phrase in e.g. fragments 1, line 1; 4, II, 15; and 13, II, 5.

Line 17

The traces after *r-bnr-n* would suit *rꜣ*, 'mouth'. Cf. writings of that noun in lines 2 and 18 of the present column. For the phrase *r-bnr-n rꜣ*, see the 13th line of Column I; fragment 19, line x+13.

Line 19

(a) The traces after *r* would suit *bnr*; the ensuing break is of sufficient size to accommodate *n* + *rꜣ*, 'mouth'. For the phrase *r-bnr-n rꜣ*, see line 17 of the present column and textual note thereon.

(b) At the end of the line, perhaps restore *wp*[*y*], 'op[en]', as in the line above. Or is the sign after *w* a *b*?

Line 20

(a) Visible after the break which follows the initial *p3* are a short vertical stroke followed by a divine determinative. This combination occurs at the end of words like *Nwn*, 'Primaeval Ocean' (cf. lines 8 and 15 above) and *Ḏḥwty*, 'Thoth' (cf. lines 6 and 11). The former would fit well in the available space here, and the preceding *p3* makes its restoration a virtual certainty.

(b) The sign at the end of the line could be *ḥ*, or possibly the initial part of a word like *ir.t*, 'eye'.

Commentary

Only the ends of the lines in the initial column of this fragment are still preserved. It is impossible to make any sort of connected sense of these. Line 1 mentions the gods on the earth (*n3 ntr.w ḥr p3 t3*), line 3 the sky, and line 5 the stars (*syw.w*). There is a reference to emerging (*pr*) in line 6. *p3 Nwn*, 'the Primaeval Ocean', appears in line 7, the divine name *Ḏḥwty*, 'Thoth', in line 8, and the noun *iṯ*, 'father', in line 9. Line 10 contains the phrase *sṯ=f t3*, 'he conducted the earth'. *sṯ* (< earlier *sṯ3*) in this idiom means 'lead in submission'.[247] Thus, a king can be said to conduct (*sṯ3*) a foreign land to a god, or a deity to lead one before a king. In the hymn to Amun preserved in P. Leiden I 344 verso, it is said to him: *in k3=k sṯ3 n=k t3*, 'It is your ka that has conducted the earth to you.'[248]

Line 12 mentions the face of a being whose name has not been preserved. Line 13 contains the phrase ⸢*r-bnr*⸣*-n r3=f*, '⸢out⸣ of his mouth'. Presumably the same being is involved in both cases. Inasmuch as line 17 of the second column of the present fragment has the phrase *r-bnr-n* ⸢*r3*⸣ *p3-Šy*, 'out of ⸢the mouth of⸣ Pshai', it could well be that deity who is involved. One may have had here a reference to the emergence of something or some beings from the mouth of Pshai, in other words, to the creative activity of that god.

In a text from the temple of Edfu, (P)shai is called *sṯni irw*, 'the one who distinguished (i.e. created) forms'.[249] In a demotic letter addressed to him and other deities, he is invoked as *p3-Šy p3 ꜥ3 i-ir qmy Kmy tr=(f)*, 'Pshai the great, who created Egypt in (its) entirety'.[250] As a result of their belief in the creative powers of the spoken word, it is frequently stated in the texts of the ancient Egyptians that everything which came into existence emerged from the creator's mouth, i.e. through the magical force of his utterance.[251] Thus, in the cosmogony of P. Bremner-Rhind (4th century BC), the

[247] Cf. *Wb.* 4, 353, 7.

[248] P. Leiden I 344 verso, XI, 1-2 (Zandee, *Der Amunhymnus des Papyrus Leiden I 344, verso*, pp. 995-6 and plate 11).

[249] Chassinat, *Edfou* 3, p. 149, line 5. Cf. Quaegebeur, *Le Dieu égyptien Shaï*, p. 116.

[250] P. Berlin P 15660, lines 3-4 (Zauzich, *Enchoria* 19/20 [1992/93], p. 167 and plate 37).

[251] Zandee in Obbink et al (eds.), *Essays on Some Aspects of the Religious Function of Words Dedicated to Dr H.W. Obbink*, pp. 33-66. Cf. Bickel, *La Cosmogonie égyptienne avant le Nouvel Empire*, pp. 100-11.

creator is made to say: *ꜥšꜣ ḫprw.w m pr m rꜣ=i*, 'Modes of being were multiplied in emerging from my mouth.'[252] Other sources, however, make a distinction, stating that gods (*nṯr.w*) came out of the mouth of the demiurge and human beings (*rmṯ.w*) from his eyes. For example, the inscription accompanying a scene depicting the rite of offering the lotus at Edfu says with respect to Re: *pr nṯr.w m rꜣ=f rmṯ.w m nṯr.ty=f*, 'The gods emerged from his mouth and men from his eyes.'[253] In a scene from the temple of Dendera depicting the same rite, it is said of Harsomtus: *pr rmṯ.w m ir.ty=f nṯr.w m rꜣ=f*, 'People came forth from his eyes and gods from his mouth.'[254] These are only two of numerous examples illustrating the same theme.[255] Since Pshai in our cosmology appears to be a solar divinity like those who figure in the pair of passages just cited,[256] it is possible that the lines under consideration originally contained a reference to the emergence of the gods from his mouth. This suggestion is supported by the fact that it is precisely at this point in the narrative that the noun *ntr.w*, 'gods', begins to be used with any regularity in our text.

Line 20 of Column I refers to an ibis, which has risen or will rise up (= Thoth). Line 21 mentions Pshai. The nature of the relationship between these two deities is elucidated in the lines immediately following.

II, 1 of our fragment preserves the phrase *iw=f in-*[*sn*]*y.w r-bnr*, 'he pa[ss]ing out'. The verb here could be predicated of a being who emanates from another one. In view of what comes next, the being in question is likely to be Thoth, who is said in other Egyptian sources to have emanated from the solar deity. According to the Neith cosmogony at Esna, for example, Thoth emerged (*pr*) from the heart of Re.[257] Similarly, in the cosmogony in the temple of Khonsu at Karnak, the god is called *pr m ib ꜥꜣ m ꜥ.wt n Rꜥ*, 'the one who came forth from the great heart in the body of Re'.[258] In a demotic spell from the London-Leiden Magical Papyrus, Thoth is invoked: *mtw=k pꜣ i-ir pr n ḥ̂ꜣṯ=f n pꜣ-Šꜥy ꜥꜣ pꜣ i̭ṯ.w* (sic) *sp-sn n nꜣ ntr.w tr=w*, 'You are the one who came forth from the heart of the great Pshai, the father of the fathers of all the gods.'[259] These parallels suggest that perhaps our text spoke of Thoth as emanating from the heart of Pshai.

[252] Faulkner, *The Papyrus Bremner-Rhind*, p. 60, lines 3-4.

[253] Chassinat, *Edfou* 4, p. 140, line 5.

[254] Chassinat, *Dendara* 2, p. 178, line 3.

[255] For others, see Ryhiner, *L'Offrande du lotus*, p. 39, note 13; Otto, *Gott und Mensch*, pp. 58 and 122-3; Assmann, *Re und Amun*, pp. 235-7; Sauneron, *Esna* 5, p. 261, note (a); Bickel, *La Cosmogonie égyptienne avant le Nouvel Empire*, pp. 93-6; Mathieu in *Hommages à François Daumas*, pp. 499-509.

[256] See commentary on fragment 4 above.

[257] Sauneron, *Esna* 3, p. 33 (206, 11). Cf. idem, *Esna* 5, pp. 266-7.

[258] Parker and Lesko in Baines et al (eds.), *Pyramid Studies and other Essays Presented to I.E.S. Edwards*, p. 172 and plate 37.

[259] London-Leiden Magical Papyrus, II, 21 (Griffith and Thompson, *The Demotic Magical Papyrus of London and Leiden* 2, plate 2).

The verb *sn(y)* used in the line under consideration occurs a number of times in the Edfu cosmogony, where it is predicated of the *ḏꜣis.w*, words or phrases uttered by the creator to bring things into existence, which are themselves hypostasised as divinities.[260] Finnestad understands the verb to refer to the emergence of the *ḏꜣis.w* (scil. from the mouth of the creator).[261] As Alliot has remarked, however, *sn(y)* could be taken instead in a metaphysical sense, to denote the passage or transmutation of the utterances into perceptible, i.e. material form.[262] Either sense would fit well in the present line of our cosmology, if my suggested reconstruction of it is actually correct. Thoth was believed to have emanated from the heart of the sun god, as the passages cited in the preceding paragraph show. However, a text from the temple of Edfu states that he was engendered 'through (or as) a thought' (*m mꜣwy*) in that heart.[263] Thus, his emanation from it necessarily involves a transmutation of something previously imagined or conceived of into perceptible reality, as was true of the *ḏꜣis.w*.

Perhaps restore *wpy=f* or *wn=f*, 'he opened', at the end of line 1, preceding the *rꜣ=f n mt* of line 2, i.e. '[He opened] his mouth to speak.' One has here the beginning of a dialogue, probably between Pshai and the newly created Thoth, both of whom are mentioned by name several times in the lines that follow. In line 3, one of the speakers says: *iw=y hb=w r tꜣ p.t*, 'I will send them to the sky.' The verb *hb* is employed deliberately to evoke the noun *hb*, 'ibis', which occurs in line 13 below and line 20 of the previous column. Similarly, in the Neith cosmogony at Esna, it is related how Thoth emerged from the heart of Re during a revolt against the sun god led by Apopis. Thereupon, the text states, *mdw=f ḥnꜥ it=f h(ꜣ)b.n=f sw r sbi*, 'He (scil. Thoth) spoke with his father and he (=Re) sent him against the rebel.'[264] Sauneron has drawn attention to the way in which the verb *h(ꜣ)b*, 'send', there serves to call to mind the ibis aspect (*hb*) of Thoth.[265] The Book of the Heavenly Cow inscribed in some royal tombs of the New Kingdom records another dialogue between that god and Re. This one takes place after mankind has rebelled against the solar deity and the uprising has been put down. Re appoints Thoth as his deputy, saying: *iw=<i> gr.t r rdit hꜣb=k wr.w r=k*, '<I> will cause you to send those who are greater than you.' The ancient commentator adds the following explanation: *ḫpr hb pw n Ḏḥwty*, 'Thus came into existence the ibis-(form) of Thoth.'[266]

[260] See Chassinat, *Edfou* 6, pp. 17, line 6, and 182, lines 9, 11-12. For the names of these divinities, see e.g. idem, *Edfou* 3, p. 315, lines 3-5.

[261] See idem, *Image of the World and Symbol of the Creator*, p. 72.

[262] See his remarks apud Barucq, *BIFAO* 64 (1966), p. 143, note (d).

[263] Rochemonteix and Chassinat, *Edfou* 1, p. 289, line 4.

[264] Sauneron, *Esna* 3, p. 33 (206, 11).

[265] Idem, *Esna* 5, p. 267.

[266] See Hornung, *Der ägyptische Mythos von der Himmelskuh*, pp. 23 and 66, note 161.

Both of these parallel instances of word play involving *h(ꜣ)b*, 'send', and *hb*, 'ibis', occur in passages recounting conversations between Thoth and the sun god. This supports, in some measure, the suggestion made above that the present column of our cosmology probably preserves the remains of a similar dialogue.[267] It is noteworthy that the conversations in the Esna cosmogony and the Book of the Heavenly Cow take place in the context of a revolt against the solar deity, the former in the middle of such an uprising, and the latter in the aftermath of one. The same may have been true of the dialogue in our text.[268]

The beginning of line 4 mentions *tꜣ sty*, 'the flame'. If the suggestion made in the preceding paragraph is correct, and our fragment is part of an account of a revolt, then perhaps this refers to the fiery eye of the sun god, here that of Pshai, who is sent against the rebels to punish them. Compare e.g. the Book of the Heavenly Cow, where Re is advised by his fellow divinities with respect to his opponents: *imi šm ir.t=k ḥw=s n=k st*, 'Send forth your eye that it might strike them for you.'[269] Cf. also fragment 30, where a flame is mentioned in conjunction with the creation and naming of Sothis. For the possibility that *tꜣ sty* in the present fragment refers to her acting as the sun god's fiery eye, see commentary ad loc.

In line 5, Pshai speaks. The following line contains the words *ḏ=w n=f Ḏḥwty n rn*, 'He was called "Thoth" by name.' One has here an account of how the god in question was named, a necessary concomitant to his creation which was described in the preceding section of the text.[270] In the Esna cosmogony, the origin of Thoth's name is explained through word play. He emerged from the heart of Re *ḫft dḥr=f*, 'when it was bitter' (scil. as a result of the ongoing revolt of Apopis), and so, the text says, he was called *Ḏḥwty*.[271]

Line 7 preserves the words *tyly n špš.w=f*, 'the raising of his arms'. If Thoth is meant, this could be an allusion to his role as uplifter of the sky, a status acquired because he was thought to participate in the creation as 'Herr des zauberkräftigen Wortes'.[272]

[267] Although such word play can occur in other contexts as well. See e.g. a text from Dendera cited in Junker, *Der Auszug der Hathor-Tefnut aus Nubien*, p. 18, where Sakhmet is addressed: *hb n=t Rꜥ ib=f m-ḫt=f m rn pfy n hb*, 'Re sent you his heart (= Thoth) after it in that name (of his) of ibis.'

[268] For a more detailed treatment of the motif of rebellion against the sun god, see commentaries on fragments 6 and 8 below; also my edition of P. Carlsberg 462, published in Frandsen and Ryholt (eds.), *The Carlsberg Papyri 3: A Miscellany of Demotic Texts and Studies*, p. 95-112. Conversations between the solar deity and Thoth, his chief adviser, are common in accounts of such uprisings. Cf. Sternberg, *Mythische Motive und Mythenbildung in den ägyptischen Tempeln und Papyri der griechisch-römischen Zeit*, pp. 156-7, 171-2, 191, and 218.

[269] Hornung, *op. cit.*, pp. 4 and 54, note 32.

[270] See discussion of the naming formula involved in the Introduction.

[271] Sauneron, *Esna* 3, p. 33 (206, 11). Cf. idem, *Esna* 5, p. 267; *Mélanges Mariette*, pp. 234-5.

[272] Kurth, *Den Himmel Stützen*, pp. 85-6.

Although some Egyptian sources treat Thoth as an autonomous primaeval creator,[273] more often he figures as the heart and/or tongue of the solar deity, through whom the thoughts and plans of the latter are expressed and given concrete form.[274] As such, he is sometimes said to be the first-begotten of the sun god.[275] It is not clear whether this was so in our text. If I, 13 of the present fragment originally recorded the emergence of the gods from the mouth of Pshai, as I have suggested in the commentary on that line, then Thoth would not seem to be reckoned as the eldest among them. He could be, however, if the said line merely predicted the ultimate emergence of the gods, but that event itself was not actually recounted until after the narrative described the creation of Thoth. Compare the Neith cosmogony at Esna, which first records the prediction of that goddess concerning the still unborn sun god to the effect that mankind will come into being from the tears of his eyes and the gods from the saliva of his lips, and only later narrates the actual occurrence of those events.[276]

Line 8 preserves the name of *pꜣ Nwn*, 'the Primaeval Ocean', as do lines 15 and 20 below. In each instance, the context is damaged. It is clear, nevertheless, that one has in the section of the text following line 8 the remains of further conversation between Pshai and Thoth. Perhaps the Primaeval Ocean was said to take part in this as well. Compare the account of the rebellion of mankind in the Book of the Heavenly Cow, where the Ocean plays an important role as an adviser to the sun god, his wise counsel being valued because of his great age.[277] Alternatively, the Primaeval Ocean could have been mentioned in this context as a possible place of refuge for the solar deity. It figures as such in the account of the revolt against Re preserved in the Book of the Fayyum.[278]

In line 13, there is a reference to an ibis (*hb*). As noted above, there is a play on words in the text between this noun and the verb *hb*, 'send', that occurs ten lines earlier. Probably, as in the Book of the Heavenly Cow, one had here an explanation of how the ibis came to be associated with Thoth.

Line 14 contains the phrase *b-ir=f w⸢ꜣ⸣y r pꜣ* [...], 'he is not/lest he be di⸢st⸣ant from the [...]'. Compare the aftermath of the rebellion of mankind in the text just cited, where Re mounts upon the back of the celestial cow and 'distances' himself from his

[273] See e.g. a text from Philae cited by Grapow in *ZÄS* 67 (1931), p. 37, where the god is called 'the one who came into existence when what exists had not come into being, the sole one who created everything'.

[274] For this conception, see Boylan, *Thoth the Hermes of Egypt*, pp. 107-23; Derchain-Urtel, *Thot à travers ses épithètes dans les scènes d'offrandes des temples d'époque gréco-romaine*, pp. 81-94.

[275] See Boylan, *op. cit.*, pp. 118-19.

[276] Sauneron, *Esna* 3, pp. 30-1 (206, 6) and 32 (206, 9).

[277] Hornung, *Der ägyptische Mythos von der Himmelskuh*, pp. 2-3, 10-11, 19, and 26.

[278] See Beinlich, *Das Buch vom Fayum*, pp. 315-16 and 318.

creation by ascending to the sky.[279] It is conceivable that Pshai expressed the intention to do something similar in our text, whereupon the other gods were filled with dismay and gave expression to the words quoted above. The next three fragments are very badly preserved, but the celestial cow is actually mentioned by name in the last of these. As suggested in the commentary on fragment 8 below, it may have been there that our text recounted the sun god's departure heavenwards.

Line 16 preserves part of an aetiological statement, [*r*]-*hn*-[*r pꜣ hr*]*w*, '[up] until [tod]ay'. On the function of such statements in the text, see Introduction. The following line reads in its entirety: *r-bnr-n* ⸢*rꜣ*⸣ *pꜣ-Šy* [...], 'out from ⸢the mouth of⸣ Pshai [...]'. Line 18 mentions the opening (*wpy*) of a mouth, while line 19 can be restored to read *r-b*[*nr-n rꜣ*]=*f r wp*[*y*(*?*)], 'o[ut of] his [mouth] to op[en(?)]'. These could be further references to the creation of the other gods through their emergence from Pshai's mouth. Cf. discussion of the phrase ⸢*r-bnr*⸣-*n rꜣ*=*f* in I, 13 of the present fragment.

FRAGMENT 6 — Plate 6

Dimensions: 5.6 × 9.6 cm. Colour: light brown. Related fragments: Nos. 22 and 23.

Transliteration

1.] *pꜣ Nwn ẖn* [
2.] *nꜣ msḥ.w* [
3.]*h nꜣ ntr.w ir Ḏḥwty* [
4.] ⌞*tr*⌟.*ṱ*=*f iw*=*n r* .[
5.] (vac) *i-ir nꜣy ẖpr r pꜣ* [
6.]... *pꜣ šꜥꜣ ꜥꜣ n H̱*[*mnw*
7. *pꜣ*]*y*=*f*(*?*) ... *pꜣy*=*f 7 n* [
8.] *iw*=*s*(*?*) *pꜣ* .[..].. *ti ẖ*[ꜥ
9.]... *r* [
10.]... ꜥ*w*[*y*
11.]....[

Translation

1.] the Primaeval Ocean in [
2.] the crocodiles [
3.]*h* the gods. Thoth made [
4.] his ⌞hand⌟. We will .[
5.] (vac) These things happened while the [

[279] Hornung, *op. cit.*, pp. 11-13. The verb employed there is *ḥr*, 'be distant'. On the solar deity as a 'distant' god, see Assmann, *Re und Amun*, pp. 101-5 and 232-3; Beinlich, *op. cit.*, p. 314.

6.]… the great sand of H[ermopolis
7. hi]s(?) … his seven [
8.] she/it(?) the .[..].. cause to ap[pear
9.]… to [
10.]… hous[e
11.]….[

Textual notes

Line 2

For the writing of *msḥ*, 'crocodile', here, see e.g. fragment 7, line x+4, P. Carlsberg 14, fragment i, 2, 3, 7, 8 (Volten, *Demotische Traumdeutung*, plate 8). The word is followed by horizontal plural strokes.

Line 3

The word before *nꜣ ntr.w* appears to end with *h*, and has no determinative. I am unable to suggest a restoration.

Line 4

For the traces of *tr.ṱ*, 'hand', at the beginning of the line, cf. *Glossar*, p. 644. Compare also the writing of the preposition *n-tr.t* in fragment 1, line 3. The latter has an extra stroke before the flesh determinative.

Line 6

The traces immediately after *ꜥꜣ n* are the beginning of *Ḫmnw*, 'Hermopolis'. Cf. better preserved writings of that toponym in P. Carlsberg 462, II, 18 (Smith in Frandsen and Ryholt [eds.], *The Carlsberg Papyri 3: A Miscellany of Demotic Texts and Studies*, plate 15), P. Carlsberg 10032, line x+5, and P. Carlsberg 10040, lines x+5 and x+10. For the significance of the *šꜥꜣ ꜥꜣ n Ḫmnw*, 'great sand of Hermopolis', mentioned here, see commentary below.

Line 7

The traces at the beginning of the line are perhaps those of *pꜣy=f*. What follows is a noun denoting a part of the body which I am unable to read with certainty. Could this be *šwy*, or perhaps a writing of *tbꜥ*, 'finger'?

Line 8

(a) At the beginning of the line, perhaps read *iw=s*. Cf. the undamaged occurrences of this combination in fragment 30, line 6. If the reading is correct, then presumably a preposition *n* has to be supplied before the following *pꜣ*.

(b) For the traces of *ḫꜥ*, 'appear', cf. the undamaged writings of that verb in fragment 12, I, 17, and fragment 13, II, 16.

Line 10

The restoration *ꜥw[y]* at the end of the line is certain. The horizontal stroke visible after the tall vertical one is part of the house determinative of the word. Cf. writings of the same determinative in words like *ist*, 'ground', in fragment 2, line 4, and *mꜣ*, 'place', in fragment 13, II, 2.

Commentary

It is difficult to make this fragment yield any connected sense. Line 1 refers to *pꜣ Nwn*, 'the Primaeval Ocean', line 2 to some crocodiles (*msḥ.w*). Line 3 mentions the gods (*nꜣ ntr.w*) and Thoth; line 4, the latter's hand. Line 6 preserves the words *pꜣ šꜥꜣ ꜥꜣ n Ḫ*[*mnw*], 'the great sand of H[ermopolis]'. At the end of line 7, there is the broken phrase *pꜣy=f 7 n* [...], 'his seven [...]'. Line 10 refers to an *ꜥwy*, 'house' or 'place'.

'The great sand of Hermopolis' is almost certainly a designation of the primaeval mound, the first land to emerge from the waters at the beginning of creation.[280] Elsewhere, at Heliopolis, this mound was called *šꜥy qꜣ*, 'the high sand'.[281] The importance of sand in the formation of the initial terrain at Thebes is stressed in the hymn to Amun preserved in P. Leiden I 350, where it is said with respect to that city: *pꜣ mw tꜣ im=s m sp tpy ii.n šꜥy r ḥnb ꜣḥ.wt r sḫpr sꜣṯw=s ḥr qꜣy ḫpr tꜣ*, 'Water and earth were in it at the first occasion. Sand came in order to delimit the fields so as to create its platform upon the high mound. Thus the land came into being.'[282] There is a similar account of the land's first emergence at Esna: *hbhb Nwn btbt nš m ꜣḫ.t m qꜣ.t tpy.t di m Nwn*, 'The Primaeval Ocean was forced back, sand was mixed with arable land, constituting the initial high mound set down amidst the Ocean.'[283] Likewise, according to the Book of the Fayyum, the primordial hill was constituted of sand.[284]

The crocodiles in line 2 are more difficult to explain. These occur in fragment 7 as well, and could be divinities who have transformed themselves into reptilian shape.[285] Alternatively, they could be familiars or servants of the Primaeval Ocean. A stela now in the Field Museum of Natural History in Chicago recounts a myth in which a goddess comes to the sea to bathe. The Primaeval Ocean sees her and desires her. He summons some crocodiles and orders them to attack her, intending to rescue her himself and

[280] Cf. commentary on fragment 1, lines 6 and 11. For Hermopolis as the *locus* for the emergence of the first land, see de Buck, *De Egyptische Voorstellingen betreffende den Oerheuvel*, pp. 35-42; Roeder, *Hermopolis 1929-1939*, pp. 37-8, 169-71; Ryhiner, *L'Offrande du lotus*, p. 198; Sauneron and Yoyotte in *La Naissance du Monde*, p. 57.

[281] De Buck, *De Egyptische Voorstellingen betreffende den Oerheuvel*, pp. 26-7. Cf. Ricke, *ZÄS* 71 (1935), pp. 107-11; Aufrère, *L'Univers minéral dans la pensée égyptienne* 2, pp. 666-7.

[282] P. Leiden I 350, II, 10-11 (Zandee, *De Hymnen aan Amon van Papyrus Leiden I 350*, pp. 25-6 and plate 2). Cf. de Buck, *op. cit.*, pp. 47-8; Drioton, *ASAE* 44 (1944), pp. 114-16.

[283] Sauneron, *Esna* 3, p. 4 (196, 1); idem, *Esna* 5, pp. 317-19.

[284] See Beinlich, *Das Buch vom Fayum*, pp. 294-6.

[285] For deities who can assume such a form, see Brunner-Traut in *Lexikon der Ägyptologie* 3 (1980), column 791 (s.v. Krokodil). In a text inscribed upon the pylon of the temple of Kom Ombo, the Ogdoad are said to have turned themselves into a crocodile (de Morgan et al, *Kom Ombos* 1, p. 59, no. 61, line 5). Cf. Gutbub, *Textes fondamentaux de la théologie de Kom Ombo*, p. 474, note (f). In some Egyptian sources, the son of the solar deity is said to assume the form of a crocodile in order to kill the enemies of his father. See e.g. Sauneron, *Esna* 2, p. 236 (127, 2); de Morgan et al, *Kom Ombos* 1, p. 127, no. 167, line 14.

thereby win the goddess's gratitude.[286] In this myth, the crocodiles are subject to the Ocean because they inhabit his watery domain.[287] Yet another possibility is that one has here the remains of a reference to the so-called 'Golden Age' in the history of the world, the reign of the Ogdoad, a time of peace and joy when *n iṯ msḥ n psḥ ḥf*, 'crocodiles did not seize and serpents did not bite'.[288] The difficulty with this last suggestion is that Thoth, who is mentioned in line 3 of the present fragment, had not yet been created at the period in question.

In the commentary on the second column of fragment 5, it was suggested that one might have preserved there part of an account concerning a rebellion against the sun god. If so, then fragment 6 could be a continuation of that account. As remarked in the aforesaid commentary, both Thoth and the Primaeval Ocean play prominent roles in the sources that record uprisings of this sort. Some of these sources describe a rebellion which took place at Hermopolis or in the vicinity of that town.[289] Spell 17 of the Book of the Dead says specifically that the sun god crushed the revolt against him and slaughtered the rebels *ḥr qꜣꜣ imy Ḫmnw*, 'on the high mound which is in Hermopolis'.[290] This is equivalent to the *šꜥꜣ ꜥꜣ n Ḫ*[*mnw*], 'great sand of H[ermopolis]', that appears in our text, which might similarly have been identified as the place where the enemies of the solar deity were punished. If so, then the *msḥ.w*, 'crocodiles', to which reference is made in line 2 could be rebels against whom Thoth and the other gods mentioned along with him are said to have contended. The crocodile can be a symbol of evil or chaos in Egyptian thought. As such, its form is frequently assumed by opponents of the divine order in mythological narratives.[291]

[286] See Ritner in Simpson (ed.), *Religion and Philosophy in Ancient Egypt*, pp. 112-13.

[287] Similarly, in the Book of the Heavenly Cow, serpents are said to be the familiars of the earth god Geb (Hornung, *Der ägyptische Mythos von der Himmelskuh*, pp. 18-19). Cf. Klasens, *A Magical Statue Base*, p. 80. Note in this connection that the name of the Primaeval Ocean is occasionally written with three crocodiles; see e.g. de Morgan et al, *Kom Ombos* 2, p. 294 (no. 941). For the Ocean itself represented as a crocodile, see Hornung, *Zwei Ramessidische Königsgräber: Ramses IV. und Ramses VII.*, pp. 66-7 and references cited there.

[288] Chassinat, *Edfou* 5, p. 85, line 14. On this period, see Kákosy, *Studia Aegyptiaca* 7 (1981), pp. 81-92; Otto in *Religions en Égypte hellénistique et romaine*, pp. 93-108; Grapow, *ZÄS* 67 (1931), pp. 34-8; Sethe, *Amun und die acht Urgötter von Hermopolis*, p. 63; Zandee, *Der Amunhymnus des Papyrus Leiden I 344, verso*, p. 957.

[289] See Roeder, *Hermopolis 1929-1939*, p. 170; Jelínková-Reymond, *Les Inscriptions de la statue guérisseuse de Djed-Ḥer-Le-Sauveur*, pp. 31, note 7, and 43, note 3; Sternberg, *Mythische Motive und Mythenbildung in den ägyptischen Tempeln und Papyri der griechisch-römischen Zeit*, p. 215, note 2; Altenmüller, *Die Apotropaia und die Götter Mittelägyptens*, pp. 101-4.

[290] Grapow, *Urkunden* 5, pp. 6-7. Cf. de Buck, *De Egyptische Voorstellingen betreffende den Oerheuvel*, pp. 37-42.

[291] See Brunner-Traut in *Lexikon der Ägyptologie* 3 (1980), columns 795-6 (s.v. Krokodil).

The Neith cosmogony at Esna recounts how that goddess made seven successive creative utterances (*ḏ3is.w*), which were subsequently hypostasised as divinities. When the rebellion against her son Re broke out, these seven beings acted as protectors for the sun god and his mother in her form as the divine cow Mehetweret.[292] Other sources, however, connect the *ḏ3is.w* with Thoth as well. In the cosmogony in the temple of Khonsu at Karnak, for instance, the seven deities are said to have appeared on (*pr ḥr*) the breast of Thoth.[293] An obscure passage in the same inscription may even refer to him as their father and creator.[294] In a scene from the chapel of the throne of Re at Edfu, it is said of this group of divine beings that they are the image of Thoth, *ir ḫ.wt nb.w ḫft wḏ=f*, 'who do all things according to his command'.[295] Perhaps these divinities were mentioned in line 7 of the present fragment as well, where there is now only the broken phrase *p3y=f 7 n* [...], 'his seven [...]'. If the pronoun 'his' refers to Thoth, one might imagine that the *ḏ3is.w* were described as exercising protection for the sun god against his enemies at the behest of that deity. Thoth is attested elsewhere fighting against the forces of chaos, aided by various groups of seven beings. Dawson has drawn attention to a papyrus in Paris (Bibliothèque Nationale 177) which depicts that god armed with two knives attacking a pig, accompanied by seven uraei, seven symbolic eyes, seven falcons, seven cats, and seven ibises.[296] It is worth noting, in this connection, that the *ḏ3is.w* are sometimes represented as falcons.[297] In Spell 71 of the Book of the Dead, they are said to slaughter foes on the Island of Fire (*iw nsrsr*), which is another name for the primaeval high mound of Hermopolis.[298]

[292] Sauneron, *Esna* 3, pp. 33-4 (206, 12-14). Cf. idem, *Esna* 2, p. 280 (163, 18), and *Esna* 5, pp. 268-9; Finnestad, *Image of the World and Symbol of the Creator*, p. 72. For comprehensive bibliography on these protectors (called *ṯs.w* in earlier texts), see Koenig, *Le Papyrus Boulaq 6*, pp. 105-6; Goyon, *Les Dieux-gardiens et la genèse des temples* 1, pp. 187-8; Ramadan el-Sayed, *La Déesse Neith de Saïs* 1, pp. 53-4 and 56-7; Bickel, *La Cosmogonie égyptienne avant le Nouvel Empire*, pp. 65-6; Wilson, *A Ptolemaic Lexikon*, pp. 1216-17.

[293] Parker and Lesko in Baines et al (eds.), *Pyramid Studies and other Essays Presented to I.E.S. Edwards*, p. 172 and plate 37. Cf. Zivie-Coche in *Hommages à Jean Leclant* 4, p. 422.

[294] Parker and Lesko, art. cit., p. 173 and plate 37.

[295] Rochemonteix and Chassinat, *Edfou* 1, p. 296, line 2. For additional references to Thoth and the *ḏ3is.w*, see *Wb.* 5, 521, 16; Wilson, loc. cit.

[296] Dawson, *Aegyptus* 8 (1927), p. 104.

[297] See Ramadan el-Sayed, *Le Déesse Neith de Saïs* 1, p. 54; Goyon, *Les Dieux-gardiens et la genèse des temples* 1, p. 187; Wilson, loc. cit.

[298] See Budge, *The Book of the Dead* 2, p. 12. On the identity of the Hermopolitan *q33*, 'high mound', and *iw nsrsr*, see Ryhiner, *L'Offrande du lotus*, pp. 197-8; Roeder, *Hermopolis 1929-1939*, p. 189; Altenmüller, *Die Apotropaia und die Götter Mittelägyptens*, pp. 88-9; Grieshammer in *Lexikon der Ägyptologie* 2 (1977), columns 258-9 (s.v. Flammeninsel); Hermsen, *Die zwei Wege des Jenseits*, pp. 153-6.

FRAGMENT 7 Plate 6

Dimensions: 6.8 × 9 cm. Colour: medium brown. Related fragments: No. 23.

Transliteration

x+1.]*ẖ*[
x+2.] *pr* ..[
x+3.]. *ẖn pꜣ mw nw=w r* ..[
x+4.] *nꜣ msḥ.w pꜣ qty* [
x+5.].*ṯ=w ḥ*[ꜣ]*ṯ=w ḥr-ꜣṯ* [
x+6. *pꜣ*]- *Šy* [..] *i-ir(?) pr n pꜣ* [
x+7.] *tr=w* [...] *ntr.w rꜣ* [
x+8.]*=f rt* ..[

Translation

x+1.]*ẖ*[
x+2.] go forth ..[
x+3.]. in the water. They looked at ..[
x+4.] the crocodiles around [
x+5.] their [...], their he[ar]ts upon [
x+6. P]shai [..] who(?) came forth from the [
x+7.] entire [...] gods, mouth [
x+8.] it grows ..[

Textual notes

Line x+3

The tall vertical stroke in front of *ẖn* at the beginning of the line could be either the plural marker or a suffix pronoun *w*.

Line x+4

With the traces of *qty* at the end of the line, cf. the better preserved instance of that word in fragment 2, line 7.

Line x+5

(a) The damaged word at the beginning of the line is determined with the flesh sign. This is followed by a *ṯ* before the suffix pronoun. One expects a feminine noun denoting a part of the body.

(b) With the traces of *ḥꜣṯ*, 'heart', here, cf. the better preserved writing of that noun in fragment 23, II, 6.

(c) With the traces of *ꜣṯ*, 'back', cf. the better preserved writings of that noun in fragments 1, line 10; 4, II, 16; and 14, II, 18.

Line x+6

(a) Traces of the serpent determinative of [*p3*]-*Šy* can be made out on the original. Whether this was followed by a divine determinative as in e.g. fragment 4, II, 20 and 22, or marked the end of the god's name as in e.g. fragment 5, II, 10, is impossible to say owing to the break in the text.

(b) The traces before *pr* could be of the past relative converter *i-ir*. Cf. writings of the latter in fragment 5, II, 14, and fragment 19, line x+4. If [*p3*]-*Šy* was determined with the serpent sign alone (see previous note), then there would be sufficient space to restore an article like *n3* before *i-ir*(?).

Line x+7

With the traces of *r3*, 'mouth', at the end of the line, cf. better preserved examples of the same noun in e.g. fragments 5, II, 2; and 21, line x+2. There are further traces of writing above the initial sign of this word and the last sign of the preceding *ntr.w*, 'gods', presumably a textual addition, which I am unable to read.

Line x+8

At the beginning of the line, perhaps restore [*iw*]=*f*. For the traces of *rt*, 'grow', after this, cf. better preserved writings of the same verb in fragment 1, lines 1 and 7, and fragment 14, I, 11.

Commentary

This is another poorly preserved fragment. The phrase *ẖn p3 mw*, 'in the water', occurs in line x+3. There is a reference to some crocodiles (*msḥ.w*) in line x+4. The following line mentions *ḥ*[*3*]*ṯ*=*w*, 'their he[ar]ts', perhaps with reference to the crocodiles. The name of the god Pshai occurs in line x+6, while line x+7 juxtaposes the nouns *ntr.w*, 'deities', and *r3*, 'mouth'.

The mention of crocodiles here indicates a connection with fragment 6, where the same reptiles figure in the narrative. For the problem of the identity of these creatures, see commentary ad loc., where the various possibilities are enumerated. I have suggested in the commentaries on fragments 5 and 6 that they may preserve part of an account concerning a rebellion against the sun god Pshai. If so, then this should be true of the present fragment as well. Unfortunately, the nature of Pshai's activity here is obscure, owing to the damaged context. The occurrence of the nouns meaning 'gods' and 'mouth' together in the penultimate line of the fragment recalls the theme of the deities' emergence from the mouth of Pshai, for which see commentary on fragment 5.

FRAGMENT 8 — Plate 6

Dimensions: 5.6 × 10.2 cm. Colour: varies from light brown (left) to medium brown (right).

Transliteration

x+1.]....[
x+2.]....[
x+3.].. *m*(?).. *r-r*=*w*(?) [

x+4.] *p3* [*b*]*l*(?) [
x+5.] *wr n Ḫmn*[*w*
x+6.]... *wpy* [
x+7. *Mḥ.t*(?)*-w*]*ly t3 iḥ.*(*t*) [
x+8.] *i-ir Ḏḥwty p3 ntr* [
x+9.] *rn=w n3.w wn-n3.w ḥm*[*sy*
x+10.]....[

Translation

x+1.]....[
x+2.]....[
x+3.].. *m*(?).. to/concerning them(?) [
x+4.] the [bli]nd(?) [
x+5.] great one of Hermopo[lis
x+6.]... open [
x+7. Mehet(?)-we]ret the cow [
x+8.] *i-ir* Thoth the god [
x+9.] their names. Those who were sit[ting
x+10.]....[

Textual notes

Line x+3

The traces at the end of the line are perhaps those of *r-r=w,* preposition *r* with third person plural suffix pronoun. Cf. the undamaged writings of that combination in fragment 5, I, 9, and fragment 12, I, 16.

Line x+4

The last preserved traces in this line appear to be the eye determinative with two round pupils below it, and the evil determinative. Perhaps restore [*b*]*l*, '[bli]nd' (*Glossar*, p. 120).

Line x+5

The determinatives of *Ḫmnw* are lost. I owe the reading of that word to Kim Ryholt, who notes that the trace preceding it would suit *n* but not *n3*. Accordingly, one must have the toponym *Ḫmnw*, 'Hermopolis', here rather than the name of the Ogdoad, since the latter, although also written *Ḫmnw*, is invariably preceded by *n3* in our text. (See note (c) on fragment 2, line 9.)

Line x+6

With the traces of *wpy* here, cf. the writing of that verb in fragment 5, II, 18.

Line x+7

The partially preserved goddess's name at the beginning of the line has *wly* (< earlier *wr.t*) as its second element. Compare the better preserved writing of that adjective in P. Carlsberg 10313, line x+8. In view of the epithet *t3 iḥ.*(*t*), 'the cow', which follows, one should probably restore either

[*Mḥ.t-w*]*ly*, as I have done above, or [*ꜣh.t-w*]*ly*, i.e. the great *Ahet*-cow. For these two alternatives, see commentary below.

Line x+9

For the traces of the semi-hieratic writing of the verb *ḥmsy*, 'sit', here, cf. fragment 1, line 20, P. Carlsberg 10029, line x+6.

Commentary

As with the two previous fragments, little connected sense can be obtained from this one. Line x+7 preserves the name of a female deity who is said to be a cow, probably either Mehetweret or Ahetweret.[299] The following line mentions Thoth. Three lines earlier, in line x+5, the epithet 'great one of Hermopolis' appears; this probably refers to Thoth as well.

The name *Mḥ.t-wr.t* means 'the great flood'. This is an appellation of the Primaeval Ocean personified as a goddess, the feminine counterpart of *pꜣ Nwn*.[300] The name also denotes the primordial cow described in some sources as swimming in the Ocean, giving rise to a secondary interpretation of it as 'the great swimmer'.[301] Both cow and flood represent the place of origin of the sun. The former gives birth to Re; that deity is said to rise from the latter.

The goddess Neith is frequently identified with Mehetweret, in which form she was supposed to have brought the cosmos into existence by means of her creative utterances.[302] In the Neith cosmogony at Esna, as stated in the commentary on fragment 6, she is said to have assumed a cow's form in order to protect her child, the sun god Re, from those who rebelled against him, being assisted in that task by her seven creative utterances (*ḏꜣis.w*).[303] Thus, another inscription from the same temple describes the goddess as *nh sꜣ=s dr btn.w=f sḥr sby.w ḥr=f ḥr mw wn=s m sꜣ=f*, 'she who protected her son, subdued his opponents, and drove away those who rebelled against him in the water, being his guard'.[304]

At Esna and elsewhere, an alternative designation for Neith-Mehetweret, the mother of Re, was *ꜣh.t*, 'the cow', or sometimes *ꜣh.t-wr.t*, 'the great cow'.[305] These seem

[299] See textual note ad loc.

[300] Kákosy in *Lexikon der Ägyptologie* 4 (1982), column 3 (s.v. Mehet-weret); Ramadan el-Sayed, *La Déesse Neith de Saïs* 1, p. 52; Bickel, *La Cosmogonie égyptienne avant le Nouvel Empire*, pp. 64-5; Beinlich, *Das Buch vom Fayum*, pp. 315-16 and 318.

[301] See el-Sayed, *op. cit.*, p. 56.

[302] Ibid., pp. 51-65.

[303] See references cited in note 292 of the commentary on fragment 6; Ramadan el-Sayed, *op. cit.*, pp. 73-4 and 106-9.

[304] Sauneron, *Esna* 3, p. 273, lines 3-4 (334, 16).

[305] Ramadan el-Sayed, *op. cit.*, pp. 24-7; Ryhiner, *L'Offrande du lotus*, pp. 58, note 5, and 159, note 8; Hornung, *Der ägyptische Mythos von der Himmelskuh*, pp. 93 and 96-101; Kákosy in *Lexikon der*

to have been the preferred forms at Hermopolis, where a stela dated in the reign of Nekhtnebef twice calls her *ꜣḥ.t-wr.t ms Rꜥ*, 'Ahetweret who bore Re',[306] and the name of the goddess's temple was *Pr-ꜣḥ.t*, 'House of the Cow'.[307] In the Fayyum, where the myth concerning the divine cow who protects the sun god from his enemies was of particular importance, both Mehetweret and Ahet are attested as names for her.[308] This last fact makes it difficult to determine which of the two designations is more likely to have been used in our text.

There is a link between the cow goddess and Thoth in that both are said to exercise control over the seven *ḏꜣis.w*, 'creative utterances'.[309] Moreover, both assist the solar deity in overcoming those who have rebelled against him.[310] Inasmuch as the pair are mentioned together in the present fragment, perhaps it too is part of the rebellion account which I have tentatively reconstructed in fragments 5-7, being concerned with the role played by these two divinities, with or without their helpers, in the protection of the sun god and the defeat of his enemies.

It is interesting to note that the motif of the sun-bearing cow goddess can be combined with that of the solar egg. This is true, for instance, of the Neith cosmogony at Esna. See commentary on fragment 4, II, 19, for the egg in that account. Similarly, in the great hymn of the Ogdoad to Amun at Hibis, it is said to that god: *s.t=k ḏr-ꜥ m qꜣy.t n Wnw sꜣḥ=k tꜣ m mr nẖꜣ.wy ẖꜥ=k m n.t m swḥ.t imn.t ꞽImn.t m-ẖt=k ẖn.n=k ḥr ꜣḥ.t ẖfꜥ=k ꜥb.wy=s smḥ.n=k ḥr Mḥ.t-wr.t*, 'Your primordial seat was the high mound of Hermopolis, you reaching land in the Sea of the Two Knives, arising from the flood within the hidden egg with Amaunet behind you. You alighted upon the Ahet-cow, grasping her two horns, and you were borne through the water on Mehetweret.'[311]

In a number of sources, the cow goddess is said to protect the solar deity from his enemies by raising him upon her back or between her horns.[312] Coffin Text Spell 647

Ägyptologie 3 (1980), columns 124-5.

306 Roeder, *ASAE* 52 (1954), pp. 403 and 413.

307 Art. cit., p. 441. Cf. Roeder, *Hermopolis 1929-1939*, p. 184; Sethe, *Amun und die acht Urgötter von Hermopolis*, p. 50.

308 See Beinlich, *Das Buch vom Fayum*, pp. 314-19, in particular 315.

309 See commentary on fragment 6.

310 For Thoth's role, see ibid. and commentary on fragment 5.

311 Davies, *The Temple of Hibis in El Khārgeh Oasis* 3, plate 33, middle, lines 22-4. On the 'Sea of the Two Knives', see Hermsen, *Die zwei Wege des Jenseits*, pp. 152-60, with references to earlier literature. The Hibis hymn is cited ibid., pp. 154-5.

312 See references cited in notes 305 and 308 above; Sauneron, *Esna* 2, pp. 170 (77, 15-16) and 280 (163, 17); idem, *Esna* 3, pp. 33 (206, 13) and 54 (216, 11); Gutbub, *Textes fondamentaux de la théologie de Kom Ombo*, pp. 71 and 119. Cf. also my discussion in Frandsen and Ryholt (eds.), *The Carlsberg Papyri 3: A Miscellany of Demotic Texts and Studies*, p. 105.

speaks of the solar egg (*swḥ.t*) as being on her vertebrae.[313] According to the Book of the Heavenly Cow, in the aftermath of the rebellion against Re, that god was borne up to the heavens on the back of his mother, who was herself then transformed into the sky.[314] A tradition current in the Fayyum describes the Ogdoad as raising their arms to elevate the divine cow and her child heavenwards.[315] With the sun thus raised into its proper place, the phase of the creation of the cosmos that is said to begin with the separation of the sky and earth comes to an end.[316] It was suggested above that this separation may have been described in fragment 4 of our cosmology.[317] Conceivably, the present fragment described the culminating stage in the process initiated by that act, the elevation of the sun into the heavens.

FRAGMENT 9 — Plate 6

Dimensions: 6 × 9.7 cm. Colour: medium brown.

Transliteration

x+1.]....
x+2. *nꜣ] Ḫmnw n tꜣ*
x+3. *nꜣ Ḫm]nw ḥp=w ẖn*
x+4.] *Nw.t m-sꜣ nꜣy*
x+5.] *n=f* [...] *r ms* .[..]
x+6.] *n ḫpr n* [*t*]*ꜣ* .. *nꜣ.w*
x+7.]..[...]..

Translation

x+1.]....
x+2. the] Ogdoad in the
x+3. the Og]doad. They were concealed in
x+4.] Thebes. After this,
x+5.] to/for him [...] in order to give birth .[..]
x+6.] They are [....] of coming into existence in [t]he ..
x+7.]..[...]..

[313] De Buck, *The Egyptian Coffin Texts* 6, p. 267f-g. Cf. Bickel, *La Cosmogonie égyptienne avant le Nouvel Empire*, pp. 66-7.
[314] Hornung, *Der ägyptische Mythos von der Himmelskuh*, pp. 11-13.
[315] See Beinlich, *Das Buch vom Fayum*, pp. 222-3 and 313.
[316] So Beinlich, *op. cit.*, p. 315: 'Damit ist die Trennung von Erde und Himmel vollzogen.'
[317] See commentary on that fragment.

Textual notes

Line x+2

For the restoration [*n3*] *Ḫmnw* at the beginning of the line, cf. fragment 2, line 12.

Line x+3

Restore [*n3 Ḫm*]*nw* at the beginning of the line. Cf. the preceding note.

Line x+5

For the writing of *ms*, 'give birth', near the end of the line, cf. better preserved examples of that verb in e.g. fragment 12, I, 12, and fragment 13, II, 8. The damaged word after *ms* appears to have begun with *s*.

Line x+6

One is tempted to read the feminine noun before *n3.w* as *ḥ.t*, 'front', but the curved part at the top of the initial sign is facing in the wrong direction. Other possibilities that I have considered are *ẖ.t* and *sn.t*, but again the shape of the initial sign is not quite right for either of these. For our scribe's writing of the former, see fragment 19, line x+14.

Commentary

n3 Ḫmnw, 'the Ogdoad', are mentioned in lines x+2 and x+3. The second reference to these deities is followed immediately by the words *ḥp=w ẖn* [....], 'They were hidden in [....].' The toponym *Nw.t*, 'Thebes', is visible in line x+4. According to local tradition, the Ogdoad were born in the region of that city, specifically in *Ꞽp.t-*(*rsy*), or Luxor.[318] Thus, a text from the propylon of the temple of Khonsu at Karnak describes them as *ms m W3s.t*, 'born in Thebes'.[319] Another Karnak inscription, this one from the propylon of the temple of Montu, speaks of *Ḫmny.w ir m W3s.t qm3.n Ṯnn m Ꞽp.t*, 'the Ogdoad who were formed in Thebes, whom (Ta)tenen created in Luxor'.[320]

A more widespread tradition, attested both at Thebes and elsewhere, identified this region as the burial place of the eight Hermopolitan deities. They were said to have been interred at Djeme (Medinet Habu), the exact site of their tomb being the small 18th Dynasty temple there.[321] The earliest evidence for this belief occurs in a private letter of 21st Dynasty date.[322] Among other sources from later periods, several demotic

[318] See Sethe, *Amun und die acht Urgötter von Hermopolis*, p. 51.

[319] Idem, *Urkunden* 8, p. 80 (95c).

[320] Ibid., p. 30 (35c).

[321] Sethe, *Amun und die acht Urgötter von Hermopolis*, pp. 53-61; Traunecker, Le Saout and Masson, *La Chapelle d'Achôris à Karnak* 2, pp. 115-16; Sternberg-El Hotabi, *Der Propylon des Month-Tempels in Karnak-Nord*, pp. 42-3, note 3; Gabolde, *BIFAO* 95 (1995), pp. 251-3.

[322] P. Bib. Nat. 198, I, lines 5-8 (Černý, *Late Ramesside Letters*, p. 66). Cf. Traunecker et al, *op. cit.*, p. 131.

graffiti from the small temple at Medinet Habu attest to it as well.[323] In the Book of the Fayyum, the members of the Ogdoad speak of themselves as resting 'in our bodies' (*m ḏ.t=n*) in the mound of Djeme.[324]

One important purpose of burial, from the Egyptian point of view, was to conceal (*ḥ3p/ḥp*) the body of the deceased (scil. from those who might seek to harm it). More than one ancient Egyptian necropolis was named *Ḥ3p-nb=s*, 'Concealer of its lord', the lord in question being the chief deity interred there.[325] A hymn to the god Sobek from the temple of Kom Ombo calls him *ḥ3p ḏ.t*, 'hidden of body', in the cemetery of that town.[326] According to a text from the temple of Edfu, the underworld was made deep (*smḏ*) for Sokar Osiris *r ḥ3p ḏ.t=f*, 'in order to conceal his body'.[327] Thus, the verb *ḥ3p* (like its causative form *sḥ3p*) could be used as a virtual synomym for words meaning 'bury' or the like.[328] In a demotic mortuary papyrus, a deceased person makes the request: *ḏ=w rn=y iw=y ḥp*, 'May they say my name when I am hidden' (= buried in the tomb).[329] Another Edfu text refers to the primordial gods of Behdet whose bodies are hidden (*ẖ3w.t=sn ḥ3p.ti*) in the necropolis there.[330] This is one of a number of inscriptions from that temple which speak explicitly of the concealment of the primordial deities after their death.[331]

Bearing this in mind, the reference to the concealment of the Ogdoad in the present fragment, occurring as it does in such close proximity to the mention of Thebes, makes it virtually certain that one has here the remains of an account of the burial of the Hermopolitan deities. After elevating the divine cow and her child the sun god into the heavens,[332] their work on earth is completed. Having grown old, they return to Thebes, die, and are laid to rest in their place of birth.[333] Presumably, the actual death of the Ogdoad was recounted in a passage now lost which stood immediately before the one under consideration.

[323] See Thissen, *Die demotischen Graffiti von Medinet Habu*, pp. 193 and 214-15.

[324] Line 1245. Cf. Beinlich, *Das Buch vom Fayum*, pp. 262-3.

[325] See e.g. Gauthier, *Dictionnaire des noms géographiques* 4, p. 10; Ray, *The Archive of Ḥor,* pp. 147-9.

[326] De Morgan et al, *Kom Ombos* 1, p. 17, no. 10, line 2; Gutbub in Vercoutter (ed.), *Hommages à la mémoire de Serge Sauneron* 1, p. 432; idem, *Textes fondamentaux de la théologie de Kom Ombo*, pp. 104, note (ac), and 266, note (l).

[327] Rochemonteix and Chassinat, *Edfou* 1, p. 376, lines 12-13.

[328] Cf. Willems, *The Coffin of Heqata*, pp. 406-7; Fischer-Elfert, *Lesefunde im literarischen Steinbruch von Deir el-Medineh*, p. 86.

[329] P. BM 10507, II, 22 (Smith, *The Mortuary Texts of Papyrus BM 10507*, p. 38 and plate 2).

[330] Chassinat, *Edfou* 4, p. 240, lines 3-4. Cf. idem, *Le Mystère d'Osiris au mois de Khoiak*, pp. 283-6.

[331] See Chassinat, *op. cit.*, pp. 279-83 and 289.

[332] See reference cited in note 315 of the commentary on fragment 8.

[333] For this sequence of events, see Sethe, *Urkunden* 8, p. 5 (6b).

FRAGMENT 10 — Plate 7

Dimensions: 10 × 11.2 cm. Colour: medium brown. Related fragments: No. 24.

Transliteration

x+1. [..] *ir=w* [

x+2. [..]…..[

x+3. [..] *r-ḏbꜣ …=f* .[

x+4. [..] *pꜣ tꜣ tr=f r-ḏ[bꜣ/nt*

x+5. *…=f r rꜣ=w smne* [

x+6. *nt pꜣy n rꜣ=f* [

x+7. *ḥp=w r-ir Ḏḥwty* ..[..]…[

x+8. *pꜣ ist n tꜣ* .[..]…[

x+9. *pꜣ i-ir ḫpr n pꜣ-Š[y]* …[

x+10. *ẖny nhsy r-ḥry [n] tꜣ ly.t ḥry.t* [

x+11. *iny nt ḥr tꜣ/nꜣ* [..].. *n pꜣ tꜣ tr=f* [

x+12. *tw n pr-imnt* [..]. *nt smne r-hn-[r pꜣ hrw*

x+13. [..] ⸤*m-sꜣ*⸥ […]… *r/nb* [

x+14. [……..]..[..] *rt* [

x+15. [……..]..[

Translation

x+1. [..] They made [

x+2. [..]…..[

x+3. [..] because of … him/it .[

x+4. [..] the entire land bec[ause of/which

x+5. … him/it, while their mouths were fixed [

x+6. which leap up from his mouth [

x+7. concealment of them which Thoth did ..[..]…[

x+8. the ground of the .[..]…[

x+9. that which happened to Psh[ai] …[

x+10. skin rise up above [

x+11. stone which is upon the [..].. of the entire land [

x+12. mountain of the west [..]. which is established up unt[il today

x+13. [..] ⸤after⸥ […]… to/every [

x+14. [……..]..[..] grow [

x+15. [……..]..[

Textual notes

Line x+3

The verb to which the suffix pronoun *f* is attached concludes with the strong arm determinative and a sign that bears some resemblance to a writing of the verb *ṯ*, 'take'. I am unable to read the two signs that stand in front of these, however. The second one might conceivably be an *s*.

Line x+4

The trace at the end of the line could be either the beginning of *r-ḏbꜣ*, 'because' (cf. the example of this preposition in the preceding line), or the relative converter *nt*.

Line x+5

(a) With the damaged occurrence of *rꜣ*, 'mouth', here, cf. the better preserved examples of that noun cited in the textual note on fragment 7, line x+7.

(b) For the traces of the verb *smne*, 'fix, establish', at the end of the line, cf. better preserved examples of that verb in fragments 11, I, 11; 21, line 5; and 24, lines x+6 and x+9.

Line x+9

Only the first half of *pꜣ-Šy* has been preserved. For more complete examples of this divine name, see e.g. fragment 1, line 2, and fragment 5, II, 10.

Line x+10

(a) For the noun *ẖny*, 'skin', see textual note on fragment 4, II, 14. With the traces of the verb *nhsy*, 'rise up', that follows, cf. the better preserved instance of that word in fragment 13, II, 7.

(b) The traces of [*n*] *tꜣ ly.t ḥry.t* can just be made out at the end of the line. Cf. better preserved examples of the same phrase in fragment 1, line 9, and P. Carlsberg 10029, line x+4.

Line x+11

(a) The definite article after the preposition *ḥr* could be either *tꜣ* or *nꜣ*. Owing to the break in the middle of the line, it is impossible to tell whether the sign above the *ꜣ* is a *t* or an *n*.

(b) With the traces of *pꜣ tꜣ tr=f* at the end of the line, cf. better preserved occurrences of the same phrase in line x+4 above, fragment 13, II, 5, and fragment 19, line x+16.

Line x+12

(a) All that is preserved of the word following *pr-imnt* is what appears to be a phallus determinative. Cf. writings of words like *sty*, 'scatter', in fragment 1, line 13, and *mtwy*, 'seed, semen', in fragment 2, line 6, both of which are determined with that sign.

(b) For better preserved examples of *smne*, 'establish', elsewhere in the text, see references cited in textual note (b) on line x+5 above.

(c) For the restoration *r-hn-*[*r pꜣ hrw*] at the end of the line, cf. better preserved examples of the same phrase in e.g. fragments 1, line 1; 4, II, 15; and 13, II, 5.

Line x+13

(a) Only the top part of *m-sꜣ* is preserved in this line. For the traces, cf. more complete examples of that preposition in e.g. fragment 1, line 1, and fragment 11, I, 11.

(b) The traces at the end of the line could be either the preposition *r* or *nb*, 'every'. The word in question is completely preserved except for a small part of its upper half. Unfortunately, this is the crucial part for determining which of the two readings given above is the correct one.

Line x+15

Only a single sign is visible here, near the beginning of the line. This could be *ḫpr*, 'exist, become', or alternatively, part of a word like *mt*, 'speak' (*Glossar*, p. 184), or *ḫrw*, 'voice' (ibid., p. 366).

Commentary

Line x+5 of this fragment refers to the mouths of some unidentifiable beings, which are said to be 'fixed' (*smne*). The following line preserves the phrase *nt p3y n r3=f*, 'which leap up from his mouth'. It is not clear to whom the third person singular masculine suffix pronoun refers here. Perhaps this is an allusion to the emergence (*pr*) of divine beings from the mouth of the solar deity, for which, see commentary on fragment 5, I, 13.[334]

Line x+7 begins with the words *ḥp=w r-ir Ḏḥwty*, 'the concealment of them which Thoth did'. This must be a reference to the burial of the Ogdoad which was narrated in the preceding fragment, q.v. Their interment is treated here as an event which has already taken place, as the use of the past relative *r-ir* shows.

The Egyptians regarded burial of parents as the duty of their offspring, in particular the eldest son, if one existed. In the case of the Ogdoad, this means the solar deity.[335] However, that god, having ascended into the sky,[336] appointed Thoth as his deputy on earth, and it is presumably in that capacity that we find the latter assuming responsibility for the interment of the Ogdoad in our text. Compare the Book of the Heavenly Cow, in which the newly raised up solar deity says to Thoth: *m=ṯn wi ꜥ3 m p.t .... iw=k m s.t=i s.ty=i*, 'See, I am here in the sky ...., while you are in my place, my deputy', and then goes on to instruct him concerning the funerary cult of the Hermopolitan deities: *iw=i gr.(t) r rdi.t dwn=k ḏr.t=k r-ḫft-ḥr-n p3wty.w ꜥ3.w r=k*, 'I will cause you to extend your hand (scil. in the offering rite) in the presence of the primordial ones who are elder than you.'[337] This is why, in an inscription from the Ptolemaic temple of Hathor at Deir el-Medineh, Thoth is called *qbḥ m T3.t-ṯ3mw.t*, 'libationer in the mound of Djeme', i.e. the burial place of the Ogdoad.[338]

Line x+9 preserves the phrase *p3 i-ir ḫpr n p3-Š[y]*, 'that which happened to Psh[ai]'. The event in question, whatever it was, is clearly envisaged as having already taken

[334] *p3y* and *pr* are sometimes used as synonyms. Cf. Smith, *The Liturgy of Opening the Mouth for Breathing*, p. 45 [notes (a) and (e) on P. Berlin 8351, II, 8].

[335] See commentaries on fragments 3 and 4; Sethe, *Amun und die acht Urgötter von Hermopolis*, pp. 50-2.

[336] See commentary on fragment 8.

[337] Hornung, *Der ägyptische Mythos von der Himmelskuh*, pp. 22-3. For *p3wty.w* as a designation of the Ogdoad, see e.g. Sethe, *op. cit.*, pp. 46-7.

[338] Piehl, *Inscriptions hiéroglyphiques* 1, plate 171 μ. The text in question is cited by Sethe, *op. cit.*, p. 61, where it is wrongly attributed to Medinet Habu.

place, i.e. not part of the ongoing narrative, as the use of the past participle *i͗-i͗r* shows. Perhaps 'that which happened to Psh[ai]' was his elevation into the sky. Cf. the words *nhsy r-ḥry* [*n*] *tꜣ ly.t ḥry.t*, 'rise up above', in the line immediately following. Conceivably, the text may have adduced that event here as the reason for Thoth's assumption of the duty of burying the Ogdoad, in line with what I have suggested in the preceding paragraph.

The words *tw n pr-imnt*, 'mountain of the west', stand at the beginning of line x+12. This could be a reference to the location of the tomb of the eight Hermopolitan deities on the Theban west bank. The site where they were buried, the mound of Djeme, is often called a **ⲧⲟⲟⲩ**, 'mountain', in Coptic sources.[339] An inscription from the propylon of the temple of Montu at Karnak calls the Ogdoad *ḥtp m ḥr.t=sn m Ḫft-ḥr-nb=s*, 'those who rest in their tomb in She-who-is-opposite-her-lord', the latter being a general designation for the rocky escarpment on the west bank, also used occasionally as a term for the mound of Djeme itself.[340] In earlier periods, the name *ḏw n imnt*, 'mountain of the west', referred specifically to the high peak north of the Valley of the Queens, now known as el-Qurn, which was worshipped as the goddess Meretseger.[341]

The end of line x+12 preserves part of an aetiological statement: *nt smne r-hn-*[*r pꜣ hrw*], 'which is established up un[til today]'. On the function of such statements, see Introduction. Perhaps one had here an explanation by the writer of how either the tomb of the Ogdoad or the mountain concealing it were made to endure up until his own time.

FRAGMENT 11 — Plate 7

Dimensions: 17 × 15 cm. Colour: medium brown.

Transliteration

Column I

1. *Nw*]*n pꜣy=w iṯ šꜥ* [........] *r-b*[*nr*(?)] *r-r=w* [........]
2.].. *ḫpr Ḏḥwty pꜣ ntr* ꜥ[ꜣ] *iw=f ꜥq r* [....]. *p.t*(?) *r ir* [........]

[339] See e.g. Winlock and Crum, *The Monastery of Epiphanius at Thebes* 1, pp. 107-8; Crum and Evelyn White, *The Monastery of Epiphanius at Thebes* 2, p. 354.

[340] Sethe, *Urkunden* 8, p. 30 (35c). Cf. Otto, *Topographie des thebanischen Gaues*, pp. 49-50; Sternberg-El Hotabi, *Der Propylon des Month-Tempels in Karnak-Nord*, p. 119, note 5; Cozi, *Göttinger Miszellen* 151 (1996), pp. 37-47; idem, *Göttinger Miszellen* 153 (1996), pp. 17-31. For the equivalence *Ḫft-ḥr-nb=s* = Djeme, see P. Rhind 1, VI d 9/h 10; P. Rhind 2, VII d 7/h 7 (Möller, *Die beiden Totenpapyrus Rhind des Museums zu Edinburg*, plates 6 and 18).

[341] See Aufrère, *L'Univers minéral dans la pensée égyptienne* 1, pp. 21-2, with references to earlier literature.

3. *ỉr]m nꜣ ntr.w mtw ...ḥ r tꜣ p.t .[..] mtw [........]*
4. *] ꜥq r tꜣ p.t n-bnr=f ..... bn-pw=w ...[........]*
5. *Pt]ḥ(?) r-r=w gm=f pꜣ ḫpr bn-pw=w ...=f ... p.[t(?) ..]..[....].[..]*
6. *s]ty=f tꜣy=f mtwy .....[........]..[..]*
7. *]...... twt ꜥnḫ ...[........]....*
8. *] mw n tꜣ ly.t ḥry.t n [........]....*
9. *r]n=f ỉw=f ꜥnw ....[..]..[....] ỉr=w(?) kꜣ s[y]m*
10. *]. šꜥꜣ .... n-ỉm=w ẖn pꜣ mw r-b-ỉr-tw*
11. *] smne ỉrm pꜣ tꜣ tr=f m-sꜣ pꜣ Nwn ...*
12. *] Ḏḥwty ỉw ...=f ..[..]⌊m⌋(?)[..] rn=f rḫ [....]*
13. *]. ḫpr .. r-ỉr=f/pꜣ [...........]*
14. *]=f Ptḥ ....[...........]*
15. *Km]y ... ḏ [...........]*
16. *p]r ...... ḫpr [..........]*
17. *] rn n pꜣ qmꜣ r-ḏbꜣ.ṱ=[s ....]*
18. *] ḏ(?) Kmy n rn [..........]*
19. *].[...........]*

Column II

Traces of two lines

Translation

Column I

1. the Primaeval] Ocean their father until [........] o[ut(?)] to them [........]
2.].. Thoth the gr[eat] god began to enter [....]. sky(?) to make [........]
3. wi]th the gods and/who ...*ḥ* to the sky .[..] and/who [........]
4.] enter the sky except him ..... They did not ...[........]
5. Pt]ah(?) to them. He discovered the fact that they did not ... him/it ... s[ky(?) ..]..[....].[..]
6.] He [sca]ttered his seed ....[........]..[..]
7.]...... living image ...[........]....
8.] water above [........]....
9.] his [na]me, he/it being fair ....[..]..[....]. They made(?) another pl[a]nt
10.]. sand .... there in the water before
11.] established together with the entire earth after the Primaeval Ocean ...
12.] Thoth came ... him/it ..[..]*⌊m⌋(?)*[..] his name, know [....]
13.]. occur .. which he/the [...] did [.........]
14.] him/it Ptah ....[...........]
15. Egy]pt ... say [...........]
16. emer]ge ...... occur [..........]

17.] name of the rush because of [it ....]
18.] said(?) ‘Egypt’ as the name [..........]
19.].[............]

Column II

Traces of two lines

Textual notes

Column I

Line 1

(a) With the writing of the phrase *p3y=w iṯ* here, cf. that in fragment 2, line 9. In the latter, the initial sign of *iṯ*, ‘father’, has been omitted through haplography.

(b) The proposed restoration *r-b*[*nr*] would fit the available space and suits the traces well. Compare writings of the same adverb in fragments 2, line 8; 5, II, 1 and 17; and 19, line x+13. For better preserved writings of the following *r-r=w*, see e.g. line 5 below; fragments 5, I, 9; and 12, I, 16.

Line 2

(a) All that is preserved of the word before *ḫpr* in this line is the pellet determinative written above plural strokes.

(b) The restoration ꜥ[*3*] after *ntr* is certain. For the traces, see the better preserved writing of that adjective in fragment 6, line 6.

(c) The three signs before *r ir* at the end of the line, *t*, ‘heaven’ sign, and divine determinative, could be a semi-hieratic writing of *p.t*, ‘sky’, as in P. Carlsberg 465, line 9, or alternatively, the last part of a longer word, the rest of which is now lost.

Line 3

(a) For the traces of [*ir*]*m* at the beginning of the line, see the better preserved writings of that word in e.g. fragment 13, II, 5, 7, 11, and 15.

(b) The partially preserved verb after *mtw* ends in *ḥ* and is determined with the strong arm sign. It is construed with the preposition *r*. Cf. perhaps *ḥlḥ*, ‘guard, protect’ (*Glossar*, pp. 326-7).

Line 4

(a) The signs immediately after *n-bnr=f* appear to be *ḥq*, followed by a tall vertical stroke, followed by *ḏ* and another vertical stroke. I am unable to think of a word or words that these traces might represent.

(b) The initial signs of the partially preserved verb after *bn-pw=w* appear to be a horizontal stroke above what is perhaps an *n*. The rest of the word is illegible.

Line 5

(a) The partially preserved divine name at the beginning of the line could be *Ptḥ*, ‘Ptah’. For the traces, see writings of that name in line 14 below, and fragment 2, lines 5, 9, and 15.

(b) With the traces that I have suggested could represent *p.t*, ‘sky’, cf. better preserved writings of the same noun in e.g. fragments 4, II, 4; 5, I, 3; and 13, II, 2 and 21.

Line 6

Restore [*s*]*ty=f* at the beginning of the line. For other examples of the phrase *sty=f t3y=f mtwy*, ‘he scattered his seed’, see fragment 1, line 13, and fragment 2, line 6.

Line 7

For the writing of *twt*, 'image', here, cf. fragment 2, line 11. The traces immediately before *twt* could be of *mw*, 'water'. Cf. the writing of that noun in line 10 below. The illegible word before those traces appears to be determined with the plant sign.

Line 8

(a) With the traces of *n tꜣ ly.t ḥry.t n*, cf. better preserved examples of the same phrase in fragment 1, line 9, P. Carlsberg 10029, line x+4.

(b) The traces at the end of the line could be the lower part of the verb *ḫpr*, 'exist, become'. Cf. writings of that word in lines 2 and 13 of the present column.

Line 9

(a) The restoration *[r]n=f* at the beginning of the line is certain. Cf. writings of *rn* in e.g. lines 9 and 12 below, fragment 2, line 10 and passim.

(b) The hieratic writing of *ꜥnw*, 'be fair, beautiful', here is not found elsewhere in the text. For the normal writing used by our scribe, see textual note (a) on fragment 3, line 13.

(c) The traces before *kꜣ* could represent *ir=w*. Cf. writings of the latter in fragment 5, II, 9, and fragment 19, line x+16 (both in the verbal prefix *i-ir̥=w*).

(d) For writings of *ky*, 'another', as *kꜣ*, see *Glossar*, p. 588. With the traces of *s[y]m* that follow, cf. the better preserved occurrence of that noun in fragment 13, II, 12.

Line 10

(a) The trace at the beginning of the line is the top part of a man with hand to mouth determinative. The remainder of the word to which it belonged has been lost.

(b) With the traces of *b-ir-tw* at the end of the line, cf. the better preserved instance of that verbal prefix in fragment 1, line 19.

Line 11

For a better preserved occurrence of the semi-hieratic writing of *Nwn*, 'Primaeval Ocean', here, see fragment 1, line 2.

Line 12

The trace visible before *rn* could be the top part of an *m*. Cf. e.g. the writing of *mtw* in line 3 above and that of *mtwy* in line 9.

Line 13

The trace after the verb *ir* at the end of the line could be either part of the suffix pronoun *f* or the definite article *pꜣ*.

Line 15

With the traces of *Kmy*, 'Egypt', at the beginning of the line, cf. the better preserved instance of that toponym in line 18 below.

Line 16

For the traces of *pr*, 'emerge', at the beginning of the line, cf. better preserved examples of the same verb in e.g. fragments 5, I, 6; 13, II, 3; and 19, lines x+12 and x+16. I am unable to read what intervenes between *[p]r* and the *ḫpr* that stands at the end of the line. The only clearly recognisable sign is the determinative used with words like *mny*, 'image' (fragment 2, line 10), and *twt*, 'statue' (ibid., line 11), roughly midway between the two.

Line 17

For the suggested restoration of the suffix pronoun *s* after *r-ḏbꜣ.ṯ* at the end of the line, cf. the occurrence of *r-ḏbꜣ.ṯ=s* in fragment 4, II, 15. In view of the preceding words *rn n pꜣ qmꜣ*, 'name of the rush', one may have had in this line an explanation of how the rush acquired a further name in addition to its original one, e.g. '[They said "X" as the] name of the rush because of [it].' See commentary below.

Line 18

The traces visible before *Kmy*, 'Egypt', here could represent the top part of *ḏ*, 'say'. Cf. better preserved writings of that verb in e.g. fragment 2, lines 9, 12, and 16.

Column II

Line x+1

The signs visible at the beginning of the line could be *nn*. For this reading, see my discussion in *Sesto Congresso Internazionale di Egittologia: Atti* 2, p. 493. Alternatively, they might represent the beginning of the obscure botanical term that occurs in P. Carlsberg 230, fragment 4+5, column x+1, lines x+6 and x+8. For the latter, see Tait in Frandsen (ed.), *The Carlsberg Papyri 1: Demotic Texts from the Collection*, pp. 69-70. Quack, *Enchoria* 21 (1994), p. 66, takes this as a writing of *sẖ.t*, 'field'.

Commentary

The initial line of Column I preserves a reference to [*Nw*]*n pꜣy=w iṯ*, '[the Primaeval] Ocean their father'. The antecedent of 'their' has been lost. Cf., however, commentary on fragment 2, note 124, for references to the Primaeval Ocean as *it nṯr.w*, 'father of the gods'. The words *ḫpr Ḏḥwty pꜣ ntr* ⸢[ꜣ] *iw=f* ⸢*q r* [....]. *p.t*(?), 'Thoth the gr[eat] god began to enter [....]. sky(?)', occur in line 2. The following line preserves the phrase *r tꜣ p.t*, 'to the sky', while line 4 begins with the words ⸢*q r tꜣ p.t n-bnr=f*, 'enter the sky except him'. I have suggested that fragment 8 may have described the initial elevation of the sun into the heavens by the celestial cow goddess, while fragments 9 and 10 are concerned with the burial of the Ogdoad by Thoth acting as the solar deity's representative on earth.[342] Perhaps lines 2-4 of the initial column of the present fragment described the subsequent entry of the ibis god into the sky as the moon.[343] Compare the Book of the Heavenly Cow, where the solar deity, having appointed Thoth as his deputy and given him instructions concerning the funerary cult of the Hermopolitan divinities, tells him further: *iw=<i> gr.t r rdi.t inḥ=k p.ty m nfr.w=k m ḥḏw.t=k*, '<I> will cause you to encompass the two heavens with your beauty and your radiance.' The ancient commentator adds: *ḫpr iꜥḥ pw n Ḏḥwty*, 'Thus came into existence the lunar form of Thoth.'[344]

[342] See commentaries on the fragments in question.

[343] Thoth was the pre-eminent lunar deity in ancient Egypt. Cf. Derchain in *La Lune: mythes et rites*, pp. 36-40; Boylan, *Thoth the Hermes of Egypt*, pp. 62-75.

[344] See Hornung, *Der ägyptische Mythos von der Himmelskuh*, pp. 23-4.

Traces of a divine name, possibly *Ptḥ*, 'Ptah', are visible at the beginning of line 5. The next line starts with the words [*s*]*ty=f tꜣy=f mtwy*, 'he [sca]ttered his seed'. This recalls the account in fragment 2 of how Ptah scattered his seed in order to fertilise the eggs or egg sacs from which the Ogdoad later emerged.[345] As it happens, the name of that god recurs in line 14 of the present fragment. Clearly, however, a different sequence of events is narrated here, even though the text's poor state of preservation makes it difficult to reconstruct what precisely is said to happen.

Apart from his role as progenitor of the Ogdoad, Ptah has another important function in our cosmology, that of creator of material abundance. Allusion is made to this function in an address to Ptah's father *pꜣ-Šy pꜣ Nwn* in lines 3-5 of fragment 1, where the younger deity is called *pꜣy=k šr i-ir ti ḫpr n=n ẖlꜣ*, 'your son who created sustenance for us', and *i-ir šꜥ ir nꜣ iꜣw.wt nhm iw=w sy*, '(the one) who first made the cattle rejoice, they being satisfied'.[345] Since this is only an allusion, one could imagine that lines 5 ff. of the column under consideration actually recounted how Ptah went about bringing the said sustenance into being. The position of this act in the narrative would then be after his begetting of the Ogdoad, with the birth of the sun god intervening. For such a sequence of events, compare the demotic cosmogony of P. Berlin 13603, the initial column of which narrates how Ptah created the eight Hermopolitan deities, while the account of how he brought material abundance into being commences in the fourth column, with the birth of the sun god being described in the intervening second column.[346]

It appears from fragment 1 of our cosmology that Ptah provided sustenance for living creatures, at least in part, by uprooting or plucking out what may have been a thicket of rushes or reeds (*qmꜣ*).[347] The words [*s*]*ty=f tꜣy=f mtwy*, 'he [sca]ttered his seed', in line 6 of the present column could refer to the fertilisation of such a growth by Ptah's semen.[348] The phrase *ir=w(?) kꜣ s*[*y*]*m*, 'they made(?) another pl[a]nt', which occurs three lines later, could describe a further consequence of the same act. The rush or reed which appears to serve as a catalyst for the creation of sustenance in fragment 1 occurs in this one as well, specifically in line 17 below, which preserves the phrase [....] *rn n qmꜣ r-ḏbꜣ.ṯ=* [*s*], '[....] name of the rush/reed because of [it]'. As remarked in the textual note on that line, one probably has here the end of an explanation telling how the plant in question received a further name in addition to its original one. Cf. the statement *ḏ=w n=f ꜣsy n rn*, 'It was called "reed" by name', in fragment 1, line 12,

[345] See commentary ad loc.

[346] See Erichsen and Schott, *Fragmente memphitischer Theologie in demotischer Schrift*, pp. 311-14 (13-16) and 316-17 (18-19) with plates 5-6.

[347] See fragment 1, line 5, and commentary ad loc.

[348] For the scattering of divine seed onto plants, see commentary on fragment 1, line 13, and references cited there.

which may refer to this plant as well.[349] In the commentary on fragment 1, line 5, it was suggested that Ptah's plucking out or uprooting of the rush or reed might be related to the fabrication of the *ḏd*-pillar, which is sometimes said to be made, at least in part, of cut off reed stems.[350] If so, then perhaps the additional name given to the rush in the present fragment was *ḏd*.

The toponym *Kmy*, 'Egypt', occurs twice in the initial column of the fragment under consideration (lines 15 and 18). The latter preserves the words *ḏ*(?) *Kmy n rn*, 'said(?) "Egypt" as the name', in an otherwise broken context. If the reading *ḏ* is correct there, then presumably one had in the present column an account of how Egypt was given its name. Cf. Introduction for the sort of naming formula involved. As we have seen, the noun *qmꜣ*, 'rush, reed', occurs only one line before this. In view of the phonetic similarity between that word and *Kmy*, 'Egypt', perhaps the name of the latter was said to have been derived from the former by paronomasia. This would be logical, since the thicket of reeds in fragment 1 appears to mark the place where the land first emerged and thus to be, in a sense, its precursor.[351] This primaeval mound is on Egyptian territory. The fact that the origins of the land *Kmy*, 'Egypt', can be traced back to a plant whose name is *qmꜣ* may have led our text's author to postulate an etymological link between toponym and botanical term.

FRAGMENT 12 — Plate 8

Dimensions: 8.6 × 16 cm. Colour: medium brown. Related fragments: No. 25.

Transliteration

Column I

1. *r*]*-hn-r p*[*ꜣ hrw*
2.] *nꜣ s-ḥm.wt* [
3.]*y* [
4.] *s*[*š*]*n* [
5.]...[
6. *s*]*rpt* [
7.]...[
8. *ẖ*]ˁ*=s*
9. *ḏw*]*f* ⸢*ir/ḏ=w*⸣ *n=s*
10. *S*]*ẖmy Tfny*
11.]. *mtw ḏwf pꜣy*

[349] See commentary ad loc.
[350] See references cited in note 56, ibid.
[351] See commentary on fragment 1, lines 11 and 18-20.

12.] *ms=w Ḥr*
13.] *p3 ḏwf iw=f*
14. *ẖ*]*n p3 ḏwf irm=f*
15.].. *n Nḫb3.t r-ḏb3*
16.]*=f r-r=w* [..] *šnb*
17.]. *p3 ḫ*ʿ [...] *p.t*

Column II

x+1. *ḏ Ḏ*[*ḥwty(?)*
x+2. *rn n p3 ḫ*[*t(?)*
x+3. *n rn=f* [
x+4. *t3 nhy* .[
x+5. *nt* ʿ*ḥ*ʿ *ẖn=f* .[
x+6. *ḫpr t3 m*[*t*
x+7. *rn=s p3* ..[
x+8. *iw=f n=s(?)* ..[
x+9. *ḫpr i-ir n3* .[
x+10. *r p3-R*ʿ ..[
x+11. [..]*=w irm=w* [
x+12. *Ḏḥwty p3 ntr* [
x+13. ...[

Translation

Column I

1. up] until t[oday
2.] the women [
3.]*y* [
4.] lo[t]us [
5.]...[
6. l]otus [
7. [...]
8.] she [app]eared
9. papy]rus. ⸢They made/said⸣ for/to her
10. S]akhmet and Tefnut
11.]. which is papyrus
12.] Horus was born
13.] the papyrus, he
14. i]n the papyrus with him
15.].. to/of Nekhbet because
16.] him/it to them [..] unite with
17.]. the appearance [...] sky

Column II

x+1. There said T[hoth(?)
x+2. name of the tr[ee
x+3. as/of its name [
x+4. the sycomore .[
x+5. which stands within it .[
x+6. The thi[ng] occurred [
x+7. her/its name, the ..[
x+8. it is hers(?) ..[
x+9. occur. *i-ir* the .[
x+10. to/concerning Pre ..[
x+11. They [..] with them [
x+12. Thoth the god [
x+13. ...[

Textual notes

Column I

Line 1

With the traces of *r-hn-r pꜣ hrw*, 'up until today', here, cf. better preserved instances of that phrase in e.g. fragments 1, line 1; 4, II, 15; and 13, II, 5.

Line 3

One has here the remains of a word which ends with *y* and the jar determinative. For the shape of the latter, cf. e.g. the determinative of the noun *ḥnqy*, 'beer', in P. Carlsberg 14, a, 2-6 (Volten, *Demotische Traumdeutung*, plate 3).

Line 4

With the writing of *sšn*, 'lotus', here, cf. that in P. Carlsberg 10040, line x+4. On this botanical term and the identification of the species which it denotes, see Ryhiner, *L'Offrande du lotus*, pp. 3 and 11-14; Germer, *Flora des pharaonischen Ägypten*, pp. 37-9; Brunner-Traut in *Lexikon der Ägyptologie* 3 (1980), columns 1091-6; Charpentier, *Recueil de matériaux épigraphiques relatifs à la botanique de l'Égypte antique*, pp. 618-21.

Line 6

With the partially preserved writing of *srpt*, 'lotus', in this line, cf. better preserved examples of the word cited in *Glossar*, p. 442. Althogh this term originally denoted the lotus leaf, it came to be used as a designation for the whole plant as well. In the latter usage, it is more or less synonymous with *sšn*. Some authorities would see in *srpt* the name of a distinct species, but their views have not won wide acceptance. See Brunner-Traut, *Lexikon der Ägyptologie* 3 (1980), column 1095; Charpentier, *op. cit.*, pp. 604-5, and references given there.

Line 8

For the writing of *ḫꜥ*, 'appear', here, cf. line 17 below, fragment 2, line 7, fragment 13, II, 16.

Line 9

With the traces of *ḏwf*, 'papyrus', here, cf. the better preserved examples of the same word in lines 13 and 14 below.

Line 11

mtw is written for *nt-iw*. The damaged word before it is determined with the plant sign.

Line 14

Only the house determinative of *ẖn* has been preserved at the beginning of the line. Cf. the undamaged occurrence of this preposition in line x+5 of the second column of the present fragment.

Line 15

The damaged word at the beginning of the line is determined with the man with hand to mouth.

Line 16

For the metathesised form *šnb* of the earlier *šbn*, 'unite' (= Coptic ϣⲱⲛⲃ), see *Glossar*, p. 499.

Line 17

The traces at the end of the line could be either a semi-hieratic writing of *p.t*, 'sky', as in P. Carlsberg 465, line 9, or alternatively, the last part of a longer word, the rest of which is now lost.

Column II

Line x+1

After *ḏ*, probably restore *Ḏ*[*ḥwty*]. Cf. better preserved occurrences of that divine name in e.g. fragment 5, II, 6, 9, and 11; also line x+12 below.

Line x+2

The word for 'tree', which I have restored as *ẖ*[*t*], could also be *šn*. Cf. fragment 13, II, 20.

Line x+6

With the traces of *mt*, 'thing', at the end of the line, cf. the undamaged example of that noun in P. Carlsberg 462, II, 9 (Smith in Frandsen and Ryholt [eds.], *The Carlsberg Papyri 3: A Miscellany of Demotic Texts and Studies*, plate 15), also the writing of the verb *mt*, 'speak', in fragment 5, II, 2.

Line x+8

With the traces that I have tentatively read as *n=s*, cf. the better preserved writing of that preposition and suffix pronoun in line 9 of the initial column of the present fragment.

Commentary

The remains of an aetiological statement, [*r*]-*hn-r p*[*ꜣ hrw*], '[up] until t[oday]', are preserved in line 1 of the initial column of this fragment On the function of such statements, see Introduction. Lotuses (*ssšn*, *srpt*) are mentioned in lines 4 and 6 of that column, while references to *pꜣ ḏwf*, 'the papyrus thicket', occur in lines 9, 13, and 14. Line 12 preserves the words *ms=w Ḥr*, 'Horus was born'. It seems fairly clear that one had here a version of the story of that god's birth and concealment by his mother Isis in the papyrus marshes of Khemmis in the delta, a myth well-known from other sources.[352]

The infant Horus was hidden so as to protect him from attack by Seth. The divinity who bore chief responsibility for guarding him was the cobra goddess Edjo.[353] However, other guardians, both divine and human, assisted in this as well. Line 2 of the present column mentions *nꜣ s-ḥm.wt*, 'the women'. Compare a late magical spell against the bite of a scorpion which relates how the 'inhabitants of Khemmis, the nurses of Pe and Dep' (*imy.w Ḫby.t mnꜥ.wt imy.w P Dp*) were instructed to protect Isis and her child.[354] Perhaps the women to whom reference is made in our text are identical with these. Line 10 of Column I mentions Sakhmet and Tefnut. Like Edjo, both of these are well suited to serve as protectresses, in view of the ferocious aspect which they share.

Some sources identify Sakhmet with Edjo.[355] Thus, in a scene from the temple of Dendera, the latter is said to have made a shelter (*ir nh*) for Horus amidst the swamp plants (*iyḥ*) and nourished him (*šd*) in the papyrus marsh (*idḥ*).[356] A comparable text from Esna associates this vegetation with the former: *ir wꜣḏ.w n Sḫm.t msḫn.t m Ḫby.t ir=sn sꜣ=k wḥm=sn mk.t=k r ḫw nṯr im=sn*, 'The papyrus plants of Sahkmet formed a resting place in Khemmis, protecting you, repeating your protection, so as to safeguard the god within them.'[357] Like Edjo, Sakhmet is sometimes called the mother of Horus as well as his guardian.[358]

One of the means by which the cobra goddess was believed to exercise protection is with the apotropaic *snb*-plant, apparently a form of papyrus.[359] In a Late Period text, Tefnut is said to employ this amulet for protective purposes as well.[360] The latter deity has a further association with the papyrus marshes of Khemmis in that, according to the New Kingdom P. Ebers, she was born there as the daughter of Isis.[361]

Line 15 of Column I preserves a reference to the goddess Nekhbet, who is well attested as a guardian and nurse of the infant Horus. In Coffin Text Spell 16, for

[352] See Gardiner, *JEA* 30 (1944), pp. 52-8; Bergman, *Ich bin Isis*, pp. 137-40; Altenmüller in *Lexikon der Ägyptologie* 1 (1975), columns 921-2; Klasens, *A Magical Statue Base*, pp. 64-78.

[353] See Gardiner, art. cit., pp. 54-6; Fischer-Elfert in *Lexikon der Ägyptologie* 6 (1986), columns 907-8 (s.v. Uto), with additional bibliography.

[354] Klasens, *A Magical Statue Base*, p. 33.

[355] See Hoenes, *Untersuchungen zu Wesen und Kult der Göttin Sachmet*, pp. 172-4.

[356] Mariette, *Dendérah* 1, plate 56a.

[357] Sauneron, *Esna* 3, p. 201 (291, 22). The deity addressed is Khnum-Re, who is identified with Horus. Cf. *Esna* 5, p. 140.

[358] Gardiner, art. cit., p. 56; Münster, *Untersuchungen zur Göttin Isis vom Alten Reiches bis zum Endes des Neuen Reiches*, pp. 114-15; Hoenes, *op. cit.*, pp. 191-2.

[359] See Koemoth, *GM* 130 (1992), pp. 33-43; Derchain, *Le Papyrus Salt 825*, pp. 168-71.

[360] P. Salt 825, VII, 8-10 (Derchain, *op. cit.*, pp. 9*-10* and plate 7).

[361] P. Ebers, XCV, 8 (Wreszinski, *Der Papyrus Ebers* 1, p. 196). Cf. Münster, *op. cit.*, pp. 146-7.

example, it is said of that divine child: *iw ms.n sw ꜣst iw šd.n sw Nḫb.t*, 'Isis bore him and Nekhbet nursed him.'[362] As the tutelary goddess of Upper Egypt, Nekhbet was supposed to have presided over the suckling of the future king Horus amidst the papyrus, along with Lower Egyptian Edjo. The two are represented in the act of doing so, for instance, in a scene from the mammisi at Philae.[363] She also serves as a midwife.[364] Some sources say of Horus that he was *ms n*, 'born to' (i.e. delivered by) Nekhbet.[365]

There may be a reference to Thoth in line x+1 of Column II.[366] He is certainly mentioned by name in line x+12 of that column, which speaks of *Ḏḥwty pꜣ nṯr*. Thoth was believed to have assisted Isis and her son during their period of concealment in Khemmis. In another late magical spell for protection against the bite of a scorpion, attested by several different copies, Isis relates how it was Thoth who originally advised her to flee to the papyrus marshes: *ḏd n=i Ḏḥwty nṯr wr ḥry-tp mꜣꜥ.t m p.t tꜣ mi r=ṯ ꜣst nṯr.t nfr gr pw sḏm ꜥnḫ wꜥ sšm s(w) ky sdg r=ṯ ḫr sꜣ nḫn iw=f n=n ḥꜥ.w=f rwḏ pḥty=f nb ḫpr di=t ḥtp=f ḥr ns.t it=f nḏ n=f iꜣ.t ḥqꜣ tꜣ.wy*, 'Thoth the great god, chief of justice in the sky and on earth, said to me: Come Isis the goddess – it is good to listen, for one lives when another guides him – conceal yourself with the infant son, that he might return to us when his body is strong and all his might has come to fullness, that you might cause him to occupy the throne of his father, the office of ruler of the two lands having been conferred upon him.'[367]

A text in the temple of Edfu gives a variant account of this episode. There, Isis tells Thoth that she is pregnant by her brother Osiris. The two of them go to inform Horus of Edfu. The latter instructs Thoth to take steps to ensure Isis' protection. These include fastening an apotropaic talisman to her body. With its aid, the goddess gives birth safely to Horus in Khemmis.[368]

According to Egyptian sources, Thoth continued to exercise protection over Horus and his mother throughout the period of their concealment in the papyrus marshes. One of the late magical spells cited previously recounts how, after Isis gave birth to her son 'in the nest in Khemmis', he was bitten by a scorpion. Isis cried out to heaven for assistance, whereupon Thoth descended from his place in the solar bark and uttered a spell to end Horus's suffering. Thereafter, at the behest of Isis, Thoth gave instructions to the gods and other inhabitants of the marshes to watch over the infant: *rs=ṯn ḥr ḫw pn ḥḥ mṯn=f m ḥr.w stnm wꜣ.wt n sby.w r=f r iṯ.n=f ns.t tꜣ.wy*, 'May you guard

[362] De Buck, *The Egyptian Coffin Texts* 1, p. 48a-b.

[363] Junker and Winter, *Das Geburtshaus des Tempels der Isis in Philä*, p. 12. Cf. ibid., pp. 376 and 380.

[364] See Heerma van Voss in *Lexikon der Ägyptologie* 4 (1982), column 367 (s.v. Nechbet).

[365] E.g. Junker and Winter, *op. cit.*, pp. 122-3.

[366] See textual note ad loc.

[367] See Klasens, *A Magical Statue Base*, pp. 9-10. The version quoted here is that of the Metternich Stela (M).

[368] Chassinat, *Edfou* 6, p. 214, lines 1-9. Cf. Alliot, *Le Culte d'Horus à Edfou au temps des Ptolémées*, p. 814; Kurth, *Studia Aegyptiaca* 14 (1992), pp. 373-4.

this child, conceal his path from men, confuse the ways of those who rebel against him until he has taken the throne of the two lands.'[369] Thoth also helped to protect Horus in Khemmis against attack by Seth and his confederates, as a further text in the temple of Edfu records.[370]

Line x+2 of this column contains the phrase *rn n pꜣ ẖ[t]* (or *š[n]*), 'name of the tr[ee]'; line x+3, the words *rn=f*, 'its name'. Line x+4 preserves a reference to *tꜣ nhy*, 'the sycomore'. Perhaps one had here an account of the creation and naming of the sycomore tree. This could have constituted an otherwise unknown episode in the story of the concealment of Isis and Horus in the marshes. A well-known representation in the tomb of Tuthmosis III in the Valley of the Kings depicts that monarch being suckled by a sycomore tree, which the accompanying inscription identifies as *mw.t=f ꜣst*, 'his mother Isis'.[371] Perhaps the goddess was described as nursing Horus in this form in our text. The sycomore, by its very name, is associated with the idea of protection or shelter (*nh.t*).[372] It is equally possible, therefore, that the tree figured in some protective capacity in the context of the present fragment.

All that can be made out in line x+10 of Column II are the words *r pꜣ-Rꜥ*, 'to Pre'. This god too, like Thoth and the other divinities mentioned above, played a role in guarding Isis and Horus in Khemmis. In the magical spell cited two paragraphs above, after Thoth has given instructions to the marshland inhabitants to watch over the infant god, he adds the statement: *Rꜥ m p.t ḥr wšb ḥr=f*, 'Re in heaven is protecting him.'[373] In a similar spell directed against the poison of scorpions, Thoth is made to say to Horus: *ii.n=i m p.t m wḏ.t Rꜥ r ir sꜣ=k*, 'I have come from heaven at the command of Re in order to protect you.'[374] Moreover, the sun god is said to have legitimised the royal status of the younger deity when he was still in Khemmis. An inscription in the mammisi at Philae preserves the text of a decree issued by Re on behalf of Horus, proclaiming the latter's inheritance of the two lands.[375]

The divine name *pꜣ-Rꜥ* appears here for the first time in our cosmology. Previously, the solar deity, if my interpretation in the commentary on fragment 4, II, 22 is correct, has been called *pꜣ-Šy pꜣ Nwn* or simply (P)shai. The former name is used to designate

[369] Klasens, *op. cit.*, pp. 22-34. For the sense of *ḥḥ* here (literally 'tread' or '[pretend to] seek'), see Egberts, *In Quest of Meaning* 1, pp. 345-58, especially p. 355.

[370] See Chassinat, *Edfou* 6, pp. 219, line 5 –220, line 3; Alliot, *op. cit.*, p. 817; Kurth, art. cit., pp. 378-9. Compare also P. Jumilhac, VI, 12-16 (Vandier, *Le Papyrus Jumilhac*, p. 117 and fourth unnumbered plate, top).

[371] Bucher, *Les Textes des tombes de Thoutmosis III et d'Amenophis II* 1, plate 24, right. For other depictions of Isis as a sycomore, see Baum, *Arbres et arbustes de l'Égypte ancienne*, p. 85, note 429.

[372] See Moftah, *Die heiligen Bäume im alten Ägypten*, pp. 121-3.

[373] Klasens, *op. cit.*, p. 34.

[374] *Op. cit.*, p. 46. Cf. the comments of Klasens, ibid., p. 69; Cauville, *BIFAO* 89 (1989), p. 54.

[375] Junker and Winter, *Das Geburtshaus des Tempels der Isis in Philä*, pp. 16-21.

that deity prior to his emergence from the Primaeval Ocean; he acquires the latter at his initial manifestation in the created cosmos.

The present fragment apparently introduces yet a third name for the sun god. This too should mark some new stage in his existence. It was suggested in the commentary on fragment 8, q.v., that one may have had in that portion of the text an account of how the sun god departed from earth and ascended to the sky, borne by his mother the celestial cow goddess. If so, then perhaps the additional name *pꜣ-Rꜥ* was given to the solar deity here to mark his ascension into the heavens. Such a view is supported by the fact that the name Pshai only occurs once in our text after fragment 8, and there clearly in a retrospective passage.[376] It is plain, moreover, from the references which I have cited two paragraphs above that the flight of Isis to Khemmis with Horus was believed to have occurred after the solar deity's ascent to the sky. Thoth says in one, for example, that Re in heaven is protecting Horus, and announces in another that he has come down from heaven, where he received a command from Re.

For *Rꜥ* as the specific name of the sun god after he has distanced himself from the earth and ascended to the sky, cf. P. Leiden I 350, III, 23-5: *imn.ti m Ỉmn m-ḥꜣ.t nṯrw* .... *ḥr=k tw m imy p.t mn.ti m Rꜥ*, '(You are) hidden as Amun at the head of the gods ...., you made yourself distant as the one who is in the sky, enduring as Re.'[377] Various Egyptian etymologies of the name of the sun god are known. According to one of these, *Rꜥ* means 'the one who ascends', which could account for its use in the present fragment with the function that I have suggested for it.[378]

Our cosmology is not unique among Egyptian works of that genre in combining material pertaining to the myth of Osiris, Isis and Horus with an account of the creation of the elements of the universe. The most notable parallel for this is the so-called Memphite Theology, which combines a detailed narrative describing Horus's contest with Seth for the royal inheritance of his father Osiris with a record of how Ptah brought the cosmos into existence.[379] There are good reasons why this should have been done. The myth of Osiris, Isis and Horus is concerned above all else with the legitimate transmission of royal authority. From an Egyptian point of view, the institution of kingship was as fundamental an element in the ordering of the cosmos as the sky, earth or underworld. Hence the incorporation of the most important myth concerned with that institution within a cosmogonical or cosmological account is understandable.

[376] See commentary on fragment 10, line x+9.

[377] Zandee, *De Hymnen aan Amon van Papyrus Leiden I 350*, plate 3.

[378] For this etymology, see Hornung, *Conceptions of God in Ancient Egypt*, p. 67, quoting idem, *Das Amduat* 1, p. 20, lines 1-2: *ꜥr=k m rn=k n Rꜥ*, 'You will ascend in your name of Re.' The divine appellation is derived from the verb by metathesis (similarly in Moret, *Le Rituel du culte divin journalier en Égypte*, p. 125).

[379] Sethe, *Dramatische Texte zu altägyptischen Mysterienspielen*, pp. 1-80.

There is a further reason that could relate to our text in particular. I have suggested that fragment 5 and the ensuing fragments may have recounted the story of a rebellion against the sun god, in which he was protected from his foes by the divine cow Mehetweret/Ahetweret.[380] In Egyptian sources, the flight of Isis to Khemmis with her son Horus, made in order to protect him from his enemies, is seen as a parallel to the story of the flight of the solar deity and his mother from those who revolted against him. The one myth is essentially a re-enactment of the other.[381] This is especially evident in another account of a rebellion against the solar deity preserved in P. Carlsberg 462.[382] There, when Pre is under threat, he is said to 'assume the form of a child again' and mount up between the cow goddess's horns.[383] The text states that Pre is called Pre-Harpocrates as a result,[384] and his mother is identified with Isis.[385]

The converse can be found in e.g. the texts relating to the birth of Horus in the mammisi at Philae. There, Harpocrates is given the epithet *ms.n ỉḥ.t-*[*.w*]*r.* [*t*], 'he whom Ahet[we]re[t] bore',[386] stressing his identification with the sun, and he is said to have been acclaimed in Khemmis as *Rʿ-Ḥr-ꜣḫ.ty m wḥm*, 'the second Re-Harakhti'.[387] Likewise, the assimilation of the newly born Horus to the sun god is a prominent theme in the second of the two Edfu narratives cited earlier.[388]

There are additional similarities between the two myths to which attention can be drawn. Some of the same divinities, Thoth for instance, appear in protective roles in both. Moreover, the seven *ḏꜣis.w*, 'creative utterances', of Mehetweret, which help to defend her and the sun god from their enemies,[389] are paralleled by seven scorpions who are said to escort Isis on her flight to Khemmis.[390] Thus, the story of the flight recapitulates the myth of the rebellion against the sun. The link between it and the latter serves further to emphasise the cosmic significance of the institution of kingship,

[380] See commentaries ad loc.

[381] The parallelism is made explicit in e.g. P. Insinger, XX, 17-18 (Lexa, *Papyrus Insinger* 1, p. 65).

[382] See my edition of the text published in Frandsen and Ryholt (eds.), *The Carlsberg Papyri 3: A Miscellany of Demotic Texts and Studies*, pp. 95-112. The equation of Horus and Isis with the sun god and the celestial cow is discussed ibid., pp. 105-7, with references to earlier literature.

[383] P. Carlsberg 462, II, 4-6. Cf. commentary ad loc.

[384] Ibid., II, 15.

[385] Ibid., II, 10 and passim.

[386] Junker and Winter, *Das Geburtshaus des Tempels der Isis in Philä*, pp. 384-5. Cf. Daumas, *Les Mammisis des temples égyptiens*, pp. 187-8 with note 3.

[387] Junker and Winter, *op. cit.*, p. 22-3.

[388] See references cited in note 370 above; Labrique, *Stylistique et théologie à Edfou*, p. 56.

[389] For these, see commentaries on fragments 6 and 8.

[390] See Klasens, *A Magical Statue Base*, pp. 10-11 and 70.

as embodied in the youthful Horus to whom the royal authority is transmitted.[391] It would not be surprising, therefore, if, as I have suggested may have been the case, the two myths concerning these events were recounted together in our cosmology, one following immediately after the other.

FRAGMENT 13 Plate 9

Dimensions: 17.1 × 22.5 cm. Colour: dark/medium brown.

Transliteration

Column I

Illegible traces of three lines

Column II

1. *iw=f ti wḏꜣ nꜣ ntr.w nt [ẖ]n=s iw=f ti rnp=w šꜥ ḏt r b[....]*
2. *ḫpr pꜣ mwt iw=f ḥl r [m]ꜣ nb nt ẖ[r] tꜣ p.t iw=f šnb irm pꜣ [ꜥnḫ i-ir(?)]*
3. *pr n pꜣ Nwn ḏ pꜣ [m]wt pꜣ b[w] pꜣ syt pꜣy pꜣ ṯꜣw pꜣ ꜥnḫ [pꜣ bw]*
4. *pꜣ Nwn pꜣy šnb=w n pꜣ 2 n wꜥ sp mtw=w nꜣ nt ti mtw=w nꜣ nt [ṯ nꜣ nt nb(?)]*
5. *nt ḥr pꜣ tꜣ tr=f r-hn-r pꜣ hrw mtw pꜣ ꜥnḫ irm pꜣ mwt nꜣ.w pꜣ ṯꜣw [pꜣ ꜥnḫ]*
6. *pꜣ nt ti ꜥnḫ nꜣ nt nb [nt] ꜥnḫ pꜣ tꜣ pꜣ ṯꜣw n pꜣ mwt* (vac) *pꜣ nt ṯ.ṱ=w pꜣ ym [ꜥꜣ n]*
7. *H̱l irm tꜣy=f ... pꜣ ṯꜣw n pꜣ Nwn [pꜣ] nt nhsy.ṱ=f [mtw=f]*
8. *pꜣ nt ms.ṱ=f pꜣ ṯꜣw n pꜣ mwt mtw=f pꜣ nt ꜥšm=f mtw nꜣ sbt.[wt] n p[ꜣ ...]*
9. *n nꜣ tww.w mtw b-ir=w šb pꜣ ṯꜣw n pꜣ Nwn pꜣ nt ti rnp[=w ...]*
10. *mny pꜣ ṯꜣw n p[ꜣ N]wn pꜣ nt wsm=w pꜣ [s]mt n [nꜣ] nt[r.w nꜣ rmt.w]*
11. *nꜣ sšm.w irm nꜣ ꜥꜥ.. nꜣ lym.w nꜣ ipt.w nꜣ ḏ[tf.w]*
12. *nꜣ sym.w pꜣ ṯꜣw n pꜣ ꜥ[n]ḫ pꜣ nt [t]i ꜥnḫ=w pꜣ ṯꜣw n pꜣ mwt pꜣ [nt ti mwt=w]*
13. *wꜥ m-qty wꜥ n-im=w pꜣ ... [mtw=f i]-ir mḥṱ n pꜣ tꜣ Ḏḥ[wty ...]*
14. *pꜣy mtw=f pꜣ nt ṯ mtw=f pꜣ nt t[i] b-ir=f sṱ=f r ti ḫpr pꜣ [....]*
15. *pꜣ mwt rḫ ṯ.ṱ=f r-ḏbꜣ ḫpr i-ir [p]ꜣ mwt irm pꜣ ꜥnḫ ḥq .[....]*
16. *mtw=f pꜣ nt ꜥq ẖn tꜣ wḏy [iw]=f ḫꜥ n itn ẖn [tꜣ p.t iw=f n]*
17. *kꜣ sty r ḏt mtw=f pꜣ nt t[i] pꜣ mw r-ḥry n pꜣ Nwn [mtw=f pꜣ nt]*
18. *sḥn pꜣ ṯꜣw n nꜣ nt plḫ [ḥ]r pꜣ [tꜣ(?)] mtw=f pꜣ nt ti pꜣy [nꜣ kꜣ.w]*
19. *šp nꜣ ꜣty.w mtw=f i-ir ti [pr(?) p]r.t nb n tꜣ sḫ.t mtw=f [pꜣ nt ..].[..]*
20. *nꜣ šn.w nꜣ sym.w ẖr ḥlly [nb(?) mtw]=f pꜣ nt ti ḫpr sḫny nb [...]..[..]*
21. *tꜣ twꜣ.t mtw=f mtw pꜣy=f(?) [....]... n tꜣ p.t [... ḏ]t m[..]*

[391] This cosmic aspect is stressed repeatedly by Daumas in his book on the Egyptian mammisis, cited in note 386 above. See ibid., pp. 124, 148, 190, 196-7, 202-3, 248-9, 290-1, 308-9, 320, 325, 346, 375, 445, and 462. See further Sternberg, *Mythische Motive und Mythenbildung in den ägyptischen Tempeln und Papyri der griechisch-römischen Zeit*, pp. 222-4; Allen, *Genesis in Egypt*, pp. 11-12.

Translation

Column I

Illegible traces of three lines

Column II

1. He makes the gods who are [i]n it well, causing them to be renewed eternally, while *b*[....]
2. Death came into existence, flying to every [pl]ace which is und[er] the sky, uniting with the [life which(?)]
3. came forth from the Primaeval Ocean, because [d]eath is the exhala[tion(?)] of the serpent. The breath of life is [the exhalation(?)]
4. of the Primaeval Ocean. The two united completely. They are the ones that give, they are the ones that [take all things(?)]
5. which are on the entire earth up until today, that is to say, life and death. The breath [of life]
6. is what gives life to everything [which] lives on earth. The breath of death (vac) is what takes them. The [great] sea [of]
7. Syria and its ..., the breath of the Primaeval Ocean is what awakens it, and [it]
8. is what gives birth to it. As for the breath of death, it is what quenches it. And the hill[s] of th[e ...]
9. of the immutable mountains, the breath of the Primaeval Ocean is what causes [them(?)] to rejuvenate [...]
10. daily. The breath of t[he Primae]val Ocean is what ruins them. [Li]kewise, for [the] go[ds, people],
11. the *sšm*-beasts and the herds(?), the fish, the birds, the ser[pents],
12. and the plants, the breath of l[i]fe is what [ca]uses them to live, and the breath of death is [what causes them to die],
13. one being just like another of them. The ..., [it is he w]ho has taken possession of the earth. Th[oth] is [...].
14. He is the one who takes. He is the one who giv[es]. He does not allow the [....] to come into existence again.
15. Death can take him, because death and life rule .[....].
16. He is the one who enters the sound eye, appearing as a disk in [the sky, as]
17. an engendering bull forever. He is the one who bri[ngs] the flood up from the Primaeval Ocean. [He is the one who]
18. ordains breath for those who are spread out [ov]er the [earth(?)]. He is the one who causes [the bulls] to ejaculate
19. and the cows to receive (it). He is the one who has caused every [se]ed of the field [to emerge(?)]. He is [the one who ..].[..]

20. the trees and the plants bearing [every(?)] flower. [H]e is the one who causes every event to occur [...]..[..]

21. the underworld. He is the one whose(?) [....]... in/of the sky [... eter]nity *m*[..]

Textual notes

Column I

Illegible traces of three lines

Column II

Line 1

(a) With the traces of *ntr.w*, 'gods', here, cf. the better preserved occurrence of that noun in e.g. fragment 5, I, 1.

(b) With the traces of *ẖn*, 'in', after *nt*, cf. the undamaged occurrences of that preposition in line 16 below.

(c) *ti* before *rnp=w* has been written above the line. The verb *rnp*, except for the determinative, is written in hieratic.

Line 2

(a) With the traces of *mꜣ*, 'place', here, cf. the better preserved occurrences of that noun in e.g. fragment 23, II, 3, P. Carlsberg 462, II, 1 (Smith in Frandsen and Ryholt [eds.], *The Carlsberg Papyri 3: A Miscellany of Demotic Texts and Studies*, plate 15), and P. Carlsberg 10040, line x+4.

(b) With the traces of *ẖr*, 'under', after *nb nt*, cf. the better preserved occurrence of the same preposition in line 20 below.

(c) For *šnb*, 'unite' (< earlier *šbn*), see note on fragment 12, I, 16.

(d) Perhaps restore *ʿnḫ i-ir* at the end of the line, which would fit the available space. Cf. the statement in line 4 below that death and life united (*šnb*).

Line 3

(a) A further instance of the partially preserved noun after [*m*]*wt* occurs in P. Carlsberg 123, line 5. The writing there is , i.e. *bw* + sail determinative, which is how the example in our text should be restored. From the context, *bw* appears to mean something like 'breath' or 'exhalation', but I have not found the word occurring elsewhere.

(b) At the end of the line, probably restore [*pꜣ bw*]. This would fill the available space and provide a good parallel with the preceding sentence.

Line 4

(a) For *šnb*, 'unite', see note on fragment 12, I, 16.

(b) At the end of the line, perhaps restore [*ṯ nꜣ nt nb*] or similar. This would fill the available space. For *ṯ*, 'taking', used in opposition to *ti*, 'giving', see line 14 below. For the former as the characteristic activity of death, see lines 6 and 15.

Line 5

(a) A trace of the walking legs determinative of *hn* is visible after that word's second consonant. Compare the writings of this element of the phrase *r-hn-r pꜣ hrw*, 'up until today', in fragment 4, II, 15, and fragment 5, II, 16.

(b) *mtw* after *pꜣ hrw* is written for *nt-iw*.

(c) At the end of the line, restore either *pꜣ ṯꜣw* [*pꜣ ʿnẖ*] as in line 3 above, or *pꜣ ṯꜣw* [*n pꜣ ʿnẖ*] as in line 12 below. Some restoration along these lines is required, in view of the obvious parallelism between this sentence and the next. If only *pꜣ ʿnẖ* has been lost, then the extent of the missing text at the end of the line is roughly 1.1 cm. If an *n* stood before *pꜣ ʿnẖ*, it will have been slightly longer.

Line 6

(a) The traces before *pꜣ ṯꜣw n pꜣ mwt* suit *ʿnẖ pꜣ tꜣ* exactly. It is difficult to see what else could be restored in front of them apart from the relative converter *nt*.

(b) It appears that nothing was ever written in the space between *pꜣ mwt* and *pꜣ nt ṯ.ṯ=w*. Presumably, the surface of the papyrus here had already been damaged in antiquity, or was thought to be unsuitable for writing upon for some other reason.

(c) The pronominal infinitive of *ṯ*, 'take', is used in a present tense relative clause, in apparent violation of the Stern-Jernstedt Rule. Similar cases occur in lines 7, 8, and 10 below. Quack, *RdE* 42 (1991), pp. 189-207, explains that such exceptions typically are found in 'non-narrative discourse', which certainly holds true for our text. Likewise, Simpson, *Demotic Grammar in the Ptolemaic Sacerdotal Decrees*, pp. 155-6, in discussing further cases of this sort, notes their regular occurrence in 'various types of general statement (habitual, gnomic, or static)'. In our text, the use of the pronominal infinitive in a present tense relative clause is restricted to such statements.

(d) At the end of the line, restore the water determinative of *ym*, 'sea', followed by [*ʿꜣ n*]. The 'great sea of Syria' is the Mediterranean. Cf. e.g. Wenamun, I, 8 (Gardiner, *Late-Egyptian Stories*, p. 61, line 10); London-Leiden Magical Papyrus, VI, 30 (Griffith and Thompson, *The Demotic Magical Papyrus of London and Leiden* 2, plate 6).

Line 7

(a) The damaged noun after *tꜣy=f* should denote a part or attribute of the Mediterranean Sea. I am unable to suggest a reading.

(b) For the use of the pronominal infinitive of *nhsy*, 'awaken', in a present tense relative clause here, see note (c) on the preceding line.

(c) The restoration of the third person singular masculine independent pronoun *mtw=f* at the end of the line yields good sense and fits the available space well.

Line 8

(a) For the use of the pronominal infinitives of the verbs *ms*, 'give birth', and *ʿšm*, 'quench', in present tense relative clauses here, see note (c) on line 6 above.

(b) It is not clear to me how the end of the line should be restored. Perhaps one had here a phrase linking the noun *sbt.*[*wt*], 'hill[s]', with the *tww.w*, 'mountains', to which reference is made in the line immediately following. For hills and mountains mentioned in conjunction elsewhere in demotic, see e.g. P. Leiden I 384, XII, 25-6 (de Cenival, *Le Mythe de l'oeil du soleil*, plate 11). The Coptic counterparts of *sbt.t* and *tw* occur frequently in parallelism with each other; cf. Crum, *CD*, p. 322b.

Line 9

(a) *mtw* before *b-ir* is a writing of *nt-iw*. For the traces of *b-ir* here, cf. the well-preserved writing of the negative aorist marker in fragment 5, II, 14. This marker occurs in line 14 of the present column as well, but there it has a vertical stroke between the first and second elements.

(b) The verb after *b-ir* is *šb*, 'change' (*Glossar*, p. 497). A writing identical with this one, 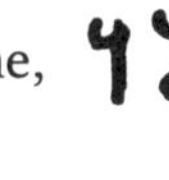, occurs in line 6 of P. Carlsberg 421, fr. 3.

(c) Perhaps restore the suffix pronoun *w* after *rnp*, the antecedent being the hills and mountains to which allusion has just been made or what is on them. I am unable to suggest what might have followed this.

Line 10

(a) For the construction *pꜣ nt wsm=w*, see note (c) on line 6 above. The antecedent of the suffix pronoun *w* is presumably identical with the subject, now lost, of the verb *rnp* at the end of the preceding line, q.v.

(b) With the traces of *ntr.w*, 'gods', here, cf. the better preserved occurrence of that noun in fragment 5, I, 1. By comparison with other lists of a similar type (see e.g. Assmann, *Ägyptische Hymnen und Gebete*, p. 567; Sauneron, *Esna* 3, pp. 128 [249, 2], 177 [275, 10], 216 [300,9], 313 [358, 31], and 348 [377, 5]), one expects *rmt.w*, 'people', to have followed *ntr.w*. It is conceivable that *nt*[*r.w irm nꜣ rmt.w*] should be restored in place of the *nt*[*r.w nꜣ rmt.w*] that I have suggested.

Line 11

(a) The term *sšm* denotes a class of supernatural beasts. Cf. e.g. P. Leiden I 384, XV, 1, where the avenging gryphon who punishes murderers is said to be the *sšm*, 'creature', of death (de Cenival, *Le Mythe de l'oeil du soleil*, plate 14). The writing there is identical with that in our text. Cf. Tait, *AcOr* 37 (1976), pp. 36, note 35, and 42, note m.

(b) The second noun in this line begins with ꜥꜥ and an animal skin determinative. I am unable to read the damaged signs which follow. On the basis of comparable lists (see references cited in note (b) on the preceding line), one expects a term denoting mammalian quadrupeds here, as distinct from the fish and birds which are mentioned subsequently. Could this be an attempt at a demotic writing of the *ꜥw.t*, 'Kleinvieh, Wild', of *Wb.* 1, 170? Compare similar writings of the noun ꜥꜥ, 'ein (heiliges) Tier', cited in *Glossar*, p. 56, which, according to the *Wb.*, is merely a specialised usage of *ꜥw.t*.

(c) *ḏ*[*tf.w*], 'ser[pents]' (or worms), seems a plausible restoration at the end of the line. One requires a noun beginning with *ḏ* that denotes the lowest class of animals. Depending upon how much space was taken up by the writing of that word, perhaps *irm*, 'and', should be supplied immediately after it as well.

Line 12

(a) For the traces of *ti*, 'cause', after *pꜣ nt*, see better preserved examples of the same verb in e.g. lines 1, 4, and 6 above.

(b) The restoration [*nt ti mwt=w*] at the end of the line is virtually certain, in view of the parallelism of this sentence with the preceding one.

Line 13

(a) I am unable to read the masculine noun . From the context, this is clearly an epithet describing the deity who is the subject of the ensuing lines as well. See commentary below.

(b) For the restoration [*mtw=f i*]-*ir*, see the better preserved occurrence of this combination in line 19 below. All that survives of *mtw=f* here are traces of the *m*. Of the following *i-ir*, only the initial stroke of the *i* has been lost. The definite article before the second element of a cleft sentence can be omitted if the initial element is a pronoun. See Johnson in Young (ed.), *Studies Presented to Hans Jakob Polotsky*, p. 420.

Line 14

(a) For the traces of *ti*, 'give', after the second occurrence of *pꜣ nt*, see better preserved examples of the same verb cited in note (a) on line 12 above, as well as the one near the end of the present line.

(b) The suffix object *f* after *sṯ* is reflexive, referring back to the subject. For the construction, cf. the usage of Coptic **ⲥⲱⲧ** in negative sentences as illustrated in Crum, *CD*, p. 360a.

(c) I am unable to suggest a restoration for the end of this line. The sense, however, is clearly that whatever perishes is not allowed to come into existence again.

Line 15

(a) For other demotic occurrences of the verb *ḥq*, 'rule' (not in *Glossar*), see Smith, *The Mortuary Texts of Papyrus BM 10507*, p. 57 [note (a) on I, 5]. Some of the examples cited there have the same serpent determinative as the present one.

(b) At the end of the line, one needs to restore the emphasised adverbial adjunct of the second tense marked by *i-ir*. Perhaps this was something like 'over everything' or 'everywhere'.

Line 16

The restoration of *tꜣ p.t* at the end of the line seems virtually certain, in view of what precedes. After *p.t* there is sufficient space to restore *iw=f n* as well.

Line 17

(a) For the traces of *ti* before *pꜣ mw*, see better preserved examples of the same verb cited in note (a) on line 12 above.

(b) Nothing appears to have been written in the blank space between *r* and *ḥry*. The restoration [*mtw=f pꜣ nt*] after *Nwn* is based upon the occurrence of those words at the beginning of the preceding and following sentences.

Line 18

(a) Perhaps restore [*tꜣ*], '[earth]', in the break after *plḫ* [*ḥ*]*r pꜣ*. What is needed is a relatively short masculine noun.

(b) For the restoration [*nꜣ kꜣ.w*], '[the bulls]', at the end of the line, see the following note.

Line 19

(a) *šp*, 'receive', is employed here with a sexual connotation. Cf. *Glossar*, p. 501; Spiegelberg, *Der ägyptische Mythus vom Sonnenauge*, p. 262 (no. 762). For the use of *ꜣty*, 'vulva', with the extended sense of 'cow', see e.g. P. BM 10508, X, 20: *hmy i-ir ꜣty.t šp pꜣy=s kꜣ*, 'Would that a cow receive its bull' (Glanville, *The Instructions of ʿOnchsheshonqy*, plate 10); P. Leiden I 384, VIII, 22: *i-ir=w ḏ r-r=k pꜣ nb n nꜣ kꜣ.w r-ir ti šp nꜣ ꜣty.w*, 'You (scil. Thoth) are called the lord of the bulls, who caused the cows to receive' (de Cenival, *Le Mythe de l'oeil du soleil*, plate 7). Just as cows are said to receive (i.e. be impregnated), so bulls are said to ejaculate (*pꜣy*), hence the restoration of *nꜣ kꜣ.w* at the end of the preceding line. Compare Chassinat, *Edfou* 4, p. 140, lines 11-12, where the members of the Ogdoad are described as *kꜣ.w pꜣy ḥm.wt šsp*, 'the bulls who ejaculate and the cows who receive'.

(b) For the construction *mtw=f i-ir ti*, see note (b) on line 13 above.

(c) For a writing of *pr.t*, 'seed', comparable with the one here, see Revillout, *Corpus Papyrorum Aegypti* 1, fascicle 4, plate 30, line 4 (left). There is just sufficient room to restore a verb like *pr*, 'emerge, come forth', in front of it. Alternatively, one could supply something like *ḫpr*, 'come into existence'.

(d) [*p3 nt*] is almost certainly to be restored after *mtw=f* at the end of the line. This was probably followed by an expression meaning 'cause to grow' or similar.

Line 20

(a) The restoration of *nb*, 'every', after *ḥlly* is only conjectural, but this would fit the available space well, making due allowance for the *mtw* which must have stood at the end of the break.

(b) The noun *sḫny* could also be translated 'constellation'. See commentary below.

Line 21

(a) *mtw* after *mtw=f* is written for *nt-iw*. For the construction involved, see note (b) on line 13 above.

(b) The traces before *n t3 p.t* are of a word that ends with the divine determinative. I am unable to suggest a reading.

(c) With the trace of *ḏt*, 'eternity', near the end of the line, cf. the undamaged occurrences of that word in lines 1 and 17 above. Perhaps restore [*r nḥḥ*] before it.

Commentary

Only illegible traces of the first column of this fragment have been preserved. The initial line of Column II begins with the statement *iw=f ti wḏ3 n3 ntr.w nt* [*ẖ*]*n=s iw=f ti rnp=w šꜥ ḏt*, 'He makes the gods who are [i]n it well, causing them to be renewed eternally.' Reference is made here to the actions of some thing or deity whose name must have been given in the closing lines of the preceding column. The antecedent of the suffix pronoun *s* after *ẖn* must have been mentioned there as well. For the possibility that this was *p.t*, 'sky', see below.

Line 2 alludes to the creation of death: *ḫpr p3 mwt iw=f ḥl r* [*m*]*3 nb nt ẖ*[*r*] *t3 p.t*, 'Death came into existence, flying to every [pl]ace which is und[er] the sky.' This implies that there was a period in the development of the cosmos when there was no death. Similarly, the much earlier Pyramid Text Spell 571 speaks of a time *n ḫprt mwt*, 'before death came to be'.[392]

Egyptian sources tend to be reticent about when and in what circumstances death was created. Beinlich has drawn attention to an obscure passage in the Memphite Theology, which he would interpret as a reference to the coming into being of life and death.[393] Even if he is correct, however, the passage tells us little about how this occurred, other than that it came about as a result of the tongue's expression of a plan conceived in the heart of the creator god Ptah. Kákosy has suggested that death was probably created as a consequence of mankind's rebellion against the sun god, described in the Book of the Heavenly Cow.[394] Certainly, some sources trace its origin

[392] Sethe, *Die altägyptischen Pyramidentexte* 2, p. 303 (1466d). Cf. Grapow, *ZÄS* 67 (1931), p. 35; Kákosy, *Studia Aegyptiaca* 7 (1981), p. 83; Bickel, *La Cosmogonie égyptienne avant le Nouvel Empire*, pp. 179-80.

[393] See *GM* 122 (1991), pp. 18-20.

[394] Kákosy, art. cit., pp. 83-86.

to that deity. For instance, in a hymn to Re preserved in a papyrus of 21st Dynasty date, now in the Louvre, it is said to him: *š3ꜥ=k ꜥnḫ snw=f mni*, 'You originated life and its fellow, death.'[395] In the demotic version of the myth of the solar eye found in P. Leiden I 384, death (*p3 mwt*) is closely associated with Pre, and both share a common epithet, *p3 ḏb3*, 'the avenger'.[396]

Our text describes death as flying (*ḥl*). A representation in another papyrus of 21st Dynasty date actually depicts a winged serpent with four legs, a human head on its neck, and a jackal's head at the end of its tail. Its wings enfold a sun disk containing a scarab. The accompanying inscription identifies this creature as *mwt p3 nṯr ꜥ3 ir nṯr.w rmṯ.w*, 'Death the great divinity who made gods and men'.[397] In this instance, death may actually figure as a form or aspect of the solar deity, rather than one of his creations,[398] as is perhaps also true in the above-cited passage from P. Leiden I 384.[399]

If my proposal to restore *ꜥnḫ i-ir* at the end of line 2 is correct,[400] then the text speaks of death, *iw=f šnb irm p3* [*ꜥnḫ i-ir*] *pr n p3 Nwn*, 'uniting with the [life which] came forth from the Primaeval Ocean'. Compare line 4 below, where it is said of death and life: *šnb=w n p3 2 n wꜥ sp*, 'The two united completely.' Elsewhere, this union is said to have been effected by the sun god Re. Thus, a hymn to that deity in P. Louvre 3292 (temp. 21st Dynasty) describes him as *twt ꜥnḫ mni*, 'uniting life and death'.[401] The two are joined together in the unending daily cycle established by Re after his ascent to the sky. As Assmann has put it, 'Durch seine eigene Bewegung über den Himmel (= Leben) und durch die Unterwelt (= Sterben) hat der Sonnengott Leben und Tod, d.h. die Vergänglichkeit des irdischen Daseins geschaffen.'[402]

That life should be said to have come forth from the Primaeval Ocean is not surprising, since the Egyptians conceived of all existence as having originated there. In recognition of this, a number of sources from the Pyramid Texts onwards employ the term *š n ꜥnḫ*, 'sea of life', as a designation for the primordial flood.[403] In a purification

[395] P. Louvre 3292, M, 4-5 (Nagel, *BIFAO* 29 [1929], p. 68 and plate 5). For the verb *š3ꜥ* here, cf. Quaegebeur, *Le Dieu égyptien Shaï*, pp. 74-5.

[396] See P. Leiden I 384, XV, 5 and 8-9 (de Cenival, *Le Mythe de l'oeil du soleil*, plate 14).

[397] The papyrus in question is BM 10018. See Schott, *Zum Weltbild der Jenseitsführer des Neuen Reiches*, NAWG 1965, nr. 11, plate 4 and p. 195 (reference courtesy of Paul Frandsen).

[398] Cf. Westendorf, *Altägyptische Darstellungen des Sonnenlaufes auf der abschüssigen Himmelsbahn*, p. 49.

[399] For a differing view, see Tait, *Acta Orientalia* 37 (1976), p. 42, note e.

[400] See textual note (d) ad loc.

[401] P. Louvre 3292, T, 5-6 (Nagel, *BIFAO* 29 [1929], p. 90 and plate 7).

[402] *Ägyptische Hymnen und Gebete*, p. 529.

[403] See Altenmüller-Kesting, *Reinigungsriten im ägyptischen Kult*, pp. 123-5; Gessler-Löhr, *Die heiligen Seen ägyptischer Tempel*, p. 349.

ritual of the Ptolemaic Period, the king is identified with Horus in the Primaeval Ocean, and it is said that death will have no power over him there.[404]

An obscure term, *bw*, is introduced in line 3. Death is called the *bw* of the serpent (*syt*), and the breath of life, if my restoration at the end of the line is correct, the *bw* of the Primaeval Ocean. I have conjectured that this noun means 'breath' or 'exhalation'.[405] Even if this is not its precise significance, *bw* clearly designates something breath-like that emanates from the being or place to which it is said to belong. In view of the association of life with the Primaeval Ocean, to which reference was made in the preceding paragraph, one can see how the breath of life could be described as the exhalation of the Ocean. More puzzling is the identification of death as the exhalation of the serpent.

There is some evidence that the Egyptians believed the exhalation of this reptile could be harmful. In the mythological account on the naos of el-Arish, for example, when Geb attempts to seize a box containing the royal uraeus so that he can place it on his brow, the text says: *pr pw ir.n s3-t3 ʿnḫ nf.w=f r ḥm n Gb m nšny=f wr sp-2*, 'The serpent emerged and its breath "lived" (i.e. was potent) against the majesty of Geb with its greatest destructiveness.' In consequence, those who were with Geb were killed and the god himself was badly burned.[406] A similar belief may underlie a maxim in the demotic P. Insinger which states: *ḥf iw=f nyf iw t3y=f mtw3.t (n) r3=f rmt ḫm t3y=f mtw3.t iw=s (n) ḥ3ṯ=f*, 'A serpent exhales, its poison (in) its mouth. As for an inferior person, his poison is (in) his heart.'[407] But to whom or what does the term *syt* in our text refer? Clearly, no ordinary serpent is meant. Does it denote a divinity or an inimical being like Apopis? In my view, it is more likely to be the former, since in Egyptian thought, as we have seen, both life and death were considered to be of divine origin. The question then arises, which god is meant? Here, one faces a more difficult problem, in view of the number of different Egyptian divinities, especially those who function as creators, that can take serpent form.[408]

[404] P. Berlin 13242, III, 2 (Schott, *Die Reinigung Pharaos in einem memphitischen Tempel*, p. 55 and plate 11).

[405] See textual note (a) on II, 3 of this fragment above.

[406] See Goyon, *Kêmi* 6 (1936), pp. 16 and 34. Compare Sambin, *Studia Aegytiaca* 11 (1988), pp. 374-5; Cauville, *BIFAO* 82 (1982), pp. 117-19. Further examples illustrating the harmful power of a serpent's *nf.w* are cited in *Wb.* 2, 250, 17; also Goyon, *Les Dieux-gardiens et la genèse des temples* 1, p. 75.

[407] P. Insinger, XXIX, 13 (Lexa, *Papyrus Insinger* 1, p. 94). For other aspects of serpents and breath in Egypt, cf. Sethe, *Amun und die acht Urgötter von Hermopolis*, pp. 27 and 92; Sauneron, *Un Traité égyptien d'ophiologie*, pp. 23-4, 29-30, 144-5, and 155-6.

[408] For lists of these, see Störk in *Lexikon der Ägyptologie* 5 (1984), columns 648-9 (s.v. Schlange); Sauneron and Yoyotte in *La Naissance du monde*, pp. 37-8; Goyon, *op. cit.*, pp. 116-17.

One possibility which merits consideration is that *syt* in our passage refers to the sun god, who can be represented as a serpent and, as such, may even be identified with death.[409] As already noted, some sources attribute the creation of death to Re.[410] Demotic *syt* derives from the earlier Egyptian word for serpent, *s3-t3*, 'son of the earth'.[411] The latter, along with variants like *s3-t3 nṯri*, 'divine serpent', and *s3-t3 wr*, 'great serpent', is well-attested as an epithet of youthful solar deities like Neferhotep and Harsomtus, and serves to emphasise their aspect as primordial creators.[412] In serpent form, these deities can be identified with Re. At Dendera, for example, Harsomtus as *s3-t3* is said to be *Rᶜ ḏs=f*, 'Re himself'.[413] Another link shared between them and him is a common association with the power of destiny which, among other things, determines when death occurs.[414] Both Re and Harsomtus, for instance, can be equated with (P)shai,[415] for whom the serpent form is characteristic.[416] Occasionally, all of these aspects are combined with one another, e.g. in a scene depicting the presentation of the lotus from the temple of Dendera, where the divine recipient of the offering, identified alternately as Re-Somtus and Harsomtus, is called both Shai and divine *s3-t3* serpent, and is said to travel in the solar barks, separating the night from the day.[417]

Lines 4-5, if my restoration of the end of the former is correct,[418] state that life and death together are responsible for giving (i.e. producing) and taking away everything on earth, a theme that is developed at greater length in the lines that follow. There

[409] See commentary on line 2 above. For the sun god in serpent form, see e.g. Schott, *Zum Weltbild der Jenseitsführer des Neuen Reiches*, pp. 193-5 and plates 2-3.

[410] See commentary on line 2.

[411] *Wb.* 3, 410, 16-17.

[412] See e.g. Ryhiner, *L'Offrande du lotus*, pp. 92 and 98; Chassinat, *Edfou* 3, p. 273, line 12; idem, *Edfou* 5, p. 194, line 1 (Neferhotep); idem, *Dendara* 1, p. 145, line 5; *Dendara* 2, p. 164, lines 4 and 9; Ryhiner, *op. cit.*, pp. 89, 95, 110, 112 (Harsomtus). On this aspect, see Kurth, *Die Dekoration der Säulen im Pronaos des Tempels von Edfu*, pp. 82-3; Ryhiner, *op. cit.*, pp. 183-4 and 189-90; el-Kordy, *BIFAO* 82 (1982), pp. 171-86.

[413] Chassinat, *Dendara* 2, p. 164, line 8. For further examples of Harsomtus equated with Re, see el-Kordy, art. cit., p. 182; Waitkus, *Die Texte in den unteren Krypten des Hathortempels von Dendera*, p. 81, note 36; Chassinat, *Dendara* 1, p. 17, line 15, p. 26, line 11, p. 35, line 12, p. 62, line 4, p. 102, line 16, and p. 117, line 14; idem, *Dendara* 5, p. 34, line 1; idem, *Edfou* 6, p. 281, lines 8 and 10.

[414] See Quaegebeur, *Le Dieu égyptien Shaï*, pp. 78-9 and 126-9; Zandee, *Death as an Enemy*, p. 51.

[415] See commentary on fragment 4, II, 22. For the possibility that *p3-Rᶜ* and *p3-Šy* are used as alternate names for the sun god in our text, see commentary on fragment 12, II, x+10.

[416] See references cited in note 70 of the commentary on fragment 1.

[417] Ryhiner, *L'Offrande du lotus*, pp. 110-11. For Harsomtus in the solar barks, see further e.g. Chassinat, *Dendara* 1, p. 48, lines 9-10; idem, *Dendara* 5, pp. 140, lines 5-6, 141, line 2, and 144, lines 10-11 and 13-14.

[418] See textual note (b) ad loc.

(lines 5-6) reference is made, for the first time in conjunction, to the breath of life (*pꜣ ṯꜣw pꜣ ꜥnḫ*) and the breath of death (*pꜣ ṯꜣw n pꜣ mwt*). The former is said to be what causes all creatures to live; the latter is what takes them away, a doctrine reaffirmed in lines 10-13 below.[419]

An earlier reference to these two kinds of breath occurs in the medical P. Ebers, which dates to the first part of the New Kingdom. The Egyptians believed that the ears were a point of entry into the human body for various sorts of vapours, fluids, and other substances.[420] In P. Ebers, the breath of life (*ṯꜣw n ꜥnḫ*) is said to enter the body through the right ear, and the breath of death (*ṯꜣw n mwt*) to enter through the left ear. A variant tradition recorded in the same manuscript identifies the right and left shoulders as the points of entry for these vapours.[421] P. Edwin Smith, roughly contemporary in date, discusses the symptoms of a person described as 'one whom something entering from outside has smitten'. A gloss in the text explains this phrase: *ir ꜥq.t m rwty ṯꜣw pw n nṯr n rwty mwt r-pw*, 'As for "something entering from outside", it means the breath of an outside god or (that of) death.'[422]

Lines 6-8 introduce the subject of 'the great sea of Syria', in other words the Mediterranean. The breath of the Primaeval Ocean (*pꜣ ṯꜣw n pꜣ Nwn*) is said to be what awakens this sea and gives birth to it, while the breath of death is what quenches it. Lines 3-4 above, according to my restoration, state that the breath of life is the exhalation (*bw*) of the Primaeval Ocean. It should follow, therefore, that the breath of life and the breath (*ṯꜣw*) of the Primaeval Ocean are closely related. Nevertheless, the two are not precisely identical, but rather complementary to each other. According to our text, the former has the function of animating or enlivening sentient creatures,[423] whereas the latter quickens or brings into being insentient objects. By contrast, the breath of death exercises its destructive power over both categories of things.

The idea that the Mediterranean Sea originates from the Primaeval Ocean can be paralleled in other Egyptian sources. In the hymn to Ptah preserved in P. Berlin 3048, for example, that god is invoked as *Nwn wr .... ngꜣ km-wr wꜣḏ-wr*, 'the great Primaeval Ocean .... who opened the Bitter Lakes and the Great Green' (= the Mediterranean).[424]

[419] For the Egyptian conception of death seizing or snatching away its victims, see Zandee, *Death as an Enemy*, pp. 85-7; Derchain, *CdE* 33 (1958), pp. 29-32.

[420] On this belief, see Smith, *The Mortuary Texts of Papyrus BM 10507*, p. 74 [note (b) on III, 6], and references cited there.

[421] P. Ebers, C, 2-5 and CIII, 15-16. See Wreszinski, *Der Papyrus Ebers* 1, pp. 207 and 215; Grapow, *Anatomie und Physiologie*, pp. 31-2.

[422] P. Edwin Smith, IV, 16 (Breasted, *The Edwin Smith Surgical Papyrus*, pp. 212-14). Cf. Caminos, *JEA* 58 (1972), p. 215, note 6.

[423] For a list of beings in this category, see lines 10-12 below.

[424] P. Berlin 3048, III, 15-18 (Wolf, *ZÄS* 64 [1929], pp. 20-1). For the pairing of *km-wr* and *wꜣḏ-wr* here, see Zandee, *Der Amunhymnus des Papyrus Leiden I 344, verso*, p. 757, note 388.

Similarly, in a hymn to Khnum from the temple of Esna, he is invoked as *Ṯnn Nwn wr ms wꜣḏ-wr*, '(Ta)tenen the great Primaeval Ocean who gave birth to the Mediterranean Sea'.[425] The verb *ms* employed there is the same as the one used in our text to describe the activity of the breath of the Primaeval Ocean in bringing the said sea into existence.

The interpretation of lines 8-10 is hampered by uncertainty over how to restore their missing portions.[426] If my suggested restorations are correct, however, then the general import of this passage is that the breath of the Primaeval Ocean (*pꜣ ṯꜣw n pꜣ Nwn*) is what causes the hills and mountains to be rejuvenated but, at the same time, can also ruin them. In the Egyptian view, both the rain that fell from the sky and the Nile inundation were manifestations of the Primaeval Ocean. *Nwn* is even used, in some sources, as an alternative name for Hapi, the god of the Nile flood.[427] In the form of rain, the Primaeval Ocean was thought to fructify hilly or mountainous regions. In a passage from the Berlin hymn to Ptah quoted in the preceding paragraph, for example, the god is called *snhp Nwn n p.t sprr mw tp ḏw.w*, 'the one who makes the Primaeval Ocean rise up to the sky, who causes water to come upon the mountains'.[428] Compare another passage from the same hymn, where Ptah is called *snṯ tꜣ.wy ḏw.w ẖꜣs.wt swꜣḏ sn m mw iw m p.t*, 'founder of the two lands, the mountains, and the hill countries, who makes them verdant with the water that comes from the sky';[429] also the following extract from the great hymn to the sun disk in the tomb of Ay at Amarna: *ẖꜣs.wt nb.wt wꜣ.wt ir=k ꜥnḫ=sn di.n=k Ḥꜥpy m p.t hꜣy=f n=sn ir=f hnw ḥr ḏw.w mi wꜣḏ-wr*, 'All distant hill countries, you make them live, having placed Hapi in the sky that he might descend to them, creating a flood upon the mountains like the Great Green.'[430]

In a slightly different sense, the inundation was thought to renew the mountains. The Egyptians, in certain contexts, conceived of mountains as living things, divine in nature.[431] As such, they could be described as bearing or giving birth to the various

[425] Sauneron, *Esna* 3, p. 152 (259, 3). For the connection between Ptah-Tatenen and the *Nwn*, see commentary on fragment 2 above.

[426] See textual notes ad loc.

[427] See van der Plas, *L'Hymne à la crue du Nil* 1, pp. 59-60, 64-5, 134; Goyon, *Le Papyrus du Louvre N. 3279*, pp. 52-3, note 2; P. Berlin 13603, IV, 3 (Erichsen and Schott, *Fragmente memphitischer Theologie in demotischer Schrift*, p. 316 [18] and plate 6). On the *tpḥ.t*, 'cavern', of *Nwn* as the source from which Hapi emerges, see van der Plas, *op. cit.* 1, pp. 60 and 177-8; Pécoil, *BSEG* 17 (1993), pp. 102-4; Gabolde, *BIFAO* 95 (1995), pp. 240-1 and 248-50.

[428] P. Berlin 3048, VIII, 4. Cf. Wolf, *ZÄS* 64 (1929), p. 31.

[429] P. Berlin 3048, III, 9 (Wolf, art. cit., p. 21). For the use of the term *swꜣḏ* here, cf. Macadam, *The Temples of Kawa* 1, p. 25, where rain is said to *sṯḥn ḏw.w*, 'make the mountains gleam like faience'.

[430] Sandman, *Texts from the Time of Akhenaten*, p. 95, lines 4-6. For the idea that the rain which falls in foreign lands derives from the same source as the Nile flood, see further van der Plas, *op. cit.* 1, pp. 70-6.

[431] See Aufrère, *L'Univers minéral dans la pensée égyptienne* 1, pp. 19-22 and 307.

minerals for which they were an important source.[432] In turn, the mining operations through which those minerals were extracted from the mountains could be characterised in terms derived from the sphere of obstetrics.[433] Just as the annual inundation fertilised the arable land, thereby allowing it to produce new crops year after year, so too was the flood believed to regenerate the mountains, permitting them to yield their mineral products in an inexhaustible supply.[434]

The inundation can be said to overflow (*ṯtf*) the mountains, with the result that *nmḥ.w nb.w s3.w*, 'all the poor are satisfied'.[435] Alternatively, it can be described as reaching the mountains (*pḥ*) or taking possession of them (*iṯ*).[436] In such cases, what is meant are the rocky massifs that delimit the Nile Valley.[437] When the flood reached these, the valley itself could be said to have been covered by the Primaeval Ocean. Thus, in a hymn to Hapi on an ostracon of New Kingdom date from Deir el-Medineh, it is said that when the inundation settles upon the high places: *hnw.w ꜥḥ3 mrw t3 m Nwn*, 'the waves contend with the desert, the land is in the Primaeval Ocean'.[438] A stela dated in the sixth year of the Kushite king Taharqa records how, in answer to his prayers for the arrival of the inundation, a particularly high Nile arose, so that *ꜥq.n=f ḏw.w t3 rsy ḥry-tp=f i3.wt t3 mḥw wnn t3 m Nwn m nnw*, 'It entered the mountains of Upper Egypt and dominated the mounds of Lower Egypt. The land was (submerged) in the Primaeval Ocean, in inertness.'[439] As van der Plas has noted, the effect of such a flood is to recreate the conditions which prevailed at the origin of the world.[440] The inundation brings everything covered by its waters into existence anew. Thus, in an inscription from the temple of Dendera, Hapi can be said to arrive in order to cause the mountains to come into being along with all the other elements that make up the cosmos.[441]

In the light of the evidence presented above, one can see why our text might speak of the breath of the Primaeval Ocean in the forms of rain and Nile flood as rejuvenating or renewing the hills and mountains. But the Egyptians knew that these natural

[432] Ibid., pp. 67-8, 101, and 319-21; Aufrère, *RdE* 34 (1982-83), p. 13.

[433] Aufrère, *L'Univers minéral dans la pensée égyptienne* 1, p. 73.

[434] Ibid., pp. 183 and 307.

[435] P. Anastasi 4, X, 7 (Gardiner, *Late-Egyptian Miscellanies*, p. 45, lines 15-16).

[436] See van der Plas, *L'Hymne à la crue du Nil* 1, pp. 30-1 and 100-1; Caminos, *Late-Egyptian Miscellanies*, p. 175.

[437] See Aufrère, *L'Univers minéral dans la pensée égyptienne* 1, p. 18.

[438] O. DeM 1675, recto, line 4 (Posener, *Ostraca hiératiques littéraires de Deir el Médineh* 3, plate 81). For a similar idea expressed in Greek, see Bonneau in Haase (ed.), *Aufstieg und Niedergang der Römischen Welt* II, 18.5, p. 3198.

[439] See Macadam, *The Temples of Kawa* 1, pp. 25 and 27. On the nature of the mounds (*i3.wt*) mentioned here, see van der Plas, *op. cit.* 1, pp. 103-4; Aufrère, *op. cit.* 1, p. 23.

[440] See first reference cited in note 436 above. For the submergence of mountains as a part of those conditions, see Aufrère, *op. cit.* 1, p. 19.

[441] Chassinat, *Dendara* 4, p. 235, line 6.

phenomena could be dangerous and destructive, as well as beneficial. The harmful effects of a torrential downpour in the mountains that occurred in the reign of Ahmose at the beginning of the 18th Dynasty are vividly described in a stela erected by that king at Karnak. According to this account, a storm arose which darkened the western part of the sky. The sound of the rain beating down on the highlands (*ḫꜣs.wt*) was louder than that to be heard where the Nile inundation emerged from its source. As a consequence, tombs and other structures were destroyed or damaged. In the words of the text, *iry.t tmm.t ir*, 'What had been created (became like) what had not been created.'[442] Similar heavy downpours are known to have fallen in Egypt in modern times, with the resultant torrents of water cascading down through the wadis from the high desert causing extensive damage on some occasions.[443]

The effects of an inundation of sufficient height to reach the mountains could be equally devastating. A stela now in Cairo records the arrival of the flood in year 29 of Amasis, which is said to have overflowed Upper and Lower Egypt and penetrated (*bs*) the mountains. The force of the water demolished a dyke to the south of Memphis.[444] A graffito dated to the third year of Osorkon III, inscribed on one of the walls of the hypostyle hall of the temple of Luxor, describes a flood which ravaged the Theban area as a great catastrophe (*wꜥꜣw wr*), saying: *iw bs Nwn m* [...] *tꜣ pn r-ḏr=f iqḥ.n=f ṯn.wy mi sp tpy* [...] *tꜣ pn n sḫm=f mi wꜣḏ-wr nn ꜥ rmṯ.w r ḫsfꜣ.t=f*, 'The Primaeval Ocean entered in [...] this entire land. It penetrated the two mountains like the first occasion [...] this land to its power like the sea. No man-made dyke could restrain its force.'[445] The devastation that could be wrought by the Primaeval Ocean when it manifested itself as the Nile inundation also provides the basis for a simile in a text addressed to Ramesses III by Amun-Re-Harakhti at Medinet Habu, in which the god tells the king: *di=i mꜣꜣ=sn ḥm=k mi pḥ.ty Nwn sḏ=f ḥdb=f niw.wt dmi.w m sꜣ n mw*, 'I cause them to see your majesty like the might of the Primaeval Ocean, invading and overthrowing towns and villages as a deluge of water.'[446]

[442] See Vandersleyen, *RdE* 19 (1967), pp. 123-59; idem, *RdE* 20 (1968), pp. 127-34. The restorations of the lacunae in this text proposed by Helck in *Historisch-biographische Texte der 2. Zwischenzeit und neue Texte der 18. Dynastie*, pp. 106-7, make the storm appear to be more catastrophic than perhaps it really was.

[443] See Bell, *AJA* 79 (1975), pp. 247-8; van der Plas, *L'Hymne à la crue du Nil* 1, pp. 126-8; Vandersleyen, *RdE* 19 (1967), p. 155, note 4; Thomas in Johnson and Wente (eds.), *Studies in Honor of George R. Hughes*, p. 215, note 28. The most recent such catastrophe occurred on 2 November, 1994; for a vivid description of its effects, see Johnson, *Chicago House Bulletin* 6, no. 1 (December 15, 1994), pp. 2-3.

[444] See Daressy, *ASAE* 23 (1923), pp. 47-8; Vandier, *La Famine dans l'Égypte ancienne*, pp. 125-6.

[445] See Daressy, *Recueil de Travaux* 18 (1896), p. 181, lines 2-4; Vandier, *op. cit.*, p. 123. Cf. Traunecker, *BIFAO* 72 (1972), p. 199, note 2.

[446] Epigraphic Survey, *Later Historical Records of Ramses III* (Medinet Habu 2), plate 102, line 2.

Inasmuch as rainfall is a phenomenon which one associates primarily with the lands outside of Egypt, whereas the Nile inundation is specific to Egypt itself, by alluding to the manifestation of the Primaeval Ocean in both forms, the author of our text draws attention to the universal influence of its power.[447] The same is done by the statement in lines 6-8 above that the breath of the Primaeval Ocean awakens and gives birth to the Mediterranean Sea.

The said breath has predicated of it both beneficial and harmful actions. One is reminded of a passage in Coffin Text Spell 319 where the deceased, identifying himself as Hapi, the god of the inundation, says: *ii=i ẖbꜣ=i šm=i grg=i*, 'When I come, I destroy. When I go, I set in order.'[448] In this respect too, the breath of the Primaeval Ocean differs from the breath of life, since the latter is a purely beneficial force.

The theme of lines 10-13 of the second column of the present fragment is the absolute dominion exercised by the breath of life and the breath of death over all living things. A long list of these is given; the order follows roughly what Fischer has termed the rule of 'diminishing progression',[449] beginning with the gods and ending with serpents and plants. That *lym.w*, 'fish', are given precedence over *ipt.w*, 'birds', may appear strange to modern eyes, but parallels for this exist in comparable lists.[450] The text affirms of all forms of existence that the breath of life is what causes them to live and the breath of death is what causes them to die.[451] In this respect, the highest form of existence is exactly like the lowest.

For the view that all living things are subject to death, compare Spell 154 of the Book of the Dead, where the deceased invokes the god Osiris: *šd=k wi m-ḫt=k kꜣ tm=i ḥwꜣ mi nw ir.n=k r nṯr nb r nṯr.t nb.t r ꜥw.t nb.t r ḏdf.wt nb.t*, 'May you save me in your following, so that I do not decay, as you have made (happen) to every god, to every goddess, to all cattle, and to all serpents.'[452] A similar view is expressed in the initial hour of the Book of Gates, which says with respect to a group of twelve beings identified as *nṯr.w smy.t*, 'the gods of the desert/necropolis': *ḫpr.w m Rꜥ m ꜣḫ.t=f pr.w m ir.t=f wḏ=f n=sn s.t imn.t stp n=s rmṯ.w nṯr.w ꜥw.t nb.t ḥrr.wt nb.t qmꜣ.n nṯr pn ꜥꜣ*, '(They are) the ones who emanated from Re, from his divine eye, those who emerged from his eye. He has ordained for them a hidden place (= the underworld) to which are carried off people, gods, all cattle and all serpents, everything that this great god has created.'[453]

[447] See second reference cited in note 430 above; also Hornung, *ZÄS* 81 (1956), p. 29; Zandee, *Der Amunhymnus des Papyrus Leiden I 344, verso*, pp. 754-61.

[448] De Buck, *The Egyptian Coffin Texts* 4, p. 143k-1. Cf. Gilula, *JEA* 64 (1978), pp. 129-30.

[449] See *JARCE* 10 (1973), pp. 5-9.

[450] See e.g. Sauneron, *Esna* 3, p. 153 (260, 5). For further cases, see references cited in note (b) on line 10 above.

[451] Cf. the similar statement in lines 5-6 of the present column.

[452] Budge, *The Chapters of Coming Forth by Day* 3, p. 10, lines 2-4. Cf. ibid., lines 9-12.

[453] Cf. Hornung, *Das Buch von den Pforten des Jenseits* 1, pp. 1-2; 2, pp. 32-5, especially p. 34 with note 4. For the dead being seized or carried off, see references cited in note 419 above.

In the middle of line 13, reference is made to a deity who is designated by an epithet which I am unable to read. He is said to have taken possession of the earth. After this statement, the divine name *Ḏḥwty*, 'Thoth', occurs in a broken sentence. From what is preserved, it is impossible to determine whether this equated Thoth with the deity mentioned previously or specified some other form of relationship between them.

Line 14 says of the earth-ruling divinity: *mtw=f pꜣ nt ṯ mtw=f pꜣ nt t*[*i*], 'He is the one who takes. He is the one who giv[es].' Lines 4-6 above identify 'taking' and 'giving' as the characteristic activities of death and life respectively. Accordingly, our deity is conceived of here as embodying that pair within himself. Cf. the hymn to Re cited earlier, where the sun god is credited with having united (*twt*) life and death.[454]

The end of line 14 says further of our divinity: *b-ir=f sṯ=f r ti ḫpr pꜣ* [....], 'He does not allow the [....] to come into existence again.' In the break, perhaps one should restore a phrase like 'one who has died' or similar. Compare the concluding statement of the New Kingdom Harper's Song preserved in P. Harris 500: *mk* {*wi*} *nn wn šm iw ꜥn*, 'Behold, there is none who departs (i.e. dies) and comes back again.'[455] Neither of these passages denies the possibility of continued existence after death, but rather that the deceased can return to their previous mode of being, since dying necessarily involves a change from that mode and regeneration into a new one.[456]

Line 15 states: 'Death can take him, because death and life rule.' The pronoun 'him' in this sentence must refer to the lost subject of the verb *ḫpr* in the preceding line. The view expressed here that death and life jointly exercise authority over all creatures recapitulates one previously elaborated in lines 4-6 and 10-13 above, q.v.

The god with whom we have been concerned is called 'he who enters the sound eye, appearing as a disk in [the sky, as] an engendering bull forever', in lines 16-17. The terms *wḏy*, 'sound eye', and *itn*, 'disk', in this context refer to the full moon.[457] The epithet *kꜣ sty*, 'engendering bull', is applied to deities noted for their procreative capabilities.[458] Here it denotes the moon. That celestial body is frequently described as a bull in Egyptian sources, inasmuch as it was thought to possess abundant powers of fecundity which it could instil in other living creatures, particularly during the period of its waxing in the first half of the lunar month.[459]

[454] See reference cited in note 401 above.

[455] P. Harris 500, VII, 2-3 (Fox, *Orientalia* 46 [1977], p. 406).

[456] See Hornung, *Conceptions of God in Ancient Egypt*, pp. 160-1.

[457] For the former term, see *Wb.* 1, 401, 12. For *itn* = 'lunar disk', cf. Smith, *OLP* 22 (1991), p. 106.

[458] See Smith, *The Mortuary Texts of Papyrus BM 10507*, p. 111 [note (b) on IX, 19].

[459] For the moon as a bull, see e.g. *Wb.* 5, 95, 16 (s.v. *kꜣ*); Herbin, *BIFAO* 82 (1982), p. 277, note 57; de Buck, *The Egyptian Coffin Texts* 7, pp. 25h and 35a; Goyon, *BIFAO* 65 (1967), p. 118, note 86; London-Leiden Magical Papyrus, IX, 3 and 12-13 (Griffith and Thompson, *The Demotic Magical Papyrus of London and Leiden* 2, plate 9); Aufrère, *L'Univers minéral dans la pensée égyptienne* 1, pp. 221-2; Leitz, *Tagewählerei*, pp. 268-9. On its supposed powers of fecundity, see Derchain in *La Lune:*

The act of entering (*ʿq*) the sound eye is normally predicated of Osiris. In a glorification ritual in the demotic mortuary papyrus BM 10507, for example, Isis tells her brother: *i Wsir ḫnṯ'Imnt iw=k mne ḫn pꜣ itm m p.t ir rʿ nby i Wsir ḫnṯ'Imnt iw=k r ʿq r wḏꜣ m ḥr hrw*, 'O Osiris, foremost in the West, you endure in the sun disk in the sky every day. O Osiris, foremost in the West, you will enter the sound eye daily.'[460] In a similar rite preserved in P. BM 10208, the same goddess says: 'O Osiris, foremost in the West, raise yourself up (twice). Do not be weary, for your son Horus overthrows your enemies so that you might rise up to the sky and unite with Re. The crew of the sun bark gives you praise, the mouths of the horizon dwellers are jubilant, expressing joy at your passage. Love of you is in their hearts, and awe of you in their breasts, when you have entered among them. May you enter the sound eye, uniting with it, and your hue cast illumination upon the earth at night. May you become Moon-Thoth, the bull which rejuvenates itself in the sky every day. I have created the Greater Ennead and equipped the Lesser Ennead for the benefit of the sound eye in the half-month festival – so it is the feast of filling the moon (*ꜣbd*)[461] - in the coming (*wꜣw*)[462] that I did. You enter the sound eye and thus comes into existence the power of Osiris, Thoth setting it in order with its necessary components. Misery is dispelled because of your appearance in the sky with Re in the half-month festival, bulls ejaculating and cows conceiving when you are seen in the sky on this day.'[463] Likewise, a text from Dendera above a representation of Osiris in a bark with Isis and Nephthys describes the god as *Wsir Wn-nfr mꜣʿ-ḫrw ʿq.tw m ꜣb.t m 15.nt*, 'Osiris Wennefer the triumphant entering the left eye (= the moon) on the day of the half-month festival'.[464]

Some sources speak of Osiris as being caused to enter the lunar eye by another divinity, notably Isis, as in the late ritual text P. BM 10209, where the goddess tells her brother: *di=i ʿq=k m ir.t iꜣb.t ḫpr=k m iʿḥ*, 'I will cause you to enter the left eye so that you become the moon.'[465] According to a text from the temple of Opet, the goddess Mut, identified with Isis, introduces (*sʿq*) Amun (= the ba of Osiris) into the left eye.[466] Other sources say that Osiris unites with (*ẖnm*) or fills (*mḥ*) the lunar eye.[467]

mythes et rites, pp. 20, 27, 33-4; Aufrère, *op. cit.* 1, pp. 247-8, 266, 272, 292, and 302; also discussion of lines 18-19 below.

[460] P. BM 10507, I, 11-12 (Smith, *The Mortuary Texts of Papyrus BM 10507*, p. 35 and plate 1).

[461] For *ꜣbd* = 'moon', see *Wb.* 1, 65, 13.

[462] For this verb, see *Wb.* 1, 246, 10-12.

[463] P. BM 10208, II, 9-15. There is a parallel to this passage in P. Louvre I. 3079, CX, 47 – CXI, 3. For both texts, see Haikal, *Two Hieratic Funerary Papyri of Nesmin* 1, pp. 61-4.

[464] Brugsch, *Thesaurus Inscriptionum Aegyptiacarum*, p. 54.

[465] P. BM 10209, III, 3 (Haikal, *op. cit.*, p. 36). Compare Rochemonteix and Chassinat, *Edfou* 1, p. 327, lines 7-8, where *sʿq* replaces *di ʿq* in a similar statement.

[466] See de Wit, *Les Inscriptions du temple d'Opet à Karnak* 1, p. 91; Gutbub, *Textes fondamentaux de la théologie de Kom Ombo*, pp. 392-3.

[467] See Gutbub, *op. cit.*, pp. 389-90; Leitz, *Tagewählerei*, p. 268.

The Egyptians had two differing traditions concerning the time when the moon became full. According to one, this happened on the sixth day of the lunar month. According to the other, it did not happen until the fifteenth day, or half-month festival, which is in fact full moon day.[468] The first tradition, of Heliopolitan origin, is based upon the myth concerning the distant eye of Re, which was said to have been reunited with him on the sixth lunar day.[469] This was sometimes conflated with the second tradition, in which the full moon is connected with the body of Osiris, which was dismembered into fourteen parts by Seth and subsequently made whole again.[470]

Osiris may be said to enter the sound eye in two distinct but related senses. According to the Egyptian view, on each of the fourteen successive days during the period of the moon's waxing, a different divinity was thought to merge with that celestial body and restore one of its missing parts.[471] These divinities constitute the Greater and Lesser Enneads which are said to have been created and equipped by Isis for the benefit of the sound eye in the passage from P. BM 10208 cited three paragraphs above.[472] The parts or components which they supply are called *dbḥ.w*, a term used indifferently of the constituent elements of the moon and those of the body of Osiris.[473] The work of these deities is completed by Thoth on the fifteenth day of the lunar month.[474] At this time, restored to a state of wholeness through their actions, Osiris may be said to enter the sound eye.[475]

[468] See Derchain in *La Lune: mythes et rites*, pp. 25-6; Aufrère, *L'Univers minéral dans la pensée égyptienne* 1, pp. 199-202; London-Leiden Magical Papyrus, X, 22, and XXIII, 22-3 and 27 (Griffith and Thompson, *The Demotic Magical Papyrus of London and Leiden* 2, plates 10 and 23).

[469] See Junker, *ZÄS* 48 (1910), pp. 101-6; Hughes, *MDAIK* 16 (1958), pp. 147-9.

[470] See Herbin, *BIFAO* 82 (1982), p. 274, note 43; Chassinat, *Le Mystère d'Osiris au mois de Khoiak* 2, pp. 495-7; Leitz, *Tagewählerei*, p. 270. For the conflation of the two traditions, see e.g. second reference cited in note 465 above; Koemoth, *CdE* 71 (1996), pp. 207-9, 216, and 218-19.

[471] For the rites associated with the waxing of the moon, see the detailed treatment in Aufrère, *L'Univers minéral dans la pensée égyptienne* 1, pp. 205-303.

[472] See Herbin, art. cit., pp. 262-3, note 8, and 269, note 24; Kaper, *JEA* 81 (1995), pp. 183-4; Koemoth, *CdE* 71 (1996), p. 207.

[473] See references cited in note 470 above; Aufrère, *op. cit.* 1, pp. 199-202.

[474] Herbin, art. cit., pp. 269, note 24, and 274, note 43. Cf. Chassinat, *Edfou* 3, p. 207, lines 11-13. For Thoth as completer of the lunar eye see further Aufrère, *op. cit.* 1, pp. 200 and 202, note 22, with references to earlier literature.

[475] The passage quoted earlier from P. BM 10507 in which Osiris is said to enter the sound eye daily (*m ẖr hrw*) presumably means daily during the period of the moon's waxing, as each part of the god is progressively reunited with the rest. The period of the moon's waning is effectively ignored there, as it is, for instance, in a text from Dendera where Shu says to Osiris: *twt ỉwn rnp rʿ nb*, 'You are the moon who is rejuvenated each day' (Chassinat, *Dendara* 2, p. 150, lines 11-12). The Egyptians evidently were reticent in discussing the second half of the lunar cycle. Cf. Derchain in *La Lune: mythes et rites*, p. 27; Barguet, *RdE* 29 (1977), p. 17.

In astronomical terms, full moon day is when the sun and the moon are in opposition. The Egyptians referred to this as the 'union of the two bulls' (*snsn kꜣ.wy*).[476] They believed that, on the day in question, the rays of the two celestial bodies mingled together. More specifically, the greater light of the sun was thought to enter and illuminate the moon. Thus, texts speak of the latter as receiving (*šsp*) the former or being supplied with (*ꜥpr m*) its rays.[477] The moon is said to be illuminated (*šsp.ti*) by the sun,[478] or the rays of the latter to pervade (*ꜣbẖ*) the former.[479] This belief is given visual expression in Egyptian paintings and reliefs which show the disk of the full moon with the right, or solar eye, representing the sun's light, inside it.[480] In the Late Period, Osiris was frequently identified with the sun as well as with the moon.[481] There is a marked emphasis upon this solar aspect in texts and representations concerned with his entry into the sound eye when the moon is full.[482] It is as a solar deity, therefore, as well as in his capacity as a resurrected god, that Osiris can be said to enter the eye on the fifteenth day of the lunar month.[483]

As Gutbub has noted, the effect of the entrance of a divinity into a celestial body is to enhance its light.[484] Accordingly, other deities apart from Osiris can be said to enter or unite with the moon if they perform such a function for it. Thus, the divinities who progressively restore that celestial body to fullness during the period of its waxing are described in some sources as entering it (*ꜥq*).[485] Thoth, who completes their work, can be said to enter the sound eye (*ꜥq r tꜣ wḏy*) and become one with the moon.[486] The same action can be predicated of sacred animals.[487] In addition, goddesses like Hathor and Sakhmet who are identified with the solar eye can be said to unite with (*ẖnm/snsn*) the

[476] See Herbin, art. cit., pp. 280-1; Derchain, *op. cit.*, p. 31; Aufrère, *op. cit.* 1, pp. 222 and 236; Berlandini, *RdE* 46 (1995), pp. 35-7; Kaper, *JEA* 81 (1995), pp. 190-3.

[477] See e.g. Herbin, art. cit., pp. 269-70, note 26; Sethe, *Urkunden* 8, p. 49 (61b). Cf. Derchain, *op. cit.*, p. 28; Aufrère, *op. cit.* 1, pp. 284 and 300, note 60.

[478] Sethe, *Urkunden* 8, p. 42 (53l).

[479] See Chassinat, *Edfou* 3, p. 208, lines 1-2.

[480] See Koemoth, *CdE* 71 (1996), pp. 205-7.

[481] For references, see Smith, *The Mortuary Texts of Papyrus BM 10507*, p. 60 [note on I, 11].

[482] See Herbin, *BIFAO* 82 (1982), pp. 265-6, notes 14-15, and 280-1; Derchain, *RdE* 15 (1963), pp. 21-5; Gutbub, *Textes fondamentaux de la théologie de Kom Ombo*, p. 390; Aufrère, *L'Univers minéral dans la pensée égyptienne* 1, pp. 286 and 296. Note also the prominent solar imagery in the passages cited earlier from P. BM 10507 and P. BM 10208 (cf. notes 460 and 463 above).

[483] Cf. Boylan, *Thoth the Hermes of Egypt*, p. 69.

[484] Gutbub, *op. cit.*, pp. 388-9.

[485] See e.g. Sethe, *Urkunden* 8, p. 41 (53g); de Morgan et al, *Kom Ombos* 1, p. 258, no. 327 (cf. Gutbub, *op. cit.*, pp. 387-8).

[486] P. Leiden I 384, VIII, 19-22 (de Cenival, *Le Mythe de l'oeil du soleil*, plate 7).

[487] See e.g. Sauneron, *Esna* 2, p. 208 (105, 3-4).

lunar disk.[488] Finally, the sun god himself, for reasons explained in the preceding paragraph, can be said to enter and become one with the full moon. In a text from the temple of Edfu describing the phases of the moon, for instance, it is said with respect to Horus-Re: *imꜣ=f iꜥḥ m snsn kꜣ.wy*, 'He merges with the lunar disk at the union of the two bulls' (= full moon day).[489] Similarly, spells in the Demotic Magical Papyrus of London and Leiden refer indifferently to Pre or the moon as filling the sound eye (*mḥ wḏꜣ.t*) on the fifteenth day of the lunar month.[490] In this particular context, 'filling' is synonymous with 'entering'.[491] The question arises, therefore, which deity is said to enter the sound eye in our cosmology? This will be treated at greater length below.

The god who enters the sound eye/full moon possesses all the powers and attributes of a lunar deity. As such, he is called 'the one who brings the flood (*mw*) up from the Primaeval Ocean' (line 17). The reference here is to the Nile inundation. According to Plutarch and other classical authors, the Egyptians believed that the course of the annual flood was linked to the phases of the moon.[492] Our text confirms this in explicit fashion.[493] In some Egyptian sources, the time of the new moon is associated with the onset of the river's rise.[494] Perhaps the full moon, in some mythological sense, was thought to be responsible for its acme. The connection between the moon and the inundation explains why, at the temple of Dendera, the rite of filling the sound eye (*mḥ wḏꜣ.t*) is set in the context of the new year, the time when the appearance of the star Sothis heralded the coming of the annual flood.[495]

Lines 17-18 describe the lunar deity as the one who 'ordains breath for those who are spread out [ov]er the [earth]'.[496] This beneficial action can be attributed to a

[488] See e.g. Chassinat, *Dendara* 4, p. 15, lines 6-7 (discussed in Herbin, art. cit., p. 268, note 21); idem, *Edfou* 3, pp. 82, lines 8-9, 16 - 83, line 2; idem, *Dendara* 3, p. 81, lines 10-11; Chassinat and Daumas, *Dendara* 6, p. 136, line 8. Compare London-Leiden Magical Papyrus, IX, 23, where Sakhmet is called [...] *n pꜣ-Rꜥ n tꜣ wḏꜣ.t r-ms iꜥḥ n tꜣ 15.nt n grḥ*, '[...] of Pre in the sound eye, whom the moon bore during the half-month festival at night' (Griffith and Thompson, *The Demotic Magical Papyrus of London and Leiden* 2, plate 9).

[489] Chassinat, *Edfou* 3, p. 208, line 1. Cf. Barguet, *RdE* 29 (1977), p. 16.

[490] For the former, see London-Leiden Magical Papyrus, XXIX, 23 and 29 (Griffith and Thompson, *op. cit.* 2, plate 29). For the latter, see ibid., XXIII, 22-3, 27 (Griffith and Thompson, *op. cit.* 2, plate 23).

[491] See Gutbub, *Textes fondamentaux de la théologie de Kom Ombo*, p. 390.

[492] See Bonneau, *La Crue du Nil*, pp. 234-6; Stricker, *De Overstroming van de Nijl*, pp. 21-2.

[493] For more equivocal attestations of such a belief, see material collected by Derchain in *La Lune: mythes et rites*, pp. 34-6. As he notes, Thoth is often described as exercising control over the inundation, but it is not always clear whether he does this as the moon or not.

[494] See Barguet, *RdE* 29 (1977), p. 20; Brugsch, *Thesaurus Inscriptionum Aegyptiacarum*, p. 390.

[495] See Herbin, *BIFAO* 82 (1982), pp. 266-7, note 18, and 281; Derchain, *RdE* 15 (1963), pp. 24-5; Aufrère, *L'Univers minéral dans la pensée égyptienne* 1, pp. 292-3.

[496] For the restoration of the noun *tꜣ* in the break, see textual note (a) on the second of those lines above.

variety of deities, most notably Amun, who is often said to grant or give breath to the living.[497] However, the moon is attested as a dispenser of this gift as well. Cf., for example, an inscription on the propylon of the temple of Khonsu at Karnak, where it is said of that god, who bears the epithets 'moon during the night', 'counterpart of the sun', 'he who fills the sound eye', and 'he who equips the left eye': *nšp ḫmm.ty m ṯ3w imy=f srq.n=f ḥty.t g3w.ti*, 'Nostrils inhale from the breath that is in him. He has caused constricted throats to breathe.'[498] Similarly, in an extract from a hymn at Dendera, the deity Geb addresses the moon: *ntk it ṯ3.w qm3 swḥ.t m-m-ʿ mḥ.(t) rdi ṯ3w i*[*m*]*=f* (sic, for *im=s*), 'You are the father of fledgelings, who created the egg by means of the north wind and placed breath th[ere]in.'[499]

Lines 18-19 state that the moon god causes bulls to ejaculate and cows to receive (scil. their semen), in other words to conceive. In the commentary on lines 16-17 above, reference was made to the procreative powers attributed by the Egyptians to the moon, and its supposed ability to induce fecundity in living creatures. These powers reached their peak when the moon was full. Other Egyptian sources agree with our cosmology in ascribing the mating activities of bulls and cows to lunar influence. Thus, in the text from the temple of Khonsu at Karnak cited in the preceding paragraph, the moon deity is called *sṯs k3.w sbq ḥm.wt*, 'the one who induces erection in bulls and causes cows to become gravid'.[500] Another text from the same monument says that he causes bulls to ejaculate (*sp3y*) and makes cows gravid (Labrique in Clarysse, Schoors and Willems [eds.], *Egyptian Religion: the Last Thousand Years* 2, pp. 884-7; with the verb *sp3y* used there, cf. *ti p3y* in our text). A glorification ritual preserved in P. BM 10208 states that when Osiris enters the sound eye and appears in the sky with Re at the time of the full moon, *p(3)y k3.w iwr ḥm.wt*, 'Bulls ejaculate and cows conceive.'[501] According to an inscription from the temple of Edfu, precisely the same thing happens when Horus appears as the moon on the fifteenth day of the lunar month.[502] In the demotic P. Leiden I 384, Thoth, by virtue of entering the sound eye and uniting with the moon, acquires the epithet *p3 nb n n3 k3.w r-ir ti šp n3 3ty.w*, 'the lord of the bulls who caused cows to conceive'.[503] Plutarch ascribes to the Egyptians a belief that the light of

[497] See Smith, *The Mortuary Texts of Papyrus BM 10507*, pp. 104-5 [note (a) on VIII, 8], and references cited there. For other divinities, see Otto, *Gott und Mensch*, pp. 157-61.

[498] Sethe, *Urkunden* 8, p. 74 (89b).

[499] Aufrère, *L'Univers minéral dans la pensée égyptienne* 1, p. 246. Cf. ibid., pp. 247-8, note (h), for Arabic and Coptic traditions associating the annual arrival of the north wind in Egypt with the waxing of the moon.

[500] Sethe, *Urkunden* 8, p. 74 (89b).

[501] P. BM 10208, II, 14-15 (Haikal, *Two Hieratic Funerary Papyri of Nesmin* 1, p. 63).

[502] See Chassinat, *Edfou* 5, p. 49, lines 5-6.

[503] See reference cited in note 486 above.

the moon possessed procreative powers conducive to the reproduction of animals.[504] This is confirmed by the sources quoted above. By contrast, there is no evidence to support the contention made by him and other classical authors that the Apis bull was engendered by the rays of the moon.[505]

According to lines 19-20, the lunar deity causes every seed to sprout and all plants to grow, from the trees to the flowers. Plutarch, in the first of the two passages cited from *De Iside et Osiride* in the preceding paragraph, mentions the germination of plants in conjunction with the reproduction of animals as phenomena which the Egyptians believed to be influenced by the moon. However, in most Egyptian sources, it is the Nile inundation which is said to produce vegetation and allow it to flourish. In the hymn to Hapi, for instance, the god of the flood is credited with *sḫpr sm.w*, 'creating plants',[506] *srd ḫt*, 'causing trees to grow',[507] and *ir it sḫpr bty*, 'making barley and bringing emmer into existence'.[508] But the same activities can be predicated of other gods who are identified with Hapi or given credit for bringing the inundation forth.[509] It is probably for this reason that the moon is said to be responsible for producing trees, plants, and flowers in our cosmology, in view of its identification above as the agent which causes the annual flood to occur. Apart from the passage under consideration, the Egyptian evidence for a direct connection between the moon and the germination of plants is sparse, and tends to be equivocal in nature.[510] Noteworthy is the suggestion that the fructifying power attributed to this celestial body is what led to the adoption of the constituent parts of the lunar symbol of the sound eye to express the fractions of the Egyptian corn-measure.[511]

Near the end of line 20, the words *[mtw]=f pꜣ nt ti ḫpr sḫny nb* [...] may be read, which I have translated '[H]e is the one who causes every event to occur [...].' However, the noun *sḫny* can mean 'constellation' as well as 'event'.[512] If the former translation is preferred, then one could render this clause: '[H]e (scil. the lunar deity) is the one who causes every constellation to come into existence [...]', seeing here an allusion to the role of the moon in regulating the movements of the stars.[513] The words

[504] De Iside et Osiride 41 (Griffiths, *Plutarch's De Iside et Osiride*, p. 182).

[505] Ibid. 43. Cf. Griffiths, *op. cit.*, pp. 186 and 462-3.

[506] See van der Plas, *L'Hymne à la crue du Nil* 1, pp. 28, 94-5, 143-5; 2, p. 38.

[507] *Op. cit.* 1, pp. 29-30 and 98-9; 2, p. 44.

[508] *Op. cit.* 1, pp. 24 and 81-2; 2, p. 20.

[509] See e.g. material collected in Zandee, *Der Amunhymnus des Papyrus Leiden I 344, verso*, pp. 768-77.

[510] See e.g. texts cited in Aufrère, *L'Univers minéral dans la pensée égyptienne* 1, p. 287; 2, pp. 526-7; idem, *Memnonia* 6 (1995), pp. 113-21.

[511] Derchain in *La Lune: mythes et rites*, p. 33.

[512] See *Glossar*, p. 455.

[513] See Assmann, *Liturgische Lieder an den Sonnengott*, p. 94; von Turajeff, *ZÄS* 33 (1895), p. 125; Boylan, *Thoth the Hermes of Egypt*, p. 83.

tꜣ twꜣ.t, 'the underworld', stand at the beginning of line 21. Due to the broken context, it is not clear what relation, if any, these have to the clause just discussed. If they are connected, and *sẖny* in line 20 actually means 'event' rather than 'constellation', one could have here a reference to a deity determining every event in the underworld. But it is equally possible that a new sentence began after *sẖny nb*. Near the end of line 21, the words *tꜣ p.t*, 'the sky', are found. Once again, the context is obscure, owing to the poor state of the manuscript's preservation at this point.

Fragment 13 of our cosmology is better preserved than any other of its constituent fragments. As will be evident from the commentary above, however, it is not without its problems of interpretation. Chief among these is the difficulty of trying to establish the identity or identities of the being or beings to whom reference is made at various points in the text solely by means of personal pronouns, without any further specification. The most problematic passages in this respect are lines 1, 13-14, and 16 of Column II. Do all refer to the same being, or is a different one involved in each case? No definitive solution can be offered for this problem. The following remarks on the matter are presented for what they may be worth.

The second column of fragment 13 begins by describing how a thing or deity, referred to with the third person singular masculine suffix pronoun, causes the gods to be well in some unspecified place and renews them eternally. The ensuing text is concerned with the origin of death and the nature of its relationship with life. The first preserved reference to *ꜥnẖ*, 'life', occurs in line 3. However, there is some reason to think that the word should be restored at the end of line 2 as well.[514] But even this is unlikely to have been its initial mention. Inasmuch as death marks the cessation of life, it is difficult to imagine that the former's creation was described without some prior allusion to the origin of the latter. It is tempting to suppose, therefore, that some account of life's origin was given at the end of the previous column. If so, then the unnamed agent of the gods' well-being and renewal in the initial line of Column II must be either life itself or the deity responsible for creating it. Although numerous divinities can be said to exercise power over life and death,[515] we have seen above that it is normally the sun god to whom their actual creation is ascribed.[516] Thus, if the third person singular masculine suffix pronouns in line 1 refer to a deity, it is most likely to be this one.

Line 13 refers to an unnamed being who is said to have gained possession of the earth. In line 14, he is called the one who takes and the one who gives. Elsewhere in the text, the powers of taking and giving are ascribed to death and life.[517] It is the sun

[514] See textual note (d) ad loc.

[515] Cf. e.g. Otto, *ZÄS* 87 (1962), pp. 150-4.

[516] Cf. discussion of line 2 of this column.

[517] See lines 4-6 above.

god who unites this pair within himself. Accordingly, I think it probable that he is the one to whom reference is made in these lines as well. The divine name Thoth occurs in a broken context at the end of line 13. In the discussion of that line above, allusion was made to the possibility that this god might be the one who is said to have taken possession of the earth. Conceivably so, but on balance I think that it is unlikely to be the case. If it were, then the sequence of sentences in lines 13-14 would require that it be Thoth as well who was called the one who takes and gives, i.e unites within himself the powers of death and life, in the latter of those two lines. But, as has been shown, this is rather a characteristic attributed to the solar deity.[518] In my view, it is more probable that the broken sentence mentioning Thoth described him as the subordinate of the sun god, since he is frequently identified in other sources as the deputy or vizier of that divinity.[519] Perhaps the text said something like '[It is he] (scil. the solar deity) who has taken possession of the earth. Th[oth] is [his deputy].'

Line 16 states, with respect to a deity whose name is not given, 'He is the one who enters the sound eye.' Here, reference must be to the god previously mentioned in lines 13-14, inasmuch as no other divinity is named in the intervening text to signal that there has been a change of subject. If my argument in the preceding paragraph is correct, and that god is the solar deity, then it follows that he is the one who is said to enter the sound eye as well. The ensuing lines describe in some detail the powers and attributes of this god in his lunar aspect. The circumstances in which he acquires such attributes have already been set out in the commentary on lines 16-17 above. It appears to me that in our cosmology the solar deity effectively absorbs his lunar counterpart. Compare the following extract from a hymn on a pyramidion of Middle Kingdom date, now in the British Museum (EA 280/1163): *Rʿ sẖʿ.ti m Rʿ rʿ nb nb ꜣḫ.t m Ḏḥwty nb Ḫmnw nswt p.t*, 'O Re, you are revealed as Re every day, the lord of the horizon, and as Thoth, lord of Hermopolis and king of the sky' (= the moon).[520] Similarly, in a hymn from the temple of Esna, *bꜣ dmḏ*, 'the composite ba', a name for the sun and moon united together, is said to be a manifestation of the force (*nẖt*) of Re.[521]

According to the interpretation suggested here, the entirety of the second column of fragment 13, and probably part of the now lost first column as well, was devoted to a description of the activities of the solar deity and the twin forces set in motion by

[518] Thoth, as reckoner of time, determines the length of lifespans (see Boylan, *Thoth the Hermes of Egypt*, pp. 85-7), but that is not really the same thing.

[519] See e.g. Boylan, *op. cit.*, pp. 81-2; Luft, *Studia Aegyptiaca* 4 (1978), pp. 131-42; Hornung, *Der ägyptische Mythos von der Himmelskuh*, pp. 66-7, notes 160 and 169. For the possibility that Thoth figures as the earthly representative of Pre in fragments 9-10 of our cosmology, see commentaries ad loc.

[520] Budge, *Hieroglyphic Texts from Stelae* 4, plate 21. Cf. Assmann, *Sonnenhymnen in Thebanischen Gräbern*, p. 7, note (a); Hermsen, *Die zwei Wege des Jenseits*, pp. 191, 199-203.

[521] Sauneron, *Esna* 3, p. 154 (260, 7). Cf. idem, *Esna* 5, p. 389, note 2.

him, those of life and death. Line 13 says that this god exercises control over the earth. The occurrence of the noun *twꜣ.t*, 'underworld', in a broken context in line 21 could have been part of a statement to the effect that the sun god governs that sphere as well. If so, then perhaps the missing antecedent of the suffix pronoun *s* in line 1, which denotes the place where our deity is said to look after the gods and renew them was *p.t*, 'sky'. In other words, the column may originally have described how the power of this deity was manifested in the sky, earth, and underworld, that is to say, the entire cosmos. Its purpose then would have been to affirm the universal character of the sun god's dominion.[522] Given the present state of the text, it is impossible to demonstrate that this interpretation is the correct one. It does, however, offer a coherent explanation of the contents of the fragment under consideration as they have been preserved to us.

FRAGMENT 14 Plate 10

Dimensions: 17.4 × 22.5 cm. Colour: dark brown. Related fragments: Nos. 26-29.

Transliteration

Column I

1.].. *nꜣ ḥf.w*
2.] *ḏ n=f pꜣ Nwn*
3.]. *tꜣ nb pꜣ šꜥꜣ*
4. *ẖ*]*n pꜣ ist*
5.].....
6.].....
7.].. *tꜣ* .. *tꜣy*
8.].. *i-ir=f ḏ* ..
9. *š*]*ꜥꜣ pꜣ sꜥ*.. *i-ir*
10.] *ḥf 2 nt ḥr pꜣ tꜣ nꜣ.w*
11.] *iw=w rt.w ḥr tꜣ* ...
12.] *sbt.wt n iny*
13.].....
14. (lost)
15.].....
16. *ḥ*]*f.w ꜥn*

[522] The sort of tripartite division postulated here is well attested in Egyptian hymns and other works of a hymn-like nature that stress the omnipotence of a particular divinity. See e.g. Sauneron, *Esna* 2, pp. 279-83 (no. 163); idem, *Esna* 3, pp. 153-6 (no. 260); Assmann, *Ägyptische Hymnen und Gebete*, p. 579.

17.].. *ẖn=w*
18.] *šꜥ pr=w*
19.] *tꜣ ꜥꜣ.t mtw=w*
20.] *rpy.w*
21. *i]ny ꜥn*
22.] *šꜥ*
23.]...*w*

Column II

1. *mtw=w ḫpr ẖn* [*nꜣ*(?) *in*]*y.w* [...]....=*f* [
2. *n nꜣ rmt.w nt* ..[... *iw*]*f.w r-r=w šꜥ* ..[
3. *ḫpr pꜣ smt hyny.w iwf.w nꜣ.w pꜣ smt n* ...[
4. *iny.w* [*n*] *nꜣ tww.w mtw* ... *hyny.w in*[*y.w*
5. *tꜣ ftyw nꜣ nkt* ... *pꜣ ḥf 2 nꜣ.w nꜣ* [
6. [*iw*]*f.w ꜥn nꜣ.w mtw nꜣ nb nꜣ.w* (*n*) *pꜣ iny mtw tꜣy=w ẖn*[*y tꜣy*
7. *ḫpr r-ḏbꜣ.ṯ=s ꜥn tꜣ ft*[*y*]*w r-bnr-n nꜣ iwf.w n* [
8. *i-ir pꜣ mw pḥ r-r=s ꜥn ḫpr tꜣ ftyw n pꜣ mw tꜣy* [
9. *tꜣ ftyw nt pr n pꜣ Nwn* [*iw*]=*f ti q*[*t*]*y*(?) [
10. *pꜣ*(?) *sḏm* ... *nt*(?) *gmꜥ* ..[
11. *r tꜣy=s ir.t nꜣ ym.w* ⸢*tꜣy*⸣ [
12. *tr=f nꜣ.w nꜣ sbt.wt n tꜣ* ..*ꜣ.t n* [
13. *nb iw=w pr n pꜣ Nwn* ...[
14. *gtg* [..].=*f n* ⸤*tꜣ*⸥ [
15. *i-ir pꜣ Nw*[*n*
16. ....[
17. .. *mḫ*(?)[
18. *ꜣt* [
19. [*pꜣ N*]*wn* [
20. [..]..[
21. ....[
22. *Ḫmnw* [
23. *n Ptḥ* .[

Translation

Column I

1.].. the serpents
2.] said to him the Primaeval Ocean
3.]. every land. The sand
4. i]n the ground
5.].....

6.].....
7.].. It is the ..
8.].. He said ..
9. s]and, the *sʿ*.. which
10.] They are the two serpents which are on the earth.
11.] they growing on the ...
12.] hills of stone
13.].....
14. (lost)
15.].....
16. ser]pents again
17.].. in them
18.] until their emergence
19.] the stone, and they
20.] temples
21. st]one again
22.] until
23.]...*w*

Column II

1. and they exist/come to be in [the(?) sto]nes [...].... him/it [
2. of the people who ..[... fle]sh to them until ..[
3. because they are likewise flesh, in the manner of ...[
4. stones [of] the mountains which ... some sto[nes
5. the efflux. They are the property ... of the two serpents. The [
6. They are [fle]sh as well, that is, the lords (of) the stone, which [is] their sk[in
7. happens because of it as well. The eff[lu]x from the flesh of [
8. the water also/still reaches it, because it is the efflux of the water [
9. the efflux that comes forth from the Primaeval Ocean, [as] it causes to cir[cu]late(?) [
10. the(?) hearing ... which(?) damages/is damaged ..[
11. ⸢it being⸣ its companion (of) the seas [
12. They are [....] entire. The hills of the ..*ꜣ.t* of [
13. all, when they emerge from the Primaeval Ocean ...[
14. hasten [..]. him/it in ⸤the⸥ [
15. *i-ir* the Primaeval [Ocean
16. ....[
17. .. *mẖ(?)*[
18. back [
19. [the Prim]aeval Ocean [

20. [..]..[
21. ....[
22. Ogdoad/Hermopolis [
23. of/for Ptah .[

Textual notes

Column I

Line 2

With the traces of *ḏ*, 'say', at the beginning of the line, cf. better preserved occurrences of that verb in e.g. fragment 1, line 7, and fragment 2, line 1 and passim.

Line 3

What stands between *tꜣ* and *pꜣ šꜥꜣ* here can only be the adjective *nb*, 'every'. An extra stroke has been added to the lower of the two signs which comprise that word, giving it the appearance of a pair of ꜥayins written one above the other, as in the noun *ꜥꜥe.w*, 'limbs' (cf. the writing of the latter in e.g. fragment 4, II, 12). This curious orthography recurs in II, 6 of the present fragment. For our scribe's normal writing of *nb*, see fragments 4, II, 2; 13, II, 2 and passim.

Line 4

Only the determinative of *ẖn*, 'in', is preserved at the beginning of the line. Cf. more complete examples of this word in fragment 2, lines 2 and 8, fragment 4, II, 10.

Line 7

The signs visible immediately before the definite article *tꜣ* are the pellet determinative above plural strokes. The noun that follows the article is determined with the flesh sign.

Line 8

A tall upright sign is visible near the beginning of the line. This could be either an *s* or the man with hand to mouth determinative.

Line 9

(a) For undamaged examples of *šꜥꜣ*, 'sand', elsewhere in the text, see fragment 1, lines 6 and 7, fragment 6, line 6.

(b) The first two consonants of the masculine noun preceding *i-ir* are *s* and ꜥ. The third is lost in a break. The word is determined with the house sign. I am unable to suggest a translation.

Line 12

(a) For the writing of *sbt.wt*, 'hills', here, see e.g. fragment 13, II, 8, and line 12 of the second column of the present fragment.

(b) With the traces of *iny*, 'stone', cf. the better preserved examples of that noun in line 6 of the second column of the present fragment, and fragment 10, line x+11.

Line 13

All that is visible in this line is a stone determinative.

Line 15

The faint traces that can be made out at the end of the line are of the sail determinative (or *ṯꜣw*, 'breath'), followed by what might be the preposition *r*.

Line 16

For the restoration [*ḥ*]*f.w*, '[ser]pents', here, see e.g. lines 1 and 10 above, line 5 of the second column of the present fragment.

Line 18

With the traces of *šꜥ*, 'until', before *pr=w*, cf. better preserved occurrences of the same word in line 22 below, fragment 11, I, 1.

Line 19

For this writing of *ꜥꜣ.t*, 'stone', see also fragment 26, line x+2.

Line 20

The *y*, *t* above house determinative, divine determinative, and plural strokes of *rpy.w*, 'temples', are reasonably clear on the photograph. The word's initial sign, which rather resembles that of *rt*, 'grow', in line 11 above, can be made out on the original.

Line 21

For the traces of [*i*]*ny*, '[s]tone', at the beginning of the line, cf. better preserved examples of that noun cited in note (b) on line 12 above.

Line 23

All that survives in this line is the end of a plural noun determined with the stone sign. The traces that precede the determinative rule out a reading as *iny*.

Column II

Line 1

For the restoration [*in*]*y.w*, '[sto]nes', here, cf. better preserved examples of that noun cited in note (b) on line 12 of the initial column of the present fragment. There appears to be sufficient room to restore the article *nꜣ* before this.

Line 2

(a) The initial signs of the word that follows *nꜣ rmt.w nt* are perhaps *ꜣy*. The remainder of it is lost in a break.

(b) With the traces of *iwf.w*, 'flesh', here, cf. the better preserved example of that word in line 7 below.

(c) For better preserved occurrences of *r-r=w* elsewhere in the text, see e.g. fragments 5, I, 9; 11, I, 5; and 12, I, 16.

Line 3

(a) For the traces of *smt* after *ḫpr pꜣ*, cf. the better preserved example of that word later on in the line.

(b) For the traces of *hyny.w*, cf. the writing of that word in the line immediately following. For those of *iwf.w*, 'flesh', see line 7 below.

(c) The copula *nꜣ.w*, although somewhat indistinct on the photograph, is clear on the original.

(d) I am unable to read the word after *pꜣ smt n* at the end of the line. Its second sign could be *ḥ*.

Line 4

(a) *iny.w*, 'stones', at the beginning of the line is clear on the original. For the traces, cf. better preserved writings of that noun cited in note (b) on line 12 of the initial column of the present fragment.

(b) For the traces of *tww.w*, 'mountains', see fragment 13, II, 9.

(c) *mtw* is written for the relative converter *nt-iw*. The initial sign of the word after this looks something like *w*. Or is the tall vertical stroke part of the following sign, e.g. an *l*?

(d) For the restoration *in*[*y.w*] at the end of the line, see reference cited in note (a) above.

Line 5

(a) The article *tȝ* at the beginning of the line is clear on the original. The feminine noun that follows it recurs in lines 7, 8, and 9 below. The best preserved instances are those in lines 7 and 8. By combining the traces of these, one obtains a word written with the consonants *f*, *t*, and *y*, followed by the water determinative and horizontal plural strokes, hence my reading *ftyw*. It is difficult to see what else this could be apart from a writing of the *fty*, 'Schweiss', of *Glossar*, p. 145, although none of the examples cited there is actually written with plural strokes. Nonetheless, the notion of plurality is inherent in any word denoting a liquid. Cf. Lacau, *BIFAO* 56 (1957), pp. 161-72; Leclant, *Orientalia* 30 (1961), p. 409, note 5.

According to *Wb.* 1, 582, 8-11, *fd.t*, the hieroglyphic ancestor of demotic *fty*, apart from its literal meaning, can be used as a figurative term for something that emanates from a divinity, particularly, but not exclusively, effluxes of a liquid nature. It is in this sense that the word is employed in our text. In line 7 below, the *ftyw* is said to come out of (*r-bnr-n*) flesh. While the nuance 'perspiration' is at first sight apt here, in fact the flesh in question is quite different to the ordinary sort. Lines 8 and 9 speak, respectively, of the *ftyw* of the water (*mw*) and the *ftyw* that comes forth from the Primaeval Ocean (*pȝ Nwn*). For the significance of the term in each instance, see commentary below.

(b) The pellet determinative and plural marker of *nkt*, 'property', are clear on the original. I am unable to read the following word, which ends with the man with hand to mouth determinative and a stroke which is either the plural marker or a suffix pronoun *w*. *tr*=*w*, 'all, entire', would fit the context well, but the traces do not suit such a reading.

Line 6

(a) For the restoration [*iw*]*f.w*, '[fle]sh', at the beginning of the line, cf. the better preserved example of that noun in the line immediately following.

(b) *mtw* here and at the end of the line is written for *nt-iw*.

(c) The writing of *nb*, 'lord', here is identical with that of *nb*, 'every', in I, 3 of the present fragment.

(d) For the restoration *ẖn*[*y*], 'sk[in]', at the end of the line, see better preserved examples of that noun in fragment 4, II, 14, and fragment 10, line x+10. It is impossible to say whether the instance here was determined with the flesh and 'leather' signs, as in the former passage, or with the leather sign alone as in the latter. The restoration of the copula *tȝy* immediately afterwards is highly probable in view of the parallelism between this relative clause and the preceding one with the copula *nȝ.w*.

Line 7

For the feminine noun *ftyw*, see note (a) on line 5 above.

Line 8

(a) *i-ir* at the beginning of the line is clear on the original. Cf. identical writings of the second tense marker in e.g. fragments 8, line x+8; 12, II, x+9; and 13, II, 15. With the sentence *i-ir pȝ mw pḥ r-r=s ꜥn*, cf. the words *mw pḥ r-r=w* which occur in a damaged context in fragment 29, line x+2. For the possible significance of this sentence, see commentary below.

(b) For the feminine noun *ftyw*, see note (a) on line 5 above.

Line 9

(a) *tꜣ ftyw* at the beginning of the line is clear on the original. For this feminine noun, see note (a) on line 5 above.

(b) For the traces of *pꜣ Nwn*, 'the Primaeval Ocean', after *pr n*, see e.g. line 13 below and the second line of the initial column of the present fragment.

(c) The top of a suffix pronoun *f* is visible immediately before *ti*. It is difficult to see what else could be restored before it apart from *iw*.

(d) The traces after *ti* would suit the verb *qty*, 'go around, circulate'. Cf. e.g. the writing of the phrase *n pꜣy=w qty*, 'around them', in fragment 2, line 7. Alternatively, the first trace could represent part of a verb like *šb*, 'change' (cf. fragment 13, II, 9, and textual note [b] ad loc.). If the latter reading is correct, then the *y* visible at the end of the line must be part of the succeeding word.

Line 10

(a) The trace visible before *sḏm* could be part of the definite article *pꜣ*, but this is by no means certain. The illegible word that follows *sḏm* appears to have a man with hand to mouth determinative.

(b) The verb *gmʿ* (cf. *Glossar*, p. 580), is certain after what I have tentatively read as *nt*. It is not clear, however, whether one has here the infinitive or the stative form of the verb, hence my provision of two alternative translations.

Line 11

(a) The possessive article *tꜣy=s* before *ir.t* is clear on the original.

(b) For the writing of *ym.w*, 'seas', see *Glossar*, p. 50. It appears that the scribe originally added the plural stroke to that word before he had written the second sign of its water determinative. This was subsequently inscribed over the stroke, and the latter written again immediately afterwards.

(c) With the traces of the copula *tꜣy* at the end of the line, cf. the better preserved writing at the end of line 8 above.

Line 12

(a) The suffix pronoun *f* after *tr* is clear on the original, as is the article *nꜣ* before *sbt.wt*. For the writing of the latter, see line 12 of the initial column of the present fragment, also fragment 13, II, 8.

(b) The feminine article *tꜣ* is clear on the original. The noun that follows begins with a sign that looks like *in* or *mn*. After this come an *ꜣ*, a flesh determinative and the feminine ending *t*. I am unable to suggest a reading or translation.

Line 13

(a) With the traces of *pr*, 'emerge', after *iw=w*, cf. better preserved occurrences of that verb in e.g. line 9 above, fragment 5, I, 6.

(b) The traces after *Nwn* are . I am unable to suggest a reading.

Line 14

All that survives of the word preceding the suffix pronoun *f* in this line is a man with hand to mouth determinative.

Line 17

The trace after *m* at the end of the line could be the lower part of *ẖ*.

Line 18

For this writing of *ꜣt*, 'back', see note (b) on fragment 4, II, 16.

Line 20

The only sign preserved in this line is either an *s* or the man with hand to mouth determinative.

Line 22

With the traces of *Ḫmnw*, cf. the writing of *Ḫmnw*, 'Ogdoad', in fragment 2, line 12, and that of *Ḫmnw*, 'Hermopolis', in fragment 24, line x+10. The former ends with horizontal plural strokes; the latter adds a toponym determinative after these. Since nothing is preserved of the present line after the horizontal plural strokes of *Ḫmnw*, it is impossible to determine which of the two words one had here.

Line 23

With the traces of *Ptḥ*, 'Ptah', here, cf. better preserved examples of that divine name in fragment 2, lines 5, 9, and 15, fragment 11, I, 14.

Commentary

The words *nꜣ ḥf.w*, 'the serpents', occur in line 1 of the initial column of this fragment. Too little is preserved there to allow one to ascertain what was said about these reptiles. The plural [*ḥ*]*f.w* recurs in line 16 below, while line 10 and the fifth line of Column II mention a pair of serpents (*ḥf 2*). It is unclear whether reference is made to the same group of creatures in all four passages. If so, then the serpents here must be distinct from those mentioned in fragment 30, which are four in number.[523]

The name of the Primaeval Ocean occurs in line 2 of Column I. The following line refers to *pꜣ šꜥꜣ*, 'the sand'; so does line 9. On the significance of this substance in Egyptian cosmological thought, see commentaries on fragment 1, line 6, and fragment 6, line 6. The words [*ẖ*]*n pꜣ ist*, '[i]n the ground', occur in line 4. For sand as the constituent material of the *ist* (< earlier *sꜣṯw*), see the first of the two commentaries just cited.

As noted above, line 10 mentions a pair of serpents. It says concerning these: *ḥf 2 nt ḥr pꜣ tꜣ nꜣ.w*, 'They are the two serpents which are on the earth.' The import of this statement is obscure. Do the snakes in question provide the basis for a rudimentary sort of zoological classification, dividing all such creatures into two groups?[524] In this case, the text could mean something like 'They represent the two (species of) serpent which are on the earth.' Or is the sense that the pair are the chief or pre-eminent members of this class of reptile on earth?

Whatever the case, it is clear from what is said about them in Column II that these are no ordinary serpents, but rather divinities in ophidian form. The serpents in

[523] See line x+7 of that fragment.

[524] Cf. the treatise on serpents preserved in P. Brooklyn 47.218.48 + 47.218.85, which seems to utilise a broad classification of all snakes into two types: *ḥfꜣ.w*, which are long and cobra-like, and *fy.w*, which are short and viper-like (Sauneron, *Un Traité égyptien d'ophiologie*, pp. 172-3).

question could be male deities, like the primordial pair *Km-ꜣt=f* and *Ỉr-tꜣ* discussed in the commentary on fragment 1, line 16. Alternatively, one could be male and the other female. Cf. the [*ḥ*]*f.t n pꜣ mḥt*, '[female se]rpent of/in the north', and *ḥf n pr-imnt*, 'male serpent of/in the west', to which reference is made in fragment 38, lines x+2 and x+5. Probably the best known pair of divine serpents in this category is that made up of (P)shai and Renenutet.[525] However, other couples, e.g. Isis and Osiris, were venerated together in serpent form as well, and so presumably could figure in the passage under consideration.[526]

Line 12 of Column I preserves the words *sbt.wt n iny*, 'hills of stone'.[527] This is the first of several references to stone in the present fragment. It is of interest that mineral-bearing hills should be mentioned in such close proximity to divine serpents as they are here. Because they lived underground and were frequently encountered by the Egyptians in the mountainous regions to which their mining and quarrying expeditions took them, snakes came to be seen as guardians or protectors of buried minerals. The patron deities of such regions, whether male or female, were often represented in serpent form.[528] Perhaps this is the function of the two serpents in the fragment under consideration.

The words *iw=w rt.w ḥr tꜣ* ... [...], 'they growing on the ... [...]', in line 11 could refer to stones or minerals. For the Egyptian conception of these as living, organic things which could be said to grow or flourish in the same way as plants, see commentary on fragment 13, II, 8-10, in particular references cited in note 434. In the demotic version of the myth of the solar eye, an elaborate comparison between the green stone of the mountains and the green papyrus plant is introduced with the words *i-ir nꜣ tw.w wtwt n wyt n mꜣꜥ.t m-qty pꜣ mꜣqr n ḏwf nt rṱ*, 'The mountains flourish with real green stone like the papyrus clump that grows.'[529] Later on in the same passage, the green stone of the mountains is explicitly said to be the 'double' (*sn.nw*) of a serpent.[530]

All that is preserved in line 18 of Column I is the phrase *šꜥ pr=w*, 'until their emergence'. This could refer to the emergence of snakes from their holes or, alternatively, to that of minerals from the places where they are concealed. In a

[525] See Quaegebeur, *Le Dieu égyptien Shaï*, pp. 152-3, with additional bibliography. For (P)shai in serpent form, see also references cited in note 70 of the commentary on fragment 1.

[526] For Isis and Osiris as serpents, see Smith, *The Mortuary Texts of Papyrus BM 10507*, p. 114 [note (c) on IX, 19]; Dunand, *BIFAO* 67 (1969), pp. 9-48.

[527] For the noun *iny* (< earlier *inr*) and its range of meanings ('stone, mineral' etc.), cf. Aufrère, *L'Univers minéral dans la pensée égyptienne* 1, pp. 95-100; Harris, *Lexicographical Studies in Ancient Egyptian Minerals*, pp. 19-20.

[528] Aufrère, *op. cit.* 1, pp. 140-1 and 317.

[529] P. Leiden I 384, VI, 3-4 (de Cenival, *Le Mythe de l'oeil du soleil*, plate 5).

[530] P. Leiden I 384, VI, 16-17 (de Cenival, loc. cit.). For the reading *sn.nw* there, see Jasnow, *Enchoria* 18 (1991), p. 208.

number of texts, minerals of various sorts are said to 'emerge' (*pr*) from the mountains or countries of their origin for the benefit of a deity or king.[531]

The noun *ꜥꜣ.t* which occurs in line 19 is often, but not exclusively, used to denote hard varieties of stone.[532] In demotic texts, it is virtually synonymous with *iny*.[533] *rpy.w*, 'temples', are mentioned in the following line. Perhaps these were identified as the destination of the *ꜥꜣ.t*.

The initial line of Column II preserves the words *mtw=w ḫpr ẖn* [*nꜣ*(?) *in*]*y.w*, 'and they exist/come to be in [the(?) sto]nes'. It is not clear to whom or what 'they' refers here. Line 2 mentions some people (*rmt.w*) and flesh (*iwf.w*) in a broken context. In line 4, there are two references to *iny.w*, 'stones'. The first calls them *iny.w* [*n*] *nꜣ tww.w*, 'stone [of] the mountains'. The noun *ftyw*, which I have interpreted to mean 'efflux, perspiration' or similar stands at the beginning of line 5.[534] Thereafter, the text continues: *nꜣ nkt* ... *pꜣ ḥf 2 nꜣ.w*, 'They are the property ... (of) the two serpents.' Since the preceding *ftyw* is feminine singular, the only possible subject for this nominal sentence is the stones mentioned in the line before. The text affirms, therefore, that the stones of the mountains belong to the two serpents. For the latter, see commentary on I, 10-12 of the present fragment. The divine nature of these reptiles is evident from the description of them given in the line that follows this one.

There (line 6) one has the enigmatic statement: [*iw*]*f.w ꜥn nꜣ.w mtw nꜣ nb nꜣ.w* (*n*) *pꜣ iny mtw tꜣy=w ẖn*[*y tꜣy*], 'They are [fle]sh as well, that is, the lords (of) the stone, which [is] their sk[in].' In other words, the owners of the stone (= the serpents) have skin composed of that material, but are flesh (i.e. living creatures) at the same time. Perhaps the partially preserved sentence *ḫpr pꜣ smt hyny.w iwf.w nꜣ.w pꜣ smt n* ...[...], 'because they are likewise flesh, in the manner of ...[...]', which occurs three lines earlier, restated the same idea only in slightly different terms.

What underlies this passage, in my opinion, is the well-attested Egyptian belief that the bodies of deities were made of metals and precious minerals. Both of the latter could be classified together as *ꜥꜣ.t*.[535] In demotic, as noted earlier, that noun was virtually synonymous with *iny*.[536]

Evidence for the Egyptian conception of the mineral nature of the gods' bodies has been assembled by Derchain-Urtel,[537] Aufrère,[538] and Zandee.[539] Only a few texts

[531] See e.g. Mariette, *Dendérah* 1, plate 70, no. 8; plate 71, nos. 3 and 13.

[532] Aufrère, *L'Univers minéral dans la pensée égyptienne* 1, pp. 101-3; Harris, *Lexicographical Studies in Ancient Egyptian Minerals*, pp. 21-2.

[533] Harris, *op. cit.*, p. 22.

[534] See textual note (a) ad loc.

[535] See e.g. Famine Stela, lines 16-17 (Barguet, *La Stèle de la famine à Séhel*, pp. 24-5 and plate 4).

[536] See commentary on I, 19 of the present fragment.

[537] *SÄK* 1 (1974), pp. 85-8.

[538] *L'Univers minéral dans la pensée égyptienne* 1, pp. 311-13; 2, pp. 433-6.

[539] *Der Amunhymnus des Papyrus Leiden I 344, verso*, pp. 349-53.

illustrating this idea will be cited here. In the great Hibis hymn to Amun, the Ogdoad say of that deity: *qs.w=f m ḥḏ inm=f m nbw ḥry-tp=f m ḫsbḏ mꜣꜥ.t*, 'His bones are silver, his skin gold, his hair is real lapis lazuli.'[540] This is one of several inscriptions where the appearance of the solar deity is described in such terms. In a hymn to Amun in the temple of Ramesses III at Karnak, the king says to that god: *ꜥꜣ.wt nb šps ꜣbḫ ḏ.t=k*, 'All precious stones are mingled in your body.'[541] In the songs of Isis and Nephthys preserved in P. Bremner-Rhind, the two goddesses address their brother Osiris thus: *šnw=k m ḫsbḏ .... inm=k ḥꜥ.w=k n=k m biꜣ Šmꜥw qs.w=k nb m ḥḏ mi ntk ẖ ṯs.w=k n=k m mfkꜣ.t*, 'Your hair is of lapis lazuli ...., your skin and flesh are yours, consisting of quartzite.[542] Your bones are fashioned of silver, according as you are a child. Your vertebrae belong to you, being made of turquoise.'[543] The bodies of deceased individuals who achieved the divine status of Osiris are likewise said to be 'mineralised' in e.g. Spell 172 of the Book of the Dead.[544] The Egyptians sought to assist this process through the application of stone or metal amulets or by gilding the mummy itself.[545]

Of divine beings with serpent form whose bodies are described as being constituted of minerals, the most familiar is that encountered by the eponymous hero of the Middle Kingdom story of the Shipwrecked Sailor, perhaps an avatar of the sun god. It is said of this creature: *ḥꜥ.w=f sḫr.w m nbw in.wy=fy m ḫsbḏ mꜣꜥ*, 'His flesh was overlaid with gold and his eyebrows (= markings?) were of real lapis lazuli.'[546] Elsewhere, the serpent demon Apopis is sometimes given the epithet *inr*, '(creature made of) stone'.[547] To these, one can now add the pair of serpents mentioned in our text.

Lines 7-8 of Column II continue: *tꜣ ft[y]w r-bnr-n nꜣ iwf.w n* [....] *i-ir pꜣ mw pḥ r-r=s ꜥn*, 'As for the eff[lu]x from the flesh of [....], the water also/still reaches it.' In view of what precedes, it seems fairly certain that the flesh in question is that of the serpents. Accordingly, restore something like *pꜣ ḥf 2* at the end of line 7. Since the serpents are themselves stone, the efflux from their flesh must take that form as well. The words *hyny.w in*[*y.w*], 'some sto[nes]', at the end of line 4 above, and *tꜣ ftyw*, 'the efflux', at

540 Davies, *The Temple of Hibis in El Khārgeh Oasis* 3, plate 33, middle, line 1. For a parallel, see Lange, *Der magische Papyrus Harris*, p. 38.

541 Epigraphic Survey, *Ramses III's Temple within the Great Enclosure of Amon* 1, plate 23, A, line 4.

542 For this translation of *biꜣ Šmꜥw*, see Aufrère, *op. cit.* 1, p. 312, note (f).

543 P. Bremner-Rhind, XV, 22-7 (Faulkner, *The Papyrus Bremner-Rhind*, p. 29).

544 Naville, *Das ägyptische Todtenbuch* 1, pp. CXCIII-CXCIV. Cf. commentary in Aufrère, *op. cit.* 1, pp. 342-5.

545 See *op. cit.* 2, pp. 379-81 and 389-90.

546 Blackman, *Middle-Egyptian Stories*, p. 43, lines 8-9.

547 See e.g. Chassinat, *Edfou* 4, p. 237, line 4 (written with stone and snake determinatives); idem, *Edfou* 6, p. 329, line 2 (with snake determinative alone). On this epithet, see Aufrère, *op. cit.* 1, pp. 98-9, with additional bibliography; Yoyotte, *RdE* 30 (1978), p. 150, note 10. (Correct the latter's citation of *Dendara* 6, p. 193, line 13 to *Dendara* 4, p. 193, line 13.)

the beginning of line 5, may be the remnants of a sentence which stated this in an explicit fashion.

A corollary of the belief that the bodies of deities were made of metals and precious minerals was the idea that these materials, in their natural state, were the efflux or emanation of the gods' bodies.[548] Thus, an inscription from the temple of Edfu calls haematite (*biꜣ*) and green stone (*wꜣḏ*) from Napata and other lands outside of Egypt *ꜥꜣ.w(t) šps pr m ḥꜥ.w nṯr*, 'the precious stones which emanate from the god's body'.[549] In a text from the treasury of the temple of Dendera, the goddess Hathor is told: *in=f n=t Tfrr ḫntš ẖr ḫsbḏ ꜥꜣ.t šps pr im=t*, 'He (scil. the king) brings you the region of *Tfrr*, rejoicing and bearing lapis lazuli, the precious stone which emanates from you.'[550] In a funerary text of Ptolemaic date, a deceased person is told: *iw n=k inr pr m Rꜥ rḏw.w nfr nw nṯr ꜥꜣ*, 'There will come to you the stone which emanated from Re, the perfect efflux of the great god.'[551] In the Embalming Ritual, it is said to a dead person: *rdi n=k Rꜥ nbw pr m Wsir*, 'Re will give you the gold which emanates from Osiris.'[552] Another passage from the same text calls orpiment (*qnw*) 'the efflux of Re' (*rḏw.w nw Rꜥ*).[553]

Thus in our text, the water is said to reach the efflux from the flesh of the two serpents, in other words, the minerals. What is meant here, I suspect, is that the waters of the inundation reach the mountains where the minerals are to be found. For the verb *pḥ*, 'reach', used of the Nile flood with this sense, see commentary on lines 8-10 of the second column of fragment 13. As noted there, the waters of the inundation were believed to have the effect of regenerating the mountains and making them fruitful in producing minerals.

This happens, according to our text, *ḫpr tꜣ ftyw n pꜣ mw tꜣy* [....] *tꜣ ftyw nt pr n pꜣ Nwn* [*iw*]=*f ti q*[*t*]*y*(?) [....], 'because it is the efflux of the water [....] the efflux that comes forth from the Primaeval Ocean, [as] it causes to cir[cu]late(?) [....]' (lines 8-9). The interpretation of these lines is hampered by the breaks in the text. If, as I have translated, the words *tꜣ ftyw n pꜣ mw tꜣy* in line 8 form a complete nominal sentence, i.e. 'it is the efflux of the water', then the subject can hardly be anything other than the antecedent of the feminine singular suffix pronoun *s* which occurs earlier in the same line, viz. the efflux of the flesh of the two serpents. In other words, what emanates from their flesh (= minerals) is said to come forth from the water as well. This could be an allusion to the fact that the mountains which bear minerals, having been covered

[548] See Aufrère, *RdE* 34 (1982-83), pp. 15-18.

[549] Chassinat, *Edfou* 6, pp. 201, line 10 - 202, line 1. Cf. Aufrère, *L'Univers minéral dans la pensée égyptienne* 2, p. 765.

[550] Chassinat, *Dendara* 4, p. 158, line 6. For *Tfrr*, see Aufrère, *op. cit.* 2, p. 464.

[551] P. Louvre N 3148 recto, X, 10 (Pierret, *Études Égyptologiques*, p. 72). For the probable identity of this stone, see Aufrère, *RdE* 34 (1982-83), p. 16.

[552] Sauneron, *Rituel de l'embaumement*, p. 27, lines 5-6.

[553] Ibid., p. 29, line 10.

by the inundation, emerge as if created anew when the flood subsides.[554] Alternatively, the copula *tꜣy* at the end of line 8 may not have been sentence final, in which case the name of whatever was identified as the efflux of the water has been lost.

The 'efflux that comes forth from the Primaeval Ocean' (*ftyw nt pr n pꜣ Nwn*) which occurs at the beginning of line 9 is another name for the Nile inundation. Cf. an inscription from the propylon of the temple of Montu at Karnak, where the god Amun-Re is called *Nwn wr ḫpr m sp tpy it it.w n Ḫmny.w pr ṯꜣw m fnḏ=f Ḥʿpy m fd.t=f*, 'the great Primaeval Ocean who came into being at the first occasion, the father of the fathers of the Ogdoad, from whose nose the wind emerges, whose efflux is the inundation'.[555] Conceivably, lines 8-9 of the column under consideration may have said that the water reaches the minerals because X (now lost) is the efflux of the water, and the water in turn is the efflux of the Primaeval Ocean. Even if the words *tꜣ ftyw nt pr n pꜣ Nwn* in the latter line and *i-ir pꜣ mw pḥ r-r=s* in the former belong to two separate sentences, they occur in such close proximity that it is difficult to imagine there was no connection between them. However one is to restore the break at the end of line 8, whether as the beginning of a new sentence or as the continuation of the preceding one, I suspect the general sense of the passage is likely to have been the same: that the Primaeval Ocean, one way or another, is ultimately responsible for causing the water to reach the minerals.

Line 11 of Column II preserves the words *r tꜣy=s ir.t nꜣ ym.w* ⸢*tꜣy*⸣, '⸢it being⸣ its companion (of) the seas'. The identity of the subject of this clause is obscure to me. It is equally unclear to whom or what the suffix pronoun *s* of *tꜣy=s* refers. Line 12 mentions some hills (*sbt.wt*). Perhaps these are the same as the hills of stone (*sbt.wt n iny*) to which reference is made in line 12 of Column I of the present fragment. The words *iw=w pr n pꜣ Nwn*, 'when they emerge from the Primaeval Ocean', occur in line 13. Perhaps they refer to the hills' emergence from the waters of the inundation after it subsides.[556]

The Primaeval Ocean is mentioned again in lines 15 and 19 of Column II. In both cases, the context is badly preserved. Either the Ogdoad or Hermopolis figures in line 22.[557] The following line mentions Ptah, once again in a broken context. Given the emphasis in the present fragment on minerals, perhaps he appeared here in his capacity as patron of artisans who worked in metal and stone.[558]

[554] See commentary on fragment 13, II, 8-10.

[555] Sethe, *Urkunden* 8, p. 2 (1b). For *fd.t* = the inundation, see also Herbin, *Le Livre de parcourir l'éternité*, p. 253; Koemoth, *Osiris et les arbres*, pp. 1-2; *Wb.* 1, 582, 10. For the equivalence of that word and the noun *ftyw* used in our cosmology, see textual note (a) on II, 5 of the present fragment.

[556] Cf. discussion of lines 8-9 two paragraphs above.

[557] See textual note ad loc.

[558] For this, see Sandman Holmberg, *The God Ptah*, pp. 45-6; Aufrère, *L'Univers minéral dans la pensée égyptienne* 1, p. 315; 2, pp. 365-6.

GROUP 2: FRAGMENTS 15-29

FRAGMENT 15 Plate 10

Dimensions: 6.9 × 6.7 cm. Colour: light brown. Related fragments: No. 2.

Transliteration

x+1.]. *mtw=f ḥwy* ..
x+2.]. *b-ir s-ḥm.t* ..
x+3.]. *Ptḥ sty*
x+4.].. [*pꜣ*]*y=f*(?) [*s*]*št*
x+5.].... *nb iw=f* ..*m*(?)
x+6.] *nꜣ swḥ*[*y*].*w r-ḥwy=f*
x+7.]........
x+8. *P*]*tḥ* [...]....[...]

Translation

x+1.]. that he might deposit ..
x+2.]. A female cannot ..
x+3.]. Ptah scatter
x+4.].. [h]is(?) [f]orm
x+5.] every .... while he ..*m*(?)
x+6.] the eg[g]s which he deposited
x+7.]........
x+8. P]tah [...]....[...]

Textual notes

Line x+1

With the traces of *ḥwy*, 'deposit', here, cf. the better preserved occurrence of the same verb in line x+6 below. My translation assumes that reference is made to the same act in both cases. For *ḥwy* meaning 'deposit, place', rather than 'throw' elsewhere in demotic, see e.g. P. Harkness, VI, 2 (Logan in Johnson and Wente [eds.], *Studies in Honor of George R. Hughes*, p. 160); P. Spiegelberg (Petubastis), IV, 16 (Spiegelberg, *Der Sagenkreis des Königs Petubastis*, plate 4); P. Vindob. D. 6165, III, 47 (Hoffmann, *Ägypter und Amazonen*, plate 3); de Cenival, *Les Associations religieuses en Égypte*, p. 128.

Line x+3

With the traces of *Ptḥ*, 'Ptah', before *sty*, cf. better preserved examples of the same divine name in fragment 1, lines 3 and 18, fragment 2, lines 5, 9 and 15.

Line x+4

(a) Visible in the middle of the line are two short strokes followed by an *f*. Since the next word is a masculine singular noun, it is tempting to restore [*pꜣ*]*y*=*f* as in e.g. fragment 6, line 7.

(b) With the traces of *sšt*, '(secret) form', here, cf. those of the same word in P. Carlsberg 462, II, 4 (Smith in Frandsen and Ryholt [eds.], *The Carlsberg Papyri 3: A Miscellany of Demotic Texts and Studies*, plate 15). The initial *s* is missing in both examples.

Line x+6

Only the *y* of *swḥy*, 'egg', has been lost. Cf. the occurrence of that noun in fragment 4, II, 19.

Line x+7

The word at the beginning of the line appears to have a divine determinative. I am unable to suggest a reading.

Line x+8

With the traces of *Ptḥ*, 'Ptah', here, cf. the better preserved occurrences of that divine name cited in the note on line x+3 above.

Commentary

Line x+1 of this fragment preserves the words *mtw*=*f ḥwy*, 'that he might deposit'. Line x+6 refers to *nꜣ swḥ*[*y*].*w r-ḥwy*=*f*, 'the eg[g]s which he deposited'. The only male being mentioned in the fragment is the god Ptah, whose name occurs in lines x+3 and x+8. It seems probable, therefore, that he is the one who is described as depositing eggs.

In line x+3, Ptah's name is followed by the verb *sty*, 'scatter'. This recalls fragment 2, line 5, where the god is said to scatter his seed (*sty tꜣy*=*f mtwy*). He does so there in order to fertilise some objects which he has fashioned, from which the members of the Hermopolitan Ogdoad subsequently emerge. The noun identifying the objects in question has been lost in a break, but the context would seem to require that they be something like eggs or egg sacs. In view of the references to Ptah in conjunction with the deposit of eggs in the present fragment, it is tempting to see this as part of the same account, even though fragment 15 does not join directly with fragment 2. If so, then perhaps the words *b-ir s-ḥm.t*, 'a female cannot', in line x+2 of the former introduced a statement affirming the uniqueness of Ptah's mode of bringing the Ogdoad into existence. Not only did he fashion the eggs from which the eight deities emerged, he fertilised them as well, something which no woman would be able to do.

FRAGMENT 16 — Plate 11

Dimensions: 4.9 × 4.4 cm. Colour: dark/medium brown. Related fragments: No. 3.

Transliteration

x+1.]..[
x+2.]*y km*

x+3. *in*]*y*
x+4.].. *km ḥ.ṯ=f*
x+5.] *r* .[..] *šy*
x+6.]. *špš.w=f*

Translation

x+1.]..[
x+2.]*y* black
x+3. sto]ne
x+4.].. black before him/it
x+5.] to .[..] nose
x+6.]. his forelegs

Textual notes

Line x+1

The tall vertical stroke in this line could be a divine determinative.

Line x+2

The partially preserved word before *km*, 'black', ends with *y*, or possibly *e*, and is determined with the flesh sign. This could be any of several different words; the possibilities are too numerous to permit a certain restoration.

Line x+3

Probably restore [*in*]*y*, '[sto]ne', in this line. Cf. better preserved writings of that noun in fragment 10, line x+11, and fragment 14, II, 6.

Line x+6

For writings of *špš.w=f* identical with the present one, see fragment 5, II, 7, and fragment 19, line x+20. For the translation 'his forelegs' here, see commentary below.

Commentary

The adjective *km*, 'black', occurs in lines x+2 and x+4 of this fragment. Various bodily parts of a male being are mentioned ibid. and in lines x+5-6. In view of the contents, one probably has here a piece that originates from the same column of text as fragment 3, which preserves an encomium of the god Amun in the form of a black bull. Hence the translation of *špš.w=f* in line x+6 as 'his forelegs', rather than 'his arms' as elsewhere in the text. For the forelegs of a bull as an attribute worthy of admiration or praise, compare an inscription from the temple of Kom Ombo in which the god Haroeris is called *kꜣ mr ꜥḥꜣ ꜥbꜥ.tw n ḫpš.wy=f*, 'bull who loves combat, of whose forelegs one boasts'.[559] Similarly, in a text on the propylon of the temple of Montu at Karnak,

[559] De Morgan et al, *Kom Ombos* 2, p. 252, no. 885.

that god is called 'bull in Medamud, young bull with sharp horns ..., the one with mighty foreleg' (*sḫm ḫpš*).[560]

FRAGMENT 17 Plate 11

Dimensions: 4.2 × 3.2 cm. Colour: dark/medium brown. Related fragments: No. 3.

Transliteration

x+1.]...[
x+2.] *n nꜣy=f ir.t* [
x+3.].. *r ḥbs* [
x+4.]. *n* [

Translation

x+1.]...[
x+2.] of/in his eyes [
x+3.].. for a jubilee [
x+4.]. of/in [

Textual notes

Line x+1
The traces here are mostly illegible. The last sign appears to be a flesh determinative.

Line x+3
ḥbs with festival determinative is a demotic writing of *ḥb-sd*, 'jubilee'. For the loss of the final consonant, see *Glossar*, p. 299; Osing, *Hieratische Papyri aus Tebtunis* 1, p. 98.

Line x+4
All that is preserved in this line is a flesh determinative followed by an *n*. This could be either genitive, dative, or a writing of the earlier preposition *m*.

Commentary

This is a fragment mentioning the eyes of a male being (line x+2) and a sed-festival or jubilee (line x+3). It too is probably part of the same column of text as fragment 3, describing Amun in taurine form. The bull is a royal symbol.[561] It is appropriate, therefore, that a sed-festival should be mentioned in connection with a god who appears in this aspect. Compare a hymn addressed to Amun by Thoth from the former's

[560] Sethe, *Urkunden* 8, p. 4 (5b).
[561] See commentary on fragment 3 above

temple at Hibis, in which that deity is called *k3 wr*, 'great bull', and is said to 'repeat millions of sed-festivals' (*wḥm ḥḥ.wt m ḥb-sd*).[562]

FRAGMENT 18 Plate 11

Dimensions: 2.2 × 4.5 cm. Colour: medium brown. Related fragments: No. 3.

Transliteration

x+1. *sḥ]t st* [
x+2. *r]3(?) p3* .[
x+3.]*lḫ* [
x+4.].[..]=*f ir.t* [
x+5. ⸢*b*].*w iw*=*f* .[

Translation

x+1. shin]ing is the tail [
x+2. mo]uth(?), the .[
x+3.]*lḫ* [
x+4.].[..] him, eye [
x+5. horn]s as he .[

Textual notes

Line x+1

With the traces of *sḥt* here, cf. better preserved examples of that word in fragment 3, lines 4, 14, and 21. For the writing of *st*, 'tail', which follows, see the second line just cited.

Line x+2

The traces at the beginning of the line could be of *r3*, 'mouth'. Cf. better preserved occurrences of that noun in fragments 5, II, 2, 18; 10, line x+5; and 19, line x+12.

Line x+3

The initial sign in this line looks more like *l* than *i*. For the combination *l*+*ḫ* here, cf. the writing of *plḫ*, 'spread out', in fragment 13, II, 18. That verb cannot be restored here, since it is determined with the strong arm sign, while the trace visible after *ḫ* in this line is of a tall vertical sign.

Line x+4

Perhaps restore [*n3y*]=*f* before *ir.t*, as in fragment 17, line x+2.

Line x+5

With the traces of ⸢*b.w*, 'horns', here, cf. better preserved writings of that noun in fragment 3, lines 8, 10 and 12.

[562] Davies, *The Temple of Hibis in El Khārgeh Oasis* 3, plate 23, south wall, lines 11-12.

Commentary

A tail (*st*) is mentioned in line x+1, horns (*ꜥb.w*) in line x+5. One has here part of a description of an animal with those attributes, most probably a bull. Fragment 3, which preserves an encomium of the god Amun as a black bull, mentions among other bodily parts both his horns and tail. This suggests that the present fragment probably belongs to that encomium as well. In at least three places, fragment 3 uses the word *sḥt*, 'shining, radiant', to describe specific features of Amun. That word can be restored at the beginning of line x+1 of the present fragment, which provides further support for the suggestion that the two pieces are related.

FRAGMENT 19 — Plate 11

Dimensions: 6.8 × 16.2 cm. Colour: medium brown. Related fragments: Nos. 4 and 5.

Transliteration

x+1.]..[
x+2.]*m*(?)[..].[
x+3. *ẖt*]*y*(?) *ḫn*[*ṱ*(?)
x+4.] *i͗-ir swsw* [
x+5.] *nꜣ* [*s*]*yw.w nꜣ nt* [
x+6.] *wꜣḥ* [..]. *ntr 4 pꜣy*(?) *hrw.w* .[
x+7. *s*]*ḥty pꜣ tꜣ t*[*r=f*
x+8.] *tꜣ n wyn* [
x+9.] *wꜣḥ=f nm*..[
x+10. *i͗*]*r=w nm ḏ/ir=f n=f tꜣ tw*[*ꜣ.t*
x+11. *ḏ*]*m šr*(?) *ṯsy=f* [
x+12.] *rꜣ=f pr=w ir*[*m*
x+13.].. *r-bnr-n rꜣ=f* .[
x+14.]. *nꜣ.w*(?) *ḥr ẖ.ṱ=f r nḥḥ ḏt* .[
x+15. *r-hn*]*-r pꜣ hrw irm pꜣ pr* ..[
x+16.] *tꜣ tr=f i-ir=w pr ḥr* [
x+17.].[..]*y nꜣ ꜥꜥe.w* [
x+18.]..*ḏy mw.w n* [
x+19. *p*]*r irm pꜣ ntr šbs* [
x+20.] *n špš* (vac) *w=f* [
x+21.]. *r-ḏb*[*ꜣ*(?)

Translation

x+1.]..[
x+2.]*m*(?)[..].[

x+3. go] downstream(?) and upstre[am(?)
x+4.] who lifted [
x+5.] the [s]tars, the ones which [
x+6.] lay down [..]. It is(?) four divinities. Days .[
x+7. il]luminate the en[tire] earth [
x+8.] earth with light [
x+9.] He has already *nm*..[
x+10.] A pole/tamarisk was [ma]de. He said to/made for him the under[world
x+11. gen]eration, a son(?). He mustered [
x+12.] his mouth. They emerged wi[th
x+13.].. out of his mouth .[
x+14.] They are(?) [..]. on his body forever and ever .[
x+15. up un]til today with the emergence ..[
x+16.] entire earth. They came forth (having) the head [
x+17.].[..]*y* the limbs [
x+18.]..*dy* water of/in [
x+19. eme]rge with the noble deity [
x+20.] of his arms [
x+21.]. becau[se of(?)

Textual notes

Line x+3

The traces near the end of the line could represent a semi-hieratic writing of *ḫnṭ*. Cf. the writing of *ḫnṭ*, 'go upstream', in P. Krall, VIII, 11 (Hoffmann, *Der Kampf um den Panzer des Inaros*, plate 8); also writings of the element *ḫnṭ* in the divine name *Ḥr-ḫnṭ-ḫty* in P. Carlsberg 465, lines 6, 11, and 12. What precedes these traces appears to be a hieratic boat determinative and, before that, a *y*. Perhaps one had here an instance of the compound expression *ẖt(y)-ḫnṭ*, 'go down- and upstream', for which see *Glossar*, p. 364.

Line x+4

For the verb *swsw*, 'raise' (= Coptic ⲥⲁⲥⲉ), elsewhere in demotic, see Vos, *The Apis Embalming Ritual*, p. 387. Compare the *swsw* with sun determinative that occurs in P. Leiden I 384, X, 20 and 34 (de Cenival, *Le Mythe de l'oeil du soleil*, plate 9). This is unlikely to be related.

Line x+5

With the traces of *syw.w*, 'stars', here, cf. the better preserved occurrence of that noun in fragment 5, I, 5.

Line x+6

The number four is written above the line. The sign immediately after *ntr* is perhaps a badly made *pꜣy*. For the writing of *hrw*, 'day', that follows, see e.g. P. Carlsberg 69, II, 7.

Line x+7

(a) With the traces of *sḥty*, 'illuminate', here, compare the better preserved example of that verb in fragment 47, line 5. This is written with the sun determinative rather than the brazier determinative found here. Cf. similar writings of *sḥt*, 'shining, radiant', in fragment 3, lines 4, 14, and 21.

(b) For better preserved examples of *tr=*, 'entire', see line x+16 below, fragment 13, II, 5.

Line x+8

Comparable writings of *wyn*, 'light', beginning with the *wn*-sign as in the example here, are cited in *Glossar*, p. 79. As Kim Ryholt has pointed out to me, the phrase *tꜣ n wyn* recurs in a damaged passage of P. Carlsberg 207 (x+I, 15). Cf. Tait in Frandsen (ed.), *The Carlsberg Papyri I: Demotic Texts from the Collection*, p. 22 and plate 3.

Line x+9

The traces after *nm* are illegible. However, this is clearly a different word to the one that occurs in the line below.

Line x+10

(a) Demotic *nm* written with wood determinative (not in *Glossar*) occurs several times in the Vienna Apis Ritual, both as the name of the tamarisk, equivalent to Coptic ⲛⲟⲙ (Crum, *CD*, p. 226a), and as a term for a pole made of the wood of that tree. See Vos, *The Apis Embalming Ritual*, pp. 80 and 360. It is not clear which of these two senses is to be preferred in the passage under consideration.

(b) The damaged verb after *nm* could be either *ḏ*, 'say', or *ir*, 'make'. If the former, then the diagonal stroke at the top of the word has been lost.

Line x+11

I am at a loss to explain the signs  which stand between *ḏm* and *ṯsy*. Could they represent *šr*, 'son'? For roughly comparable writings of that noun, see *Glossar*, p. 517. Alternatively, could one have here a crude dittography of the determinative of the preceding *ḏm*? The verb *ṯsy*, 'muster', is frequently used in connection with the latter. See Smith, *The Mortuary Texts of Papyrus BM 10507*, pp. 88-9 [note (b) on V, 18], and references cited there.

Line x+12

For the traces of *irm*, 'with', at the end of the line, cf. better preserved occurrences of that preposition in e.g. fragment 13, II, 2, 5, 7, 11, and 15.

Line x+13

The traces before *r-bnr-n*, 'out of', are of the plural determinative. For the signs in question, see e.g. the writing of *syw.w*, 'stars', in fragment 5, I, 5.

Line x+14

The two signs immediately before *ḥr ẖ.ṯ=f* are perhaps *nꜣ.w*, the plural copula. Cf. examples of that word in fragment 2, lines 10, 13, 14, and fragment 13, II, 5.

Line x+15

(a) For the restoration [*r-hn*]-*r pꜣ hrw*, '[up un]til today', here, cf. better preserved occurrences of that phrase in e.g. fragment 4, II, 15, and fragment 13, II, 5.

(b) For the traces of *pr*, 'emergence', cf. the better preserved examples of that word in line 12 above and line 16 below.

Line x+16

For the traces of *ḥr*, 'face' (here = 'head'), at the end of the line, cf. the better preserved example of that noun in fragment 2, line 8.

Line x+17

The word before *n3 ʿʿe.w*, 'the limbs', ends with *y* and the flesh determinative. Various restorations are possible, e.g. *ḫny*, 'skin' (cf. fragment 10, line x+10), or *šy*, 'nose' (cf. fragment 16, line x+5).

Line x+18

The word before *mw.w*, 'waters', ends with the consonants *ḏy* and a tall vertical stroke. The latter could be either a determinative or the plural marker.

Line x+19

(a) For the traces of *pr*, 'emerge', at the beginning of the line, see better preserved occurrences of that verb in lines x+12 and x+16 above.

(b) *šbs* (not in *Glossar*) is a variant writing of the adjective *šps*, 'noble', cited ibid., p. 503. For the substitution of *b* for *p* in this word, see *Wb.* 4, 445; Quack in Gestermann and Sternberg-El Hotabi (eds.), *Per Aspera ad Astra: Wolfgang Schenkel zum neunundfünfzigsten Geburtstag*, p. 114. Further demotic examples of *šbs* occur in fragment 41, line x+6, and P. Louvre E 3452, XI, 3 and 8, all written with divine determinative as in the present fragment. Elsewhere in the same Louvre papyrus, the noun *šps*, 'Art Krug', of *Wb.* 4, 451, 12, is written three times as *šbs*. Cf. ibid., X, 4, 7, and 9 (Legrain, *Le Livre des Transformations*, plate 10).

Line x+20

For writings of *špš.w=f*, 'his arms', identical with the one here, see fragments 5, II, 7, and 16, line x+6. The sign above the *f* is the plural marker. The scribe, for reasons that are unclear, left a blank space between this and the preceding flesh determinative.

Line x+21

The traces visible here appear to be a stroke followed by *r-ḏb3* or, alternatively, a stroke, an *h*, and another stroke. If the former, then perhaps restore [*r-hn-r p3 hr*]*w r-ḏb*[*3.ṯ=s*], '[up until toda]y becau[se of it]', or similar, as in fragment 4, II, 15.

Commentary

Line x+3 of this fragment preserves traces which could represent the end of the verb *ḫty*, 'go downstream', followed by the beginning of *ḫnṯ*, 'go upstream'. See textual note ad loc. The same pair of verbs occurs in a text on the propylon of the temple of Khonsu at Karnak which describes how the Ogdoad, having given birth to the solar deity in the *š3ʿ ʿ3* of Hermopolis, fared upstream (*ḫnt*) with him to Thebes so that he could receive the kingship of Amun, and then downstream (*ḫd*) to Memphis again, so that he could assume the throne of *Ỉr-t3*, 'Creator of the earth'.[563] Perhaps our text described a journey made by a primordial deity or deities as well. On the other hand, the verbs *ḫnt*

[563] See Sethe, *Urkunden* 8, p. 80 (95c); idem, *Amun und die acht Urgötter von Hermopolis*, pp. 52-3.

and *ḫd* are sometimes used to denote the movements of the sun as perceived by the Egyptians.[564] It is conceivable that our text also employed them in this manner.

The phrase *i-ir swsw*, 'who lifted', is visible in line x+4. This calls to mind the words *ṯsy=f*, 'he raised', which occur in fragment 4, II, 17. The latter, as I have suggested in the commentary ad loc., could refer to the act of elevating the sky to separate it from the earth, thereby bringing the cosmos into existence. Perhaps *i-ir swsw* in the present fragment has similar reference.

Line x+5 preserves the noun *syw.w*, 'stars'. These are mentioned in fragment 5, I, 5 as well, again in a damaged context. Perhaps the reference is to the creation of the stars, which necessarily follows that of the firmament.[565] Line x+6 mentions four gods whose identity is obscure. Compare the four serpents mentioned in fragment 30, line 7. The illumination of the earth, presumably by the sun, is the subject of lines x+7-8.

A *nm*, 'tamarisk tree', or possibly a pole of tamarisk wood, is said to have been made by some unspecified beings in line x+10. Its function is obscure. The tamarisk occupied a place of importance in the religious thought of the Fayyum, which attributed apotropaic powers to it.[566] There follows a sentence which could mean either 'He made the under[world] for him' or 'He said "the under[world]" to him.'[567] If the former is the correct rendering, then one can compare fragment 4, II, 18, which speaks of the creation (*ir*) of the sky, earth, and underworld, i.e. the coming into existence of the cosmos.

Line x+11 mentions a generation (*ḏm*) in conjunction with the words *ṯsy=f*, 'he mustered'. Perhaps the sense is that the being to whom the suffix pronoun *f* refers began to produce offspring.[568] In line x+12, *rꜣ=f*, 'his mouth', is preserved, followed by the phrase *pr=w ir*[*m*], 'they emerged wi[th]'. The following line has the words *r-bnr-n rꜣ=f*, 'out from his mouth'. One is reminded of fragment 5, I, 13 and II, 17, where there occur the phrases ⸢*r-bnr*⸣*-n rꜣ=f*, '⸢out⸣ from his mouth', and *r-bnr-n* ⸢*rꜣ*⸣ *pꜣ-Šy*, 'out from ⸢the mouth⸣ of Pshai'. (The former recurs ibid., II, 19.) In the commentary on those lines above, it was suggested that they could refer to the emergence of the gods from the mouth of Pshai, who as solar deity would be their creator. The same could be true of lines x+12-13 of the present fragment. If so, then perhaps line x+11 alludes to this event as well, with the generation that is said to be mustered there denoting the newly-created gods, and the masculine singular suffix pronoun *f* after *ṯsy* referring to their creator.

[564] See Derchain, *CdE* 64 (1989), p. 78; Thomas, *JEA* 42 (1956), pp. 77-8.

[565] See e.g. a hymn to Amun-Re from the temple of that god in Hibis, which describes how he elevated the sky above the earth and set the stars on their course (Davies, *The Temple of Hibis in El Khārgeh Oasis* 3, plate 32, middle, line 5).

[566] See Beinlich, *Das Buch vom Fayum*, p. 298. On this tree in general, see Chassinat, *Le Mystère d'Osiris au mois de Khoiak* 2, pp. 539-51; Germer in *Lexikon der Ägyptologie* 6 (1986), columns 193-4 (s.v. Tamariske).

[567] See textual note (b) on line x+10 above.

[568] See second reference cited in the textual note ad loc.

Line x+15 preserves part of an aetiological statement: [*r-hn*]*-r p3 hrw*, '[up un]til today'. On the function of such statements in the text, see Introduction. This is followed by the phrase *irm p3 pr*, 'with the emergence', perhaps a further reference to the emergence of the gods from the mouth of the solar deity, as discussed in the preceding paragraph. The same could be true of *i-ir=w pr ḥr* [...], 'they came forth (having) the head [...]', in the following line. This may have described the appearance of the newly emergent deities. The damaged noun ending in *y* and the words *n3 ʿʿe.w*, 'the limbs', in line x+17 could be part of such a description as well.

Line x+18 refers to some *mw.w*, 'waters', in an obscure context. The phrase [*p*]*r irm p3 ntr šbs*, '[eme]rge with the noble deity', occurs in line x+19. Once again, this could refer to the emergence of the gods from the mouth of the sun god. Cf. the phrase *pr=w ir*[*m*] that occurs in line x+12 above. It is of interest that at Hermopolis one of the names of the solar deity was *Šps*, 'the noble one'.[569] The words *n špš.w=f*, 'of his arms', may be read in line x+20. Cf. discussion of the phrase *tyly n špš.w=f*, 'raising of his arms', in the commentary on fragment 5, II, 7 above.

Attempting to fit the present fragment into the sequence established for the other fragments of our cosmology is complicated by the fact that no divine names are preserved in it. For reasons explained in the Intrduction, these can be of use in assigning a rough position to a fragment even when little of the context in which they occur has survived. Such words and phrases as can be read with certainty in fragment 19, however, are often paralleled by, or bear a close resemblance to, words and phrases in fragments 4 and 5, which suggests that there may be some connection between its contents and theirs. One might even consider, with all due reserve, assigning this fragment a position in between the other two, although it is clear from an examination of the papyrus fibres that it formed no part either of the second column of fragment 4 or of the first column of fragment 5. In the commentary on the latter fragment, the question was considered whether or not Thoth, whose creation and naming are described there, was said to have been the first-born offspring of the sun god by the author of our text. If fragment 19 actually preceded that fragment, and the reconstruction of its contents suggested above is correct, then it would seem that this question has to be answered in the negative.

FRAGMENT 20 Plate 11

Dimensions: 8.4 × 6.9 cm. Colour: light brown. Related fragments: No. 5.

Transliteration

1. *p3-Š*]*y*(?) [..].*=f tr=w w.*[
2.] *p3y* ... *r hb* [..]. *ḏ* .. *ḥ*[*f*].*w*(?) *h*[*b*

[569] See Westendorf in *Lexikon der Ägyptologie* 5 (1984), column 584; Sethe, *Amun und die acht Urgötter von Hermopolis*, pp. 50-1; Roeder, *Hermopolis 1929-1939*, p. 171.

3.] *tw3.t irm n3 nt ẖ[n(?)* ......]..[
4.]. *hb* [
5.]....[

Translation

1. Psh]ai(?) all of his [..]. *w*.[
2.] this ... to send [..]. say .. ser[pe]nts(?) se[nd(?)
3.] underworld together with those who are i[n(?)
4.]. ibis [
5.]....[

Textual notes

Line 1

(a) The traces near the beginning of the line could be of *p3-Šy*, 'Pshai'. Cf. writings of that divine name in e.g. fragments 4, II, 20, 22; and 5, I, 21, and II, 10.

(b) For the traces of *tr=w* here, cf. fragment 4, II, 12, fragment 7, line x+7.

(c) Near the end of the line read *w* and not *ir=w*. In writings of the latter, the plural stroke normally extends below the bottom of the line. Cf. e.g. fragment 5, II, 9, fragment 19, line x+16.

Line 2

(a) I am unable to read the word after *p3y*. For the traces of *hb*, 'send', later on in the line, cf. the better preserved instance of that verb in fragment 5, II, 3.

(b) Near the end of the line, part of a serpent determinative and a plural stroke are visible. The trace before these, on the other side of the break, could represent *ḥ*. Perhaps restore *ḥ[f].w*, 'ser[pe]nts'. Cf. writings of that noun in e.g. fragments 14, I, 1, and II, 5; 26, line x+5; and 30, line 3.

(c) Only the *h* and part of the *b* of *hb* are preserved at the end of the line. The determinative is lost. It is not clear whether one had here the verb *hb*, 'send', as in the first part of the present line, or the noun *hb*, 'ibis', as in line 4 below.

Line 3

The traces after *nt* could represent *ẖn*, 'in'. Cf. better preserved examples of that preposition in e.g. fragment 2, line 8, fragment 4, II, 10, and fragment 5, I, 4.

Line 4

With the writing of *hb*, 'ibis', here, cf. that in fragment 5, I, 20 and II, 13. For the use of this noun in conjunction with the verb *hb*, 'send', in line 2, see commentary below.

Commentary

The divine name *p3-Šy*, 'Pshai', is perhaps to be read at the beginning of line 1; see textual note (a) ad loc. I have suggested that, in our text, this appellation serves to denote the solar deity during a specific period of his existence, starting with his initial manifestation in the created cosmos, and ending with his withdrawal from the earth to the sky.[570] If so, and if the restoration of the name Pshai in the present line is correct, then the fragment under consideration must have been concerned with an event or events which took place during that period. The verb *hb*, 'send', occurs in line 2; the noun *hb*, 'ibis', two lines later. The occurrence of these two homonyms in conjunction is paralleled in the columns of fragment 5. There, they are used in an account of the creation of the god Thoth by Pshai, probably in a play on words to explain how the ibis form of Thoth came into existence.[571] Perhaps fragment 20 preserves an additional part of the same account.

FRAGMENT 21 — Plate 11

Dimensions: 3.4 × 5.7 cm. Colour: medium brown. Related fragments: No. 5.

Transliteration

x+1.] *n=f mw r* [
x+2.]. *r3=f n* .[
x+3.].. *iw=f* ..[
x+4.] *r3 p3 ḥ*[*f*
x+5.]. *r3 2* .[
x+6.].. *p3* .[
x+7.]..[

Translation

x+1.] to/for him water at [
x+2.]. his mouth to .[
x+3.].. he ..[
x+4.] mouth of the serp[ent
x+5.]. two mouths .[
x+6.].. the .[
x+7.]..[

Textual notes

Line x+2

The damaged word after *r3=f n* at the end of this line begins with a long horizontal stroke. This could be part of the initial sign of a word like *mw*, 'water', or *in*, 'bring'.

[570] See commentary on fragment 12, II, x+10

[571] See commentary on fragment 5, II, 3 and 13.

Line x+3

The first two signs preserved here appear to be the divine and serpent determinatives. These are unlikely to represent the end of the name of the god *p3-Šy*, since the divine determinative of that name, when one is present, invariably follows its serpent determinative. See note (a) on fragment 7, line x+6.

Line x+4

(a) The initial sign of the writing of *r3*, 'mouth', in this line is different to the one in the writing that occurs in line x+2 above and elsewhere in the text. For parallels, see *Glossar*, p. 240. A third, hieratic, writing of this sign occurs in line x+5 below.

(b) For the restoration *ḥ*[*f*], 'serp[ent]', at the end of the line, cf. undamaged examples of that noun in e.g. fragments 14, I, 1, and II, 5; 30, lines 3 and 7; and 38, line x+5.

Line x+5

The first sign visible in this line is the flesh determinative of a word now lost. This is followed by a semi-hieratic writing of the noun *r3*, 'mouth'. Cf. note (a) on the preceding line. The tall upright sign after *r3*, which I have read as the numeral two, could also be an *s*.

Commentary

r3=f, 'his mouth', occurs in line x+2 of this fragment. Line x+4 mentions *r3 p3 ḥ*[*f*], 'the mouth of the serp[ent]'. In the following line, the noun *r3*, 'mouth', recurs, possibly followed by the number two.[572] The frequent references to this part of the body in the present fragment suggest that it might be related to fragment 5 which, as I have already indicated, may preserve part of an account of how the gods first emerged from the mouth of the solar deity Pshai.[573] The suffix pronoun *f* after *r3* in line x+2 could have reference to that god. Equally, the words *r3 p3 ḥ*[*f*] in line x+4 could be another way of saying 'the mouth of Pshai', who is regularly represented in serpent form.[574]

FRAGMENT 22 — Plate 11

Dimensions: 7.9 × 12.1 cm. Colour: medium brown. Related fragments: No. 6.

Transliteration

1.]=*f ẖn n p3 Nwn*
2.] *ntr 7 mtw p3y=w rn* ..
3.] *r b-ir rḫ=w p3 g3y*
4.] *t3 s*[*t*]*y ir* .[..]
5.]=*f smne=f t*[*3/n3* ..].

[572] See textual note ad loc.

[573] See commentary on fragment 5, I, 13; also that on fragment 19, lines x+11-13.

[574] See references cited in note 70 of the commentary on fragment 1.

6.] *ir tꜣ p.t* .[..].
7.]. *nw r-* (vac) *r=w šꜥ*
8.]..*ly* [..]..
9.]*y ir*[*m/iw*(?)]
10.] *nꜣ ḥ*[*f.w*]
11.] *r-ḏb*[*ꜣ* ..]
12.]..[...]
13.]..[...]

Translation

1.] him/it in the Primaeval Ocean
2.] seven gods whose names ..
3.] they not knowing the manner
4.] the fl[am]e, make .[..]
5.] him/it. He fixed th[e ..].
6.] create the sky .[..].
7.]. look at them until
8.]..*ly* [..]..
9.]*y* wi[th/come(?)]
10.] the ser[pents]
11.] becau[se of ..]
12.]..[...]
13.]..[...]

Textual notes

Line 1

With the traces of *ẖn*, 'in', here, cf. better preserved instances of that preposition in e.g. fragment 2, lines 2 and 8, fragment 4, II, 10. The combination *ẖn n* does not occur elsewhere in the papyrus.

Line 2

Better preserved examples of *rn*, 'name', occur in fragment 1, lines 11, 12, and fragment 2, lines 12, 13, 14, and 16, q.v.

Line 3

For the traces of *gꜣy*, 'manner', at the end of the line, see fragment 44, line x+9.

Line 4

Better preserved examples of *sty*, 'flame', occur in fragments 5, II, 4; and 30, line 5, q.v.

Line 5

The traces after *smne=f* could be either of the feminine singular definite article *tꜣ* or the plural article *nꜣ*.

Line 7

There is a blank space between the initial and final elements of *r-r=w*. Perhaps the scribe found the intervening papyrus surface unsuitable for writing.

Line 8

The partially preserved word at the beginning of the line ends with the consonants *ly* and is determined with the sail sign. Perhaps restore a verb like the *tyly*, 'rise up, raise', that occurs in fragment 5, II, 7. The sign visible at the end of the line is not the plant determinative, although one's initial impression might suggest otherwise.

Line 9

The damaged word at the beginning of the line ends in *y* and is determined with the house sign. Perhaps restore an adverb like *ẖly*, 'downwards' (cf. fragment 24, line x+7). The traces that follow could be part of the preposition *irm*, 'and', or the verb *iw*, 'come'. Cf. undamaged examples of the former in e.g. fragment 13, II, 5 and passim; and of the latter in fragment 1, line 10, and fragment 5, II, 6.

Line 10

With the traces of *ḥf.w*, 'serpents', here, cf. undamaged examples of the same noun in e.g. fragments 14, I, 1, and II, 10; 30, lines 3 and 7.

Line 11

For the restoration *r-ḏb*[*ꜣ*], 'becau[se of]', cf. undamaged writings of the same preposition in fragments 2, line 11; 4, II, 15; 12, I, 15; and 13, II, 15.

Commentary

The initial line of this fragment preserves the third person singular masculine suffix pronoun *f* followed by the phrase *ẖn n pꜣ Nwn*, 'in the Primaeval Ocean'. Line 2 mentions seven deities (*ntr 7*) whose names have not been preserved. The words *pꜣy=f 7*, 'his seven', occur in a broken context in line 7 of fragment 6. I have suggested that this might be a reference to the seven *ḏꜣis.w*, 'creative utterances', of the cow goddess Mehetweret, who are also said to be under the authority of Thoth. Their role there is perhaps that of protectors of the sun god from his enemies who have rebelled against him.[575] The words *ntr 7* in the fragment under consideration could refer to the seven *ḏꜣis.w* as well, in which case there is a good possibility that it and fragment 6 are related.

Line 4 mentions *tꜣ s*[*t*]*y*, 'the fl[am]e', again in a broken context. This could refer to the fiery eye of the sun god, who is sent against the rebels to punish them. Cf. commentary on fragment 5, II, 4. Line 5 speaks of a male being who establishes (*smne*) something, the name of which has not been preserved. Line 6 contains a reference to the creation (*ir*) of the sky, a subject which has already been treated in fragment 4, II, 18. Cf. commentary ad loc.

[575] See commentary ad loc.

Line 10 mentions some serpents (*ḥf.w*). For the occurrence of these in conjunction with a flame, see also fragments 30 and 31. If the present fragment is, in fact, part of an account narrating a rebellion against the sun god, then the reptiles in question could be either protectors of the solar deity or his enemies. Cf. the crocodiles (*msḥ.w*) which occur in fragments 6, line 2, and 7, line x+4. Just as the latter are familiars of the Primaeval Ocean, so serpents are familiars of the earth god Geb.[576]

FRAGMENT 23 Plate 12

Dimensions: 3.6 × 9 cm. Colour: medium brown. Related fragments: Nos. 6 and 7.

Transliteration

Column I

Trace of one line

Column II

1. *tyꜣ.*[
2. *tw=s* [
3. *pꜣ mꜣ* [
4. *r-r=w* [
5. *nꜣ l*[*y.t*(?)
6. *ḥr ḥꜣt* [
7. *n rn* [
8. *iw=f ḏ nṯr* [
9. ...[

Translation

Column I

Trace of one line

Column II

1. *tyꜣ.*[
2. behold [
3. the place [
4. to/concerning them [

[576] See references cited in note 287 of the commentary on fragment 6. For serpents as inimical beings, see commentary on fragment 1, line 16.

5. the s[ides(?)
6. on the heart [
7. as a name [
8. he says: god [
9. ...[

Textual notes

Column I

All that is visible in this column is a sign resembling the determinative of *msḥ*, 'crocodile', written opposite the fifth line of the adjoining column. For the determinative in question, cf. the occurrences of *msḥ* in fragments 6, line 2, and 7, line x+4.

Column II

Line 1

The traces in this line are of *ty3*, with the final *3* written over another low, broad sign. Compare similar writings of *3* in words like *3by*, 'thirst', in P. Carlsberg 68, leftmost fragment, line 3, and *3bẖ*, 'forget', in P. Carlsberg 69, II, 8. I am unable to suggest a translation.

Line 4

With the traces of *r-r=w*, 'to them', here, cf. better preserved examples of that preposition and suffix pronoun in fragments 5, I, 9, and 12, I, 16.

Line 5

The feminine noun at the beginning of the line has *l* as its initial consonant. Perhaps restore *l*[*y.t*], 's[ide]'. Cf. fragment 11, I, 8, P. Carlsberg 10029, line x+4.

Commentary

If the partially preserved sign in the initial column of this fragment is actually the determinative of *msḥ*, 'crocodile',[577] then perhaps there is some connection between it and fragments 6 and 7, which mention crocodiles as well. For the role of these reptiles in the narrative, cf. commentaries on the fragments in question.

FRAGMENT 24 — Plate 12

Dimensions: 6.1 × 11 cm. Colour: light brown. Related fragments: No. 10.

Transliteration

x+1.]. *p3 Nwn* [
x+2. *Nw*]*n* *ˁn i-ir=w* [
x+3.]... *ẖr-r=w* ...[

[577] See textual note ad loc. above.

x+4.].... ꜥ*n* ...[
x+5.]...=*f* ...[
x+6.] *smne=f pꜣ(?)* [..]..[
x+7.]. *ẖly n pꜣ ist* [
x+8.]. *iw pꜣy=w sk r nḥḥ ḏ*[*t*
x+9.].=*f smne pꜣ tw 2* ..[
x+10. *Ḫ*]*mnw irm S*..[
x+11.] *iw=f* .....[

Translation

x+1.]. the Primaeval Ocean [
x+2. the Primaeval] Ocean again. They [
x+3.]... under them ...[
x+4.].... again ...[
x+5.]... him/it ...[
x+6.] He established the(?) [..]..[
x+7.]. down to the ground [
x+8.]. their collection/assemblage being forever and ev[er
x+9.]. he establishing the two mountains ..[
x+10. H]ermopolis and *S*..[
x+11.] he .....[

Textual notes

Line x+1

For better preserved examples of *Nwn*, 'Primaeval Ocean', see e.g. fragment 2, line 13, and fragment 13, II, 4 and 7. The second half of the same divine name is preserved at the beginning of the line immediately following.

Line x+3

The damaged word immediately before *ẖr-r=w*, 'under them', is determined with the wood sign. I am unable to suggest a reading. The long horizontal stroke which passes through the determinative and extends below *ẖr-r=w* itself appears to be inadvertent.

Line x+4

The partially preserved divine name at the beginning of the line is determined with serpent and god signs. What stands before this appears to be a horizontal line bisected by a diagonal stroke. No reading suggests itself to me. At the end of the line is a sign shaped like an elongated *t*.

Line x+6

What I have read as *pꜣ* with a query could also be the top of an *f*. The traces near the end of the line bear some resemblance to *ḥry*, 'upper part'. Cf. writings of that word in the compound adverb *r-ḥry*, 'upwards', in e.g. fragment 10, line x+11, fragment 13, II, 17.

Line x+8

sk written with strong arm determinative would appear to be a variant of the *sq*, 'zusammenfügen, sammeln', of *Glossar*, p. 466. With the phrase *iw pꜣy=w sk nḥḥ ḏ*[*t*], cf. the [..] *pꜣy=w sk r Rꜥ iꜥḥ* which occurs in fragment 44, line x+11. For the use of the words *r Rꜥ iꜥḥ* there with the same sense as *r nḥḥ ḏt*, see textual note ad loc.

Line x+10

For the traces of *Ḫmnw*, 'Hermopolis', at the beginning of the line, cf. undamaged examples of that toponym in P. Carlsberg 10040, lines x+5, x+8, and x+10. All that can be read of the word that follows [*Ḫ*]*mnw irm* is its initial sign, an *s*. The coordination with [*Ḫ*]*mnw* suggests that it may have been a toponym as well.

Line x+11

The verb after *iw=f* is determined with the man with hand to mouth. Its initial sign is oval-shaped with two short projections at the right end. On the photograph, this appears to resemble the initial sign of *mr*, 'love'; on the original, however, it is clearly different. I am unable to suggest a reading.

Commentary

The name of the Primaeval Ocean occurs in lines x+1 and x+2. A male being whose name has not been preserved is said to have established (*smne*) something in line x+6. Perhaps this is the same being who is described as establishing the two mountains in line x+9. The latter expression (*pꜣ tw 2*) refers to the twin ranges which delimit the Nile valley to the east and west.[578] In non-demotic sources, these are generally designated with the dual form *ḏw.wy*.[579] Fragments 9 and 10 are concerned with the burial of the Ogdoad at Thebes. Line x+12 of the latter mentions the *tw n imnt*, 'mountain of the west', which is probably a designation for the specific site where those deities were interred.[580] The end of that line preserves part of an aetiological statement: *nt smne r-hn-*[*r pꜣ hrw*], 'which is established up un[til today]'. This could have reference either to the tomb of the Ogdoad or the mountain which concealed it. If the latter, then perhaps there is a connection with the present fragment, which describes the establishment of the western and eastern mountains in conjunction. The phrase *iw pꜣy=w sk r nḥḥ ḏ*[*t*], 'their collection/assemblage being forever and ev[er]', occurs in line x+8. The antecedent of the third person plural suffix pronoun has not been preserved. Could 'their' here refer to the eight buried deities? It is of interest that the toponym *Ḫmnw*, 'Hermopolis', occurs in line x+10, since this is the city most closely associated with the Ogdoad.

[578] See e.g. P. Krall, IX, 17-18 (Hoffmann, *Der Kampf um den Panzer des Inaros*, plate 9); P. Cairo 50142, line 22 (Spiegelberg, *CGC: Demotische Inschriften und Papyri*, plate 59).

[579] *Wb.* 5, 544, 14-15.

[580] See commentary ad loc.

FRAGMENT 25 Plate 12

Dimensions: 3.4 × 5 cm. Colour: medium brown. Related fragments: No. 12.

Transliteration

x+1. *n* ..[
x+2. *tꜣ iḫ*[*y.t*(?)
x+3. *tꜣy iw*(?) [
x+4. *i-ir nꜣ* .[
x+5. *n ꜣst* [
x+6. [..]*s*(?) [

Translation

x+1. of/to ..[
x+2. the hori[zon(?)
x+3. this, while(?) [
x+4. *i-ir* the .[
x+5. of/to Isis [
x+6. [..]*s*(?) [

Textual notes

Line x+2

The traces after *tꜣ* could be of *iḫy.t*, 'horizon'. Cf. the better preserved occurrence of that word in P. Carlsberg 462, II, 14 (Smith in Frandsen and Ryholt [eds.], *The Carlsberg Papyri 3: A Miscellany of Demotic Texts and Studies*, plate 15).

Line x+6

What I have read as *s* here could also be the man with hand to mouth determinative.

Commentary

Line x+2 may preserve the beginning of the noun *iḫy.t*, 'horizon'. Cf. textual note ad loc. The goddess Isis is mentioned in line x+5. The subject matter of this fragment may have been related to that of fragment 12, which concerns the birth of that goddess's son Horus. For Isis and the horizon mentioned in conjunction, see P. Carlsberg 462, II, 13-15 (Smith, loc. cit.).

FRAGMENT 26 Plate 12

Dimensions: 4.6 × 5 cm. Colour: light brown. Related fragments: No. 14.

Transliteration

x+1.]...[
x+2.] *ꜥꜣ.t nt ḫꜥ* [
x+3. *wt*]*ḥ*(?) *nt p*[*r*
x+4. *w*]*tḥ nꜣ m*[
x+5. *ỉw*]*f.w n nꜣ ḥf.*[*w*

Translation

x+1.]...[
x+2.] stone which appears [
x+3. vess]el(?) which comes [forth
x+4. sm]elt the *m*[
x+5. fle]sh of the serpent[s

Textual notes

Line x+3

(a) Visible at the beginning of the line are an *ḥ* and a metal determinative. Perhaps restore here the *wtḥ*, 'Gefäss, Schale', of *Glossar*, p. 107. This would provide an assonance with the verb *wtḥ* that occurs in the line immediately following.

(b) Only the first half of *pr*, 'come forth', is preserved. Cf. undamaged examples of that verb in e.g. fragments 5, I, 6; 13, II, 3; and 14, II, 9.

Line x+4

The traces at the beginning of the line are of the verb *wtḥ*, 'smelt' (*Glossar*, p. 107). Of the *w*, only the left-hand tip has been preserved.

Line x+5

Only the *f*, flesh determinative, and plural stroke of *ỉwf.w*, 'flesh', have been preserved. Cf. the undamaged writing of that noun in fragment 14, II, 7.

Commentary

Line x+2 refers to *ꜥꜣ.t nt ḫꜥ*, 'stone which appears'. Line x+5 mentions [*ỉw*]*f.w n nꜣ ḥf.*[*w*], '[fle]sh of the serpent[s]'. The fact that minerals and serpents' flesh occur here in conjunction makes it probable that this fragment should be associated with fragment 14, in particular the second column, which describes a pair of serpents whose flesh is said to be made of stone and from whose bodies minerals emanate. See commentary on the column in question. The reference to smelting (*wtḥ*) in line x+4 of the present fragment presumably relates to the treatment of the minerals after they have been mined or quarried.

FRAGMENT 27 Plate 12

Dimensions: 2.2 × 4.4 cm. Colour: light-medium brown. Related fragments: No. 14.

Transliteration

x+1.] *šꜥ* [
x+2.] *pꜣ mw ir* [
x+3.].=*f iw*=*f* .[
x+4.]. *sbt*.[*t*
x+5.] *mw.w*(?) *nt ẖ*[*n*
x+6.] *ms* [

Translation

x+1.] until [
x+2.] the water makes [
x+3.]. him/it, while he .[
x+4.]. hil[l
x+5.] waters(?) which are i[n
x+6.] give birth [

Textual notes

Line x+1

For the writing of *šꜥ*, 'until', here, cf. better preserved examples of that preposition in fragments 2, line 17; 11, I, 1; and 14, I, 22.

Line x+3

All that is preserved of the word before the suffix pronoun *f* at the beginning of the line is a man with hand to mouth determinative.

Line x+4

With the traces of *sbt.t*, 'hill', cf. better preserved occurrences of that noun in fragments 13, II, 8; 14, I, 12, and II, 12.

Line x+5

(a) The signs at the beginning of the line could be *mw.w*, 'waters', or, alternatively, the determinative of a longer word denoting a liquid of some type followed by a plural stroke.

(b) For the traces of *ẖn*, 'in', at the end of the line, cf. better preserved occurrences of that preposition in fragments 2, line 8; 4, II, 10; and 5, I, 4.

Commentary

mw, 'water', is mentioned in line x+2 of this fragment, possibly in line x+5 as well. See textual note (a) on the latter. The noun *sbt.t*, 'hill', occurs in a broken context in line x+4. Cf. fragment 14, I, 12, and II, 12 for further references to hills. Column II, lines 7-9 of that fragment describe how water reaches the hills and mountains. For the regenerative or fructifying effect which the water was believed to have on them, see commentary ad loc.; also commentary on fragment 13, II, 8-10. Perhaps the present fragment was originally part of this description, or another one like it.

FRAGMENT 28 Plate 12

Dimensions: 9 × 9 cm. Colour: light brown. Related fragments: No. 14.

Transliteration

x+1.] *n-im=w* [
x+2. *N*]*wn* [...] *tꜣ* .[
x+3.].[..]..[... *N*]*wn* [
x+4.]... *pꜣ N*[*w*]*n ꜥn* [..] *nꜣ wn*[*my*(?)
x+5.]*y ḥty=f i-ir* [...] *wnmy* [
x+6.]. *i-ir=w*(?) *ḏ n=f šꜥꜣ n r*[*n*
x+7. *N*]*wn* [...]... *ꜥn* [
x+8.]..[...].. *n rn ḏ* [
x+9. *Nw*]*n*(?) [*pꜣ*] *Nwn* .[
x+10.]... *pꜣ iny* [
x+11.]...[

Translation

x+1.] in them/there [
x+2. Prim]aeval Ocean [...] land .[
x+3.].[..]..[... Prim]aeval Ocean [
x+4.]... the Prim[aeval] Ocean as well/again [..] the *wn*[*my*(?)
x+5.]*y* his *ḥty i-ir* [...] *wnmy* [
x+6.]. They(?) called it 'sand' by na[me
x+7. Prim]aeval Ocean [...]... as well/again [
x+8.]..[...].. by name, saying [
x+9. Prim]aeval Ocean(?). [The] Primaeval Ocean .[
x+10.]... the stone [
x+11.]...[

Textual notes

Line x+1

n-im=w could be either the preposition *n* (< earlier *m*) + third person plural suffix pronoun or the adverb meaning 'there'. For the traces, cf. e.g. fragment 5, II, 9, fragment 11, I, 10.

Line x+2

For the traces of *Nwn*, 'Primaeval Ocean', in this and the next line, see better preserved examples of that divine name in fragment 2, line 13, and fragment 13, II, 7 and 9.

Line x+4

(a) For the traces of *Nwn* in the middle of the line, see references cited in the preceding note.

(b) Visible at the end of the line are the plural definite article *nꜣ* followed by a *w* and an *n*. This could be the beginning of the substantive *wnmy* which occurs in the next line, q.v.

Line x+5

(a) The partially preserved noun before *ḥty=f* ends in *y* and is determined with the flesh sign. Various restorations are possible, e.g. *ẖny*, 'skin' (cf. fragment 10, line x+10), or *šy*, 'nose' (cf. fragment 16, line x+5). *ḥty* itself has no determinative; I am unable to suggest a translation.

(b) *i-ir* after *ḥty=f* could be either the past relative converter or the marker of the second tense. Too little of the context can be made out to determine which with certainty.

(c) The determinative of *wnmy* at the end of the line has been lost, rendering its identification uncertain. Cf. perhaps the *wnm(y).t*, 'Festessen, Nahrung', of *Glossar*, p. 91.

Line x+6

(a) The traces before *ḏ* could be *i-ir=w*. Cf. writings of that verbal prefix in fragment 5, II, 9, and fragment 19, line x+16. Immediately before this, a man with hand to mouth determinative is clear on the original. The rest of the word to which it belongs has been lost.

(b) For the traces of *šʿꜣ*, 'sand', cf. better preserved writings of that noun in e.g. fragments 1, line 7; 6, line 6; 11, I, 10; and 14, I, 3.

(c) Restore *n r[n]*, 'by na[me]', at the end of the line. Cf. similar phrases like *ḏ=w n=f ꜣsy n rn* in fragment 1, line 12; *iw=* [*w*] *ḏ Šy n rn* in fragment 4, II, 22; and *ḏ=w n=f Ḏḥwty n rn* in fragment 5, II, 6.

Line x+7

For the traces of *Nwn*, 'Primaeval Ocean', near the beginning of the line, see better preserved examples of that divine name cited in the note on line x+2 above. I am unable to read what stands between this and *ʿn*, 'as well/again', at the end of the line.

Line x+9

(a) The traces immediately after the break near the middle of the line could represent the end of *Nwn*, 'Primaeval Ocean'. Cf. examples of that divine name cited in the note on line x+2 above. Alternatively, they could be the end of *p.t*, 'sky'. Compare better preserved instances of that noun in e.g. fragment 13, II, 2 and 21.

(b) The initial *n*, *w* and water determinative of *Nwn* near the end of the line are clearly visible even on the photograph. The rest of the word can be made out on the original. There is sufficient space to restore the definite article *pꜣ* before this divine name.

Line x+10

For the traces of *iny*, 'stone', here, cf. better preserved occurrences of that noun in fragment 10, line x+11, and fragment 14, II, 6. In the latter passage, *iny* is preceded by the definite article *pꜣ*, as here.

Commentary

This fragment preserves several instances of the name of the Primaeval Ocean, unfortunately all in a broken context. In line x+6 the phrase *i-ir=w(?) ḏ n=f šꜥꜣ n r[n]*, 'They(?) called it "sand" by name', occurs. Elsewhere in the text, sand figures as the material of which the primaeval mound, the first land to emerge from the water, is constituted.[581] Earlier, this had been the platform in the midst of the primordial flood on which the creator stood up in order to initiate the process of bringing the cosmos into existence.[582] The primaeval mound is specifically called *pꜣ šꜥꜣ ꜥꜣ*, 'the great sand', in our cosmology.[581] Therefore, the recipient of the simple name *šꜥꜣ*, 'sand', in the line under consideration is probably not that mound but something else, although what this may have been is uncertain. The words *pꜣ iny*, 'the stone', occur in line x+10. Stones figure very prominently in fragment 14, q.v. Perhaps the present fragment should be associated with that one.

FRAGMENT 29 — Plate 12

Dimensions: 2.3 × 1.7 cm. Colour: light brown. Related fragments: No. 14.

Transliteration

x+1.] *mtw=w gm.ṯ=w* [
x+2.] *mw pḥ r-r=w* [

Translation

x+1.] and they find them [
x+2.] water reaches them [

Textual notes

Line x+1

(a) For the traces of *mtw=w* at the beginning of the line, cf. the undamaged occurrence of the conjunctive marker + third person plural suffix pronoun in fragment 14, II, 1. Compare also the similarly written third person plural independent pronoun *mtw=w* in fragment 13, II, 4.

(b) For the writing of *gm*, 'find', here, see fragment 1, line 10, and fragment 11, I, 5. For the infinitival ending *ṯ* and following suffix pronoun *w*, cf. the writing of *ṯ.ṯ=w* in fragment 13, II, 6.

Line x+2

For the traces of *r-r=w*, see better preserved examples of that preposition and suffix pronoun in fragments 5, I, 9; 11, I, 5; and 12, I, 16. Perhaps restore something like [*i-ir pꜣ*] *mw pḥ r-r=w*. Cf. the similar phrase *i-ir pꜣ mw pḥ r-r=s* in fragment 14, II, 8.

[581] See commentary on fragment 6, line 6.
[582] See commentary on fragment 1, line 6.

Commentary

The words *mw pḥ r-r=w*, 'water reaches them', occur in a broken context in line x+2. As remarked in the textual note ad loc., this recalls the sentence *i-ir p3 mw pḥ r-r=s* in fragment 14, II, 8. There, the text is describing how water reaches the efflux of the flesh of the two serpents, which is another way of saying stones or minerals.[583] It seems probable, on the basis of the similarity between that sentence and the broken one in the line under consideration, that fragment 29 should be associated with fragment 14, more specifically with its second column. The plural object of the preposition *r* in the phrase *mw pḥ r-r=w* would then refer to stones or minerals, while the words *mtw=w gm.ṯ=w*, 'and they find them', in the preceding line could have reference to their discovery.

GROUP 3: FRAGMENTS 30-48

FRAGMENT 30 Plate 12

Dimensions: 6.7 × 8.6 cm. Colour: dark-medium brown. Related fragments: No. 31.

Transliteration

1.].. *r p3 ḥry* .....[
2.] *ir.ṯ=w n p3 nt pr* ..[
3.] *r wn n3 ḥf.w ir.*[*ṯ=w*
4.] *ḏ n=s Spt.t n rn r* [
5.]. *t3 sty p3y m-qty* [
6.].=*s iw=s mḥ iw=s ḫpr*(?) [
7.] *4 ḥf.w* [
8.]..*y*[

Translation

1.].. upwards .....[
2.] their eyes to the emergent one ..[
3.] when the serpents opened [their] eye[s
4.] called her Sothis by name *r* [
5.] It is [..]. the flame like [
6.]. her/it, burning and becoming(?) [

583 See commentary on the line in question.

7.] four serpents [
8.]..*y*[

Textual notes

Line 2

For the traces of *pr*, 'emerge', cf. better preserved examples of that verb in e.g. fragment 5, I, 6, and fragment 13, II, 3.

Line 3

For the restoration *ir.*[*ṯ=w*], '[their] eye[s]', see the better preserved occurrence of that noun and suffix pronoun in the preceding line.

Line 4

For the traces of *rn*, 'name', here, see e.g. fragment 2, lines 11, 12, 13, 14, 15, 16, fragment 4, II, 6 and 22. For the partially preserved naming formula [...] *ḏ n=s Spt.t n rn*, see discussion on the ordering of the fragments in the Introduction.

Line 5

The trace before *tꜣ sty*, 'the flame', could be the end of *n=s*, 'to/for her'. Cf. the writing of that preposition and suffix pronoun in the line above.

Line 6

As Kim Ryholt has pointed out to me, the sign which I have read as *ḫpr* with a query could also be the initial sign of *sṯ*, 'draw, conduct'. Cf. the writing of that verb in fragment 13, II, 14.

Line 7

The semi-hieratic writing of the numeral four here is different to the ordinary one that occurs in fragment 2, lines 8, 14 and 15. For the use of the construction *4 ḥf.w* rather than *4 ḥf*, see Smith, *The Liturgy of Opening the Mouth for Breathing*, Appendix, p. 78 [note on P. Strasbourg 3 verso, x+IV, x+5]. On the possible identity of the serpents in question, see commentary below.

Commentary

The words *ir.ṯ=w*, 'their eyes', occur in line 2 of this fragment. In the following line, reference is made to some serpents who are said to have opened their eyes. Presumably it is their eyes that are meant in line 2 as well. As we learn from line 7, the serpents in question are four in number. They are mentioned again in fragment 31, which almost certainly comes from the same column of text as the present fragment. Line x+2 of that piece probably contained a further reference to the serpents' opening their eyes. See textual note ad loc.

The identity of these reptiles is difficult to establish. In Egyptian religious thought, the serpent form is a general characteristic of primordial creator gods.[584] According to

[584] See references cited in note 408 of the commentary on fragment 13.

Sethe, the Ogdoad is said to be composed of four horned vipers, one embodying each constituent pair of that corporation, in a text on the propylon of the temple of Khonsu at Karnak.[585] In my view, this is incorrect, since the signs , which he translates 'die vier Hornvipern', are probably to be interpreted rather as a writing of *it it.w*, 'father of fathers', an epithet of the creator of the Hermopolitan deities. Likewise, the writing of the word Ogdoad with a hieroglyph depicting four rams' heads linked together, which Sethe sees as further evidence for a conception of these gods as a group of four serpents, serves rather to illustrate the idea that the Hermopolitan deities are an embodiment of Amun.[586] The solar deity was supposed to have been protected by four fiery serpents.[587] These could be the ones mentioned in our text, but there are no doubt other possibilities as well. The four serpents that figure in the present fragment could be identical with the four gods mentioned in line x+4 of fragment 19, q.v.

Line 4 describes the naming of the goddess Sothis: [...] *ḏ n=s Spt.t n rn*, '[...] called her "Sothis" by name'. In our text, naming is a concomitant of creation. Accordingly, the present fragment must have recounted how Sothis came into existence.[588] Other Egyptian sources provide little information regarding the circumstances in which this event occurred. I have found no other text that describes it in detail. According to Erichsen, there is an account of the birth of Sothis in the demotic cosmogony preserved in P. Berlin 13603.[589] However, Zauzich has demonstrated that what he reads as *ms Spt.t* and translates as 'Sothis-Geburt' is, in fact, a toponym, *Ms-ty*, the name of a locality north of Athribis in the delta.[590]

Line 5 preserves a reference to a flame (*sty*). In the following line, this is said to burn (*mḥ*). There is a flame mentioned in line x+3 of fragment 31 as well. Reference in both cases is almost certainly to Sothis, this goddess being noted for her fiery appearance in the sky.[591] One has a play on words here between *sty* and *Spt.t*, the latter having lost its medial *p* by the Graeco-Roman Period.[592] It might be imagined that the

[585] Sethe, *Amun und die acht Urgötter von Hermopolis*, pp. 43-4. For the text in question, see idem, *Urkunden* 8, p. 72 (87b).

[586] See commentary on fragment 3 above.

[587] See Assmann, *Ägyptische Hymnen und Gebete*, pp. 101-2.

[588] See discussion of the naming formula used here and others like it in the Introduction.

[589] The supposed references to this event occur in P. Berlin 13603, IV, 6 and 8. See Erichsen and Schott, *Fragmente memphitischer Theologie in demotischer Schrift*, pp. 316 (18) - 317 (19) and plate 5.

[590] Idem, unpublished lecture given at the Sixth International Congress of Demotic Studies, Cairo, 1996. For the toponym in question, see Grimal, *La Stèle triomphale de Pi(ʿankh)y au Musée du Caire, JE 48862 et 47086-47089*, p. 164, note 484.

[591] See Kákosy in *Lexikon der Ägyptologie* 5 (1984), column 1111 (s.v. Sothis).

[592] Kákosy, art. cit., column 1110; Bergman, *Isis-Seele und Osiris-Ei*, pp. 46-7, note 4.

text described how a bright light emanated from Sothis at her creation and the four serpents mentioned in lines 3 and 7 opened their eyes to behold it.

It is not easy to determine at what point in the narrative of our cosmology this event was actually recounted. The creation of Sothis could have been described in conjunction with that of the other stars, whose ruler she is sometimes said to be.[593] Alternatively, it could have been mentioned in the context of an account of the origin of the Nile inundation, inasmuch as the appearance of Sothis in the sky at the beginning of each year was believed to portend the annual flood.[594] Note that serpents (*ḥf.w*) and a flame (*sty*) are mentioned in conjunction in fragment 22 as well, which I have suggested might be part of a narrative describing a rebellion against the sun god.[595] The creation of Sothis may have been set in that context with the fiery goddess coming into existence in order to punish the solar deity's foes.[596] Normally this is done by deities like Sakhmet or Tefnut. Cf., however, an inscription from the pylon of the temple of Isis at Philae which recounts how the former burned the enemies of the sun god with her flame and rose up to the sky as *Nsr.t*, 'the fiery one'. The text concludes: *ḫpr rn=s pw n Spd.t*, 'Thus came into existence her name of Sothis.'[597]

FRAGMENT 31 — Plate 13

Dimensions: 3.4 × 3.6 cm. Colour: dark-medium brown. Related fragments: No. 30.

Transliteration

x+1.]..=*w* ⌊*n3*⌋ *ḥf.*[*w*
x+2. *w*]*n n3 ḥf.w* [*ir.ṯ=w*(?)
x+3. *s*]*ty w3ḥ r* .[
x+4.].. *r wn* [
x+5.] *n pḥ.ṯ* [

Translation

x+1.].. them ⌊the⌋ serpent[s
x+2.] The serpents [open]ed [their eyes(?)

[593] See e.g. Chassinat, *Dendara* 3, p. 35, line 9; Sauneron, *Esna* 2, p. 281 (163, 19); Gutbub, *Textes fondamentaux de la théologie de Kom Ombo*, p. 39, note (f). Elsewhere in the present text, the stars (*syw.w*) are mentioned in fragment 5, I, 5, and fragment 19, line x+5.

[594] See references cited in note 495 of the commentary on fragment 13. For further literature, cf. Sambin, *Studia Aegyptiaca* 11 (1988), p. 306, note 2. (The title of the first book cited there should be *Isis-Seele und Osiris-Ei*, not *Ich bin Isis*.)

[595] See commentary on that fragment.

[596] See commentary on fragment 5, II, 4.

[597] Junker, *Der grosse Pylon des Tempels der Isis in Philä*, Abb. 35 und p. 69, lines 6-9.

x+3. fl]ame set down to .[
x+4.].. while there opened [
x+5.] of/in the hindpart of [

Textual notes

Line x+1

The partially preserved word at the beginning of the line appears to end in *e* or *y*, and is determined with the man with hand to mouth.

Line x+2

For the trace of *wn*, 'open', before *nꜣ h̤f.w* (all that is preserved is the strong arm determinative), see fragment 30, line 3. Probably restore *ỉr.ṱ=w*, 'their eyes', after *h̤f.w* here, as in the fragment just cited.

Line x+3

With the traces of *sty*, 'flame', here, cf. the better preserved instance of the same word in fragment 30, line 5.

Line x+4

(a) The partially preserved word at the beginning of the line is determined with the pellet sign written above plural strokes. I am unable to suggest a restoration.

(b) The combination *r wn* also occurs in fragment 30, line 3, q.v. Perhaps restore [*nꜣ h̤f.w ỉr.ṱ=w*] after *wn* here, as in that fragment.

Line x+5

With the traces of *pḥ.ṱ*, 'hindpart', here, cf. the better preserved occurrences of the same noun in fragment 3, lines 3 and 5. It is impossible to be certain which suffix pronoun should be restored after this word.

Commentary

References to serpents (*h̤f.w*) occur in lines x+1 and x+2. A flame (*sty*) is mentioned in line x+3. This fragment is probably from the same column of text as fragment 30. Its contents have been discussed in the commentary on the latter, q.v.

FRAGMENT 32 Plate 13

Dimensions: 6.7 × 3.9 cm. Colour: dark-medium brown.

Transliteration

x+1.]..[
x+2.].[..].. *pꜣ mw* ..[
x+3.]. *nꜣ nt ẖn=w r-hn-*[*r pꜣ*] *hrw* [*r*]-*ḏb*[*ꜣ*
x+4.]. *ỉr.ṱ=w* ..[..]..*šn/št*.[

x+5.]*t3/n3/s* ....[
x+6.]. *r-ḏb*[*3* ....[
x+7.]..[

Translation

x+1.]..[
x+2.].[..].. the water ..[
x+3.]. those who are in them up un[til to]day [be]cau[se of
x+4.]. their eyes ..[..]..*šn/št.*[
x+5.] the/her/it ....[
x+6.]. becau[se of ....[
x+7.]..[

Textual notes

Line x+3

For the restoration *r-hn-*[*r p3*] *hrw* [*r*]-*ḏb*[*3*], 'up un[til to]day [be]cau[se of]', see the better preserved occurrence of the same phrase in fragment 4, II, 15. The *ṯ* and suffix pronoun *s* which follow *r-ḏb3* in that passage should probably be supplied here as well. Cf. note on line x+6 below.

Line x+4

(a) The damaged word after *ir.ṯ=w* appears to begin with *n, q, in*, or another low, broad sign written above a *b*.

(b) *š* written above *t* or *n* can be made out near the end of the line. It is not clear whether this is the beginning of a new word or the latter part of the same word as the illegible traces that precede.

Line x+5

The sign at the very beginning of this line could be an article, either *t3* or *n3*, or a pronoun *s*. The word that follows begins with either a *b* or *i* and is determined with the strong arm sign.

Line x+6

The vertical stroke visible before *r-ḏb*[*3*], 'becau[se of]', here could be the last sign of *hrw*, 'day'. Perhaps one has here the remains of the locution *r-hn-r p3 hrw r-ḏb3*, for which see note on line x+3 above.

Commentary

No coherent context can be recovered from this fragment. Aetiological statements occur in line x+3 and (possibly) line x+6. See textual note on the latter above. On the function of such statements, see Introduction. *ir.ṯ=w*, 'their eyes', is legible in line x+4. Cf. fragment 30, which speaks of serpents opening their eyes at the fiery creation of Sothis.

FRAGMENT 33 Plate 13

Dimensions: 5.1 × 3.3 cm. Colour: medium brown.

Transliteration

x+1.] *mḥt ḥr tꜣy=f ꜣꜥ.*[*t*

x+2. *w*]*sẖ* [..].*m*.[

x+3.].. *nb* [

Translation

x+1.] north upon his bac[k

x+2. h]all [..].*m*.[

x+3.].. every/lord [

Textual notes

Line x+1

(a) Writings of *mḥt*, 'north', identical with the one here occur in P. Carlsberg 462, II, 5 (Smith in Frandsen and Ryholt [eds.], *The Carlsberg Papyri 3: A Miscellany of Demotic Texts and Studies*, plate 15); and P. Carlsberg 10287, line x+2.

(b) For other writings of *ꜣt*, 'back', with ꜥ after the initial *ꜣ*, see *Glossar*, p. 12.

Line x+2

The damaged word at the beginning of the line is determined with the house sign. Cf. note on fragment 6, line 10. It is difficult to imagine any other suitable restoration apart from [*w*]*sẖ*, '[h]all'.

Line x+3

nb here could be either the adjective 'every' or the noun 'lord'.

Commentary

The noun *mḥt*, 'north', occurs in line x+1 of this fragment. The north plays an important role in some accounts which describe a rebellion against the sun god, where it serves as a place of refuge for that deity and his mother.[598] It is conceivable, therefore, that the present fragment might be linked in some way with the group of fragments (nos. 5-8) which I have suggested may be concerned with such a revolt. Another possible connection is with fragment 38, line x+2 of which mentions a female serpent of or in the north (*mḥt*) in an obscure context.

[598] See my discussion in Frandsen and Ryholt (eds.), *The Carlsberg Papyri 3: A Miscellany of Demotic Texts and Studies*, p. 105.

FRAGMENT 34 — Plate 13

Dimensions: 3 × 7.8 cm. Colour: dark-medium brown.

Transliteration

x+1.]...[
x+2.] *r* ...[
x+3. *pr*]-*ʿnḫ pꜣ sn*(?) [
x+4.] *pr r ir* ..[
x+5. *tꜣ*] *p.t pꜣ tꜣ tꜣ* [*twꜣ.t*
x+6.]*dw*.[
x+7.].. *nb*(?) [
x+8.] *ḥr* .[
x+9.]..[

Translation

x+1.]...[
x+2.] to ...[
x+3. house of] life, the brother(?) [
x+4.] come forth to do/make ..[
x+5. the] sky, the earth, and the [underworld
x+6.]*dw*.[
x+7.].. every/lord(?) [
x+8.] upon .[
x+9.]..[

Textual notes

Line x+3

For comparable writings of *pr-ʿnḫ*, 'house of life', see Spiegelberg, *Der demotische Text der Priesterdekrete von Kanopus und Memphis*, p. 125. All that is preserved of the present instance is the hieratic *ʿnḫ*-sign, and the house and writing determinatives.

Line x+4

(a) With the traces of *pr*, 'come forth', here, compare the better preserved occurrences of that verb in fragment 5, I, 6, and fragment 13, II, 3.

(b) All that is preserved of the word that follows *ir*, 'do, make', is a long horizontal stroke.

Line x+5

The restoration of *twꜣ.t*, 'underworld', at the end of the line is virtually certain, in view of the preceding references to the sky and earth followed by the feminine singular definite article *tꜣ*.

Line x+7

nb at the end of the line could be either the noun 'lord' or the adjective 'every'. I am unable to read the preceding traces.

Line x+8

All that is visible in this line is the preposition *ḥr*, 'upon', followed by a short diagonal stroke. It is conceivable that the latter extends downwards from the line immediately above.

Commentary

There is a reference to a *pr-ʿnḫ*, 'house of life', in line x+3 of this fragment, but the context is obscure. Line x+5 speaks of the sky, earth and (probably) the underworld. One could imagine that the text mentioned a work describing the creation of those elements of the cosmos which was said to be kept in the library of the aforesaid institution, perhaps even our cosmology itself or an earlier original from which it was copied.[599] One would expect to find a statement of this sort either in an introduction or a colophon.

FRAGMENT 35 — Plate 13

Dimensions: 6 × 6.3 cm. Colour: medium brown.

Transliteration

1.].*y pꜣ*(?) *ẖ*[
2.]. *nꜣy*=*w* .[
3.] *tꜣ twꜣ.t* [
4.] *ir*(?) ...[
5.].[...] *ḏ* [

Translation

1.].*y* the(?) *ẖ*[
2.]. their .[
3.] the underworld [
4.]make/do(?) ...[
5.].[...] say [

Textual notes

Line 1

The traces at the beginning of the line are a tall vertical sign and a *y* followed by the determinative used in words like *ẖꜣy*, 'high, long' (cf. fragment 3, lines 8, 10 and 14), and *qꜣ*, 'be

[599] On the nature of the house of life and the sorts of writings deposited there, see Weber in *Lexikon der Ägyptologie* 3 (1980), columns 954-7 (s.v. Lebenshaus).

exalted' (cf. ibid., line 5). I am unable to suggest a restoration. The initial trace would suit a word like *ṯsy*, 'high, exalted' (*Glossar*, pp. 670-1), but that is never written with this determinative, so far as I am aware.

Line 2

All that is preserved of the word before *n3y=w* is the walking legs determinative.

Line 4

Perhaps read *ir*, 'make, do', at the beginning of the line. The word which follows has the flesh determinative.

Commentary

This fragment mentions the underworld (*tw3.t*) in a context which is obscure.

FRAGMENT 36 — Plate 13

Dimensions: 4.3 × 5.2 cm. Colour: medium brown.

Transliteration

1. *r*]*ṯ̂=f fy* [
2. *p*]*r=f n-im=f mtw=* ⌊*f*⌋ [
3.].. *bnr m*[*t*]*w=f* [*p3 i*]*-ir ḏ* [
4.]. *r-r=w* .[
5.]..[

Translation

1.] his [fe]et bear [
2.] he will [co]me forth from it and ⌊he⌋ [
3.].. out. He is [the one w]ho said [
4.]. to/concerning them .[
5.]..[

Textual notes

Line x+2

With the traces of *pr*, 'come forth', here, compare the better preserved writings of the same verb in e.g. fragments 5, I, 6, and 13, II, 3.

Line x+3

(a) For the semi-hieratic writing of *bnr* at the beginning of the line, see *Glossar*, p. 118.

(b) The restoration [*p3 i*]*-ir ḏ* is conditioned more by the available space than by any other factor. On purely grammatical grounds, [*i*]*-ir ḏ* alone would be equally possible (see, for example, fragment 13, II, 19), but that would be insufficient to fill up the space after *m*[*t*]*w=f*.

Line x+4

With the traces of *r-r=w* in this line, compare the better preserved examples of this preposition and suffix pronoun in fragment 5, I, 9, and fragment 12, I, 16.

Commentary

Three of the five preserved lines of this fragment refer to an unidentifiable male being. The context is obscure.

FRAGMENT 37 Plate 13

Dimensions: 2 × 2 cm. Colour: dark-medium brown.

Transliteration

x+1.] *ir.t* [
x+2.]*yw* ..[

Translation

x+1.] eye [
x+2.]*yw* ..[

Commentary

The word *ir.t*, 'eye', is visible in line x+1. Nothing else can be read with certainty. Cf. fragments 3, 18, 30 and 32, all of which mention eyes as well.

FRAGMENT 38 Plate 13

Dimensions: 6.1 × 13.4 cm. Colour: medium brown.

Transliteration

x+1. *i*]*-ir=s/tꜣ/nꜣ* ..[..]
x+2. *ḥ*]*f.t n pꜣ mḥt* ..
x+3. *kk*]*y rn=f tꜣ ḥf.t*
x+4. *i*]*-ir=f ms pꜣ itn 2*
x+5.] *nw r pꜣ ḥf n pr-imnt*
x+6. *i*]*tn 2 n tꜣ twꜣ.t irm*
x+7.] *ḫpr ꜥn irm nꜣ.w qmꜣ*
x+8.] *ḏ rn=f ḫpr*
x+9.]. *wḫꜣ=f s*
x+10.].. *ꜥq=f*

x+11.].. *tw=f* ..
x+12.] *ẖr(?) lwš*
x+13.] *ḫpr ḏ i-ir(?)*
x+14. *p*].*t mḥt.t*
x+15. *p*].*t iw=f ti*
x+16.]....

Translation

x+1. *i*]-*ir* she/it/the ..[..]
x+2. female se]rpent of/in the north ..
x+3. dark]ness is his name. The female serpent
x+4.] He fashioned/brought forth the two disks
x+5.] saw the male serpent of/in the west
x+6.] two [di]sks of/in the underworld with
x+7.] happened again with those ones. Create
x+8.] said his name. There occurred
x+9.]. He desired her/it
x+10.].. He entered
x+11.].. He gave/caused ..
x+12.] filled with(?) concern
x+13.] happened, saying: *i-ir(?)*
x+14.] northern [sk]y
x+15. sk]y, he giving/causing
x+16.]....

Textual notes

Line x+1

[*i*]-*ir* at the beginning of the line is the marker of the second tense. What follows could be either the suffix pronoun *s*, or a definite article, *tꜣ* or *nꜣ*. The initial sign of the word after this appears to be ꜥ.

Line x+2

For the restoration [*ḥ*]*f.t*, '[female se]rpent', here, see the undamaged example of that noun in the line immediately below.

Line x+3

Only the *y* and determinatives of *kky*, 'darkness', are preserved at the beginning of the line. Cf. the undamaged example of that noun in fragment 4, II, 10.

Line x+4

The traces before *ms*, 'give birth, bear', are of *ir=f*. Cf. writings of that verb and suffix pronoun in fragment 4, II, 15 and 16. This must represent the last part of the second tense *i-ir=f*, since the simple past 'he bore' would have been expressed with *ms=f* rather than *ir=f ms*. The scribe wrote *i-ir=f* somewhat differently in e.g. fragment 2, line 1, and fragment 4, II, 3.

Line x+5

(a) With the traces of *nw*, 'see', at the beginning of the line, cf. better preserved instances of that verb in fragments 1, line 12; 7, line x+3; and 22, line 7.

(b) *pr* here is written for *pꜣ*; cf. *Glossar*, p. 127. For the traces of *pr-imnt*, 'west', see fragment 10, line x+12.

Line x+6

Restore [*i*]*tn*, '[di]sk', at the beginning of the line. Cf. line x+4 above.

Line x+9

Visible before *wḫꜣ=f* is a tall vertical stroke. This could be either a plural marker or the third person plural suffix pronoun.

Line x+12

With the sign that I have suggested might be *ẖr*, 'under, filled with', here, cf. the writing of that preposition in fragment 13, II, 20.

Line x+14

For the restoration [*p*].*t*, '[sk]y', here and in the following line, see undamaged writings of that noun in e.g. fragments 4, II, 4: 5, I, 3; and 13, II, 2 and 21.

Line x+16

The only word visible in this line is determined with the strong arm sign. I am unable to suggest a reading.

Commentary

A female serpent of or in the north is mentioned in lines x+2 and x+3 of this fragment. Cf. fragment 33, line x+1, where the same cardinal point figures in a broken context. Line x+3 also refers to a male being whose name is said to be *kky*, 'darkness'. This is clearly distinct from *Kk*, the member of the Hermopolitan Ogdoad, whose name has a quite different form in fragment 2, line 13. Cf. the distinction made elsewhere in the text between *pꜣ Nwn* 'the Primaeval Ocean', and *Nwn*, the member of the Ogdoad, as discussed in the commentary on fragment 2, line 16.

It is presumably this being of whom it is said in line x+4: [*i*]-*ir=f ms pꜣ itn 2*, 'He fashioned/brought forth the two disks.' The disks in question are not, as one might think, the sun and the moon, since in a subsequent line (x+6), they are described as the [*i*]*tn 2 n tꜣ twꜣ.t*, 'two [di]sks of the underworld'.

Line x+5 speaks of someone or something looking at a male serpent of or in the west. Perhaps it is to this reptile that the third person singular masculine suffix pronouns in lines x+3-4 refer. The mention of a male serpent in conjunction with a female serpent here recalls the two serpents who appear in fragment 14, I, 10, and II, 5, where they are described as the owners of the stones of the mountains. Cf. commentary ad loc. Possibly, the two serpents mentioned in the fragment under consideration are identical with that pair.

Further references to a male being occur in lines x+8-11 and x+15. Whether this is the serpent of or in the west is unclear. Line x+14 mentions the northern sky, line x+15 the sky, both in a context which is obscure. Conceivably, the female serpent of or in the north who appears in line x+2 above may have been connected in some way with the northern sky.

FRAGMENT 39 Plate 13

Dimensions: 5 × 6.3 cm. Colour: dark brown.

Transliteration

x+1.] *p3 mw* [
x+2.].. *n p3 N*[*wn*
x+3.] *sḏr ḥr* [
x+4.] *iw=f ṯ.ṯ=f ꜥn* [
x+5.]=*f ḥr n3 ẖ*(?)..[

Translation

x+1.] the water [
x+2.].. of/in the Pri[maeval Ocean
x+3.] lie upon [
x+4.] He will/if he takes it again [
x+5.] him/it on the *ẖ*(?)..[

Textual notes

Line x+2

The first sign visible is the serpent determinative. Probably one had here the noun *ḥf*, 'serpent', or a divine name written with the sign in question. For the restoration *N*[*wn*], 'Pri[maeval Ocean]', at the end of the line, see better preserved examples of that divine name in fragment 2, line 13, and fragment 13, II, 4 and 7.

Line x+3

With the traces of *ḥr*, 'upon', after *sḏr*, cf. better preserved examples of that preposition in e.g. line x+5 below, fragment 2, lines 1 and 4.

Line x+4

iw=f before *ṯ.ṯ=f* in this line could be either future or conditional. If it is the former, then probably restore *r* in the space before *ṯ.ṯ=f*.

Line x+5

I am unable to read the noun which follows *n3*. Its initial sign could be an *ẖ*.

Commentary

Too little of this fragment is preserved to permit the nature of its contents to be ascertained. The words *p3 mw*, 'the water', occur in line x+1. The following line mentions the Primaeval Ocean. Lines x+4-5 refer to an unidentifiable male being.

FRAGMENT 40 Plate 14

Dimensions: 2.9 × 6 cm. Colour: medium brown.

Transliteration

1.].. *p3-Šy(?)* [
2.] *n(?) n3 k3.w m*[
3.] *g3y n ḫpr* [
4.]*nw* .[
5.]...[

Translation

1.].. Pshai(?) [
2.] of/to(?) the bulls *m*[
3.] manner of coming into being [
4.]*nw* .[
5.]...[

Textual notes

Line 1

(a) The damaged word before *p3* appears to end with the divine determinative. I am unable to suggest a reading.

(b) A *š* and *y* are legible after *p3*. This could be part of the divine name *p3-Šy*, 'Pshai', as in e.g. fragments 4, II, 20, 22; and 5, II, 10. Alternatively one could have here the remains of a masculine noun like the *šym*, 'Stall, Gang', of *Glossar*, p. 486, or the *šyḫ*, 'Staub', of ibid., p. 487.

Line 2

k3, 'bull', is written here with a hieratic *k3*-sign. The sign in question has a slightly different appearance in writings of that noun in fragment 3, line 9, and fragment 13, II, 17, q.v. The following word, now lost, began with *m*.

Line 3

For the traces of *g3y*, 'manner', see the better preserved example of that noun in fragment 44, line x+9.

Line 4

The partially preserved word at the beginning of this line ends with *nw* and the man with hand to mouth determinative.

Line 5

A tall vertical stroke is visible in the middle of the line. The following sign is either a strong arm determinative or the beginning of the verb *tw*, 'cause, give'.

Commentary

The divine name *p3-Šy*, 'Pshai', is perhaps to be read in the initial line of this fragment. Cf. textual note (b) ad loc. If so, then it should belong with the other fragments in which the solar deity is designated by that name. See commentary on fragment 12, II, x+10. Some bulls (*k3.w*) are mentioned in line x+2. Compare the use of this term to denote the four male members of the Ogdoad, discussed in the commentary on fragment 3.

FRAGMENT 41 Plate 14

Dimensions: 4.3 × 7.7 cm. Colour: medium brown.

Transliteration

x+1.].. *r p3* ..[
x+2.]....[
x+3.]. *n3 mt.*[*w*] *ḫpr*(?) .[
x+4. *r-w*]*n-n3.w ḫn t3 sw*[
x+5.] *p3y i-ir=f ḏ.ṯ=s r-*[*ḏb3*(?)
x+6.] *p3 šbs i-ir s.*[
x+7.].. *trš m-qty* [

Translation

x+1.].. to the ..[
x+2.]....[
x+3.]. the thing[s] occur(?) .[
x+4. which] was in the *sw*[
x+5. He/it is [...]. He said it con[cerning(?)
x+6.] the noble one who *s.*[
x+7.].. red like [

Textual notes

Line x+2

This line consists mostly of illegible traces. The low horizontal sign immediately after the vertical stroke near the middle of the line is perhaps *wr*, 'great'.

Line x+3

For the traces of *mt*, 'thing', here, cf. the better preserved example of that word in fragment 5, II, 2. What follows could be *ḫpr*, 'become, occur'. Cf. writings of that verb in e.g. fragments 2, line 7; 4, II, 8; and 13, II, 2.

Line x+4

(a) For the traces of *wn-nꜣ.w* at the beginning of the line, cf. fragment 8, line x+9, P. Carlsberg 143, line x+7. In view of the following preposition *ẖn*, this must be part of the imperfect relative converter *r-wn-nꜣ.w*. Although I have translated '[which] was', alternative renderings like '[who] was' or '[which] were' are equally possible.

(b) At the end of the line, perhaps restore *sw[ḥy]*, 'eg[g]'. Cf. fragment 4, II, 19, and fragment 15, line x+6.

Line x+5

The trace at the end of the line could be part of *r-ḏbꜣ*, 'concerning'. Cf. undamaged occurrences of that preposition in e.g. fragments 2, line 11; 4, II, 15; 12, I, 15; and 13, II, 15.

Line x+6

For *šps*, 'noble', written *šbs*, as here, see note (b) on fragment 19, line x+19.

Line x+7

The partially preserved word before *trš* ends in *e* or *y* and is determined with the wood sign. For *trš* as a writing of *tšr*, 'red', see *Glossar*, p. 658.

Commentary

Reference is made to an unidentifiable male being in line x+5 of this fragment. Perhaps this is identical with *pꜣ šbs*, 'the noble one', who is mentioned in the following line. For *šbs* (< *šps*) as a name for the solar deity at Hermopolis, see commentary on fragment 19, line x+19. If the suggested restoration of *swḥy*, 'egg', at the end of line x+4 is correct (cf. textual note [b] ad loc.), then perhaps this fragment should be associated with fragment 4 which, I have argued, is likely to preserve part of a narrative concerning the emergence of the sun god from an egg. See commentary on II, 20-22 of the latter.

FRAGMENT 42 — Plate 14

Dimensions: 2.2 × 4.7 cm. Colour: medium brown.

Transliteration

x+1.]...[
x+2. *ir*]*m pꜣ kꜣ* ..[
x+3.]. *imnt* [
x+4.]....[
x+5.].[

Translation

x+1.]...[
x+2. wit]h the other ..[
x+3.]. west [

x+4.]....[
x+5.].[

Textual notes

Line x+2

For the traces of *irm*, 'with', at the beginning of the line, see e.g. fragment 13, II, 2, 5, 7, 11, and 15. For demotic writings of *ky*, 'another', as *kꜣ*, see fragment 11, I, 9; *Glossar*, p. 588.

Line x+3

With the traces of *imnt*, 'west', cf. those of the same noun in fragment 10, line x+12, and fragment 38, line x+5.

Line x+4

The first two traces in this line are totally illegible. What comes after them might conceivably be the verb *ḫpr*, 'become', followed by *ntr*, 'god'.

Commentary

This fragment mentions the west (*imnt*) in a context which is obscure.

FRAGMENT 43 — Plate 14

Dimensions: 4.4 × 5.3 cm. Colour: medium brown.

Transliteration

1.]... *ḏ tꜣ* [
2.]*y pꜣ wr*(?) [
3.]..[

Translation

1.]... say the [
2.]*y* the chief(?) [
3.]..[

Textual notes

Line 2

(a) The partially preserved word at the beginning of the line ends in *y* and is determined with the knife and metal signs. Perhaps restore a noun like *bnpy*, 'iron' (*Glossar*, p. 117) or *sfy*, 'sword' (ibid., p. 429).

(b) At the end of the line, after *pꜣ*, perhaps read *wr*, 'chief, great one'. Cf. writings of that noun cited in *Glossar*, p. 92; also P. Carlsberg 10287, line x+10, P. Carlsberg 10314, line x+3.

Commentary

No coherent context can be recovered from this fragment.

FRAGMENT 44 Plate 14

Dimensions: 7 × 9 cm. Colour: medium brown.

Transliteration

x+1.] *tꜣ ms[ḫny*
x+2.] *sty* ....[
x+3.]........[
x+4.]... *r-tw pꜣ* ..[
x+5.]..... *šp=s* .[
x+6.] *pꜣ m*[..].=*w n* ..[..].[
x+7.] *iw*(?) [..] *ṯs*[*y* ...] *r ḥry r* [
x+8.] *n sw* ...[...] *swty* [
x+9.]. *pꜣ gꜣy n* [...] *ḥ.t* .[
x+10.] *tꜣ msḫny n tꜣ/nꜣ* [
x+11.] *pꜣy=w sk r Rꜥ iꜥḥ* .[
x+12. *p*].*t*(?) *mtw=s ṯs*[*y*

Translation

x+1.] the birth [brick
x+2.] fire ....[
x+3.]........[
x+4.]... which there gave the ..[
x+5.]..... She received .[
x+6.] the *m*[..]. them in ..[..].[
x+7.] *iw*(?) [..] elev[ate ...] up to [
x+8.] of/as a star ...[...] withdraw(?) [
x+9.]. the manner of [...] front .[
x+10.] the birth brick of the [
x+11.] their collection/assemblage forever and ever .[
x+12. sk]y(?) and it elev[ates/is elevated

Textual notes

Line x+1

For the restoration *ms*[*ḫny*], 'birth [brick]', after *tꜣ*, cf. the undamaged example of that noun in line x+10 below. The word does not appear in *Glossar*. Cf., however, *Wb.* 2, 148, 6-14. As shown by the examples cited there, the range of its meanings is somewhat wider than the 'birth brick' given in my translation. *msḫny* can also signify the resting place of a deity, a temple, or even birth itself. The context of the occurrences in the present fragment is too poorly preserved to permit one to determine which of these meanings is the most appropriate here.

Line x+2

I am unable to read the traces which follow *sty*, 'fire'. The tall sign just before the end of the line is either an *s* or the man with hand to mouth determinative.

Line x+4

Visible before *r-tw* are the sun sign and a divine determinative. On their own, these could represent the name of the god Re. Alternatively, they could be part of another divine name incorporating those signs, e.g. *Ptḥ*, 'Ptah' (cf. fragment 2, lines 5, 9, and 15).

Line x+6

The verb to which the suffix pronoun *w* is attached is determined with the walking legs sign. Nothing else of it survives to permit an identification.

Line x+7

(a) The traces visible at the beginning of the line could be of the circumstantial converter *iw*. For the traces of *ṯsy*, 'elevate, raise', that follow, see line x+12 below, fragment 4, II, 17.

(b) For the writing of the adverb *r-ḥry* here, see e.g. fragment 10, line x+10, fragment 13, II, 17.

Line x+8

(a) The traces after *sw*, 'star', appear to be . The last sign looks like an *n*; the one before that could be *wn*. What precedes that sign looks something like a *ṯ*.

(b) The determinative of *swty* at the end of the line is lost. Compare the verb *swty* which occurs in P. Krall, XXII, 30 (Hoffmann, *Der Kampf um den Panzer des Inaros*, p. 369 and plate 22). There too the determinative is lost. Hoffmann (ibid., note 2261) interprets this as a writing of *sṯ* (< earlier *sṯꜣ*), 'turn, withdraw'. This verb is written in the normal fashion in fragment 13, II, 14. Alternatively, *swty* might be a variant writing of the verb *swṯe*, 'travel', which occurs in BM mummy board 35464, line 6. For that word, see Vittmann, *ZÄS* 117 (1990), p. 84 and plate 4.

Line x+9

(a) The tall vertical stroke before *pꜣ gꜣy* could be a suffix pronoun *w*, a plural marker, or part of a sign like the divine determinative.

(b) The trace after *ḥ.t* might be the *ṯ* which is attached to that word when it is followed by a suffix pronoun. If so, the pronoun in question is no longer preserved.

Line x+10

For *msḫny*, 'birth brick', see note on line x+1 above. Only the lower half of the article that follows *msḫny n* has been preserved. This could be either *tꜣ* or *nꜣ*.

Line x+11

For *sk*, 'collection, assemblage', see note on fragment 24, line x+8. With the phrase [..] *pꜣy=w sk r Rʿ iʿḥ* here, cf. the [..] *iw pꜣy=w sk r nḥḥ ḏ[t]* which occurs in that line. The parallelism of the two phrases suggests that *r Rʿ iʿḥ* is equivalent in meaning to *r nḥḥ ḏt*. Compare the hieroglyphic expression 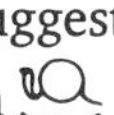 cited in *Wb.* 2, 402. Although this generally signifies 'every day', it can mean 'forever and ever' as well. Cf. Baines in Minas and Zeidler (eds.), *Aspekte spätägyptischer Kultur*, p. 25, note (d). The pair of signs in question is customarily read as *rꜥ nb*. Our text suggests that at least some Egyptian scribes interpreted them as *Rꜥ iꜥḥ* instead. Such an interpretation could have arisen from the Egyptian conception of eternity as a ceaseless cycle in which the sun and moon follow one another through the sky. Compare the statement of Horapollo: αἰῶνα σημαίνοντες ἥλιον καὶ σελήνην γράφουσι διὰ τὸ αἰώνια εἶναι στοιχεῖα, 'When they (scil. the Egyptians) represent eternity, they write the sun and the moon because they are the eternal elements', discussed in Drioton, *BSAC* 10 (1944), pp. 72-3; and Janssen, van der Walle, and Vergote, *CdE* 22 (1947), pp. 251-2.

Line x+12

(a) The traces at the beginning of this line could represent the last part of *p.t*, 'sky'. Cf. better preserved examples of that noun in e.g. fragments 4, II, 4; 5, I, 3; 13, II, 2 and 21. Another possibility is that one has here the end of *Nwn*, 'Primaeval Ocean'. For better preserved instances of that divine name, see fragments 2, line 13; and 13, II, 4 and 7.

(b) For the traces of *ṯsy*, 'elevate', here, cf. the better preserved example of that verb in fragment 4, II, 17.

Commentary

It is difficult to distinguish a single consistent theme in this fragment. A *msḫny*, 'birth brick', is mentioned in lines x+1 and x+10. Line x+2 refers to a *sty*, 'flame'; line x+8 to a star (*sw*). Flame and stars occur together in fragment 30 as well. There, the context is the fiery birth of the star Sothis.

Line x+11 describes the assemblage or collection of some beings or things as existing eternally. What these may have been is impossible to say, owing to the damaged state of the text. As mentioned in the textual note on that line, a similar description occurs in fragment 24, line x+8. In the commentary ad loc., I have suggested that the assemblage in question there might be that of the bodies of the members of the Ogdoad.

If the restoration [*p*].*t*, '[sk]y', at the beginning of line x+12 is correct (see textual note [a] on that line), then one could have in what follows a reference to the elevation (*ṯsy*) of the sky. For this event, see commentary on fragment 4, II, 17-18.

FRAGMENT 45 — Plate 14

Dimensions: 4.5 × 4.4 cm. Colour: medium brown.

Transliteration

1.].. *ỉhy*
2.]. [*tꜣ*] *p.t*
3.]..

Translation

1.]‥ *ỉhy*
2.]. [the] sky
3.]‥

Textual notes

Line 1

The initial *ỉ* of *ỉhy* is written twice, the second time as a simple vertical stroke. This curious orthography can be paralleled in other demotic texts from the Fayyum. Cf. e.g. , a writing of the divine name *Ỉḫs(s)f*, in P. Berlin 6750, VI, 16-18 (Spiegelberg, *Demotische Papyrus aus den königlichen Museen zu Berlin*, plate 79). For more normal writings of that name, see Beinlich, *Das Buch vom Fayum*, pp. 319-22. I am unable to suggest a translation for *ỉhy*. Its determinative resembles the seated person as it appears in some demotic texts of the Roman Period (cf. e.g. writings of *hy*, 'husband', cited in *Glossar*, p. 267). The seated person determinative in our text, however, looks quite different. Cf. the writing of *rmt*, 'man', in fragment 2, line 11. In my opinion, the determinative of *ỉhy* is more likely to be the hair sign which appears e.g. at the end of *km*, 'black', in fragment 3, lines 1 and 6.

Line 2

The tall vertical stroke at the beginning of the line could be either the suffix pronoun *w* or a plural marker.

Commentary

Too little of this fragment is preserved to permit the nature of its contents to be ascertained. The noun *p.t*, 'sky', occurs in line 2.

FRAGMENT 46 — Plate 14

Dimensions: 3.4 × 3.8 cm. Colour: medium brown.

Transliteration

x+1.].....[
x+2.]..*wg* .[..]..[
x+3.] *nꜣ* ..*y* [
x+4.]....[
x+5.]. *pꜣ*(?) ..[

Translation

x+1.].....[
x+2.]..*wg* .[..]..[

x+3.] the ..*y* [
x+4.]....[
x+5.]. the(?) ..[

Textual notes

Line x+2

The only signs which I can make out in this line are a *w* and a *g* followed by an evil determinative. It is tempting to restore a word like *swg*, 'töricht, dumm' (*Glossar*, p. 417), but the preceding traces do not really suit *s*.

Line x+3

The damaged noun after *nꜣ* ends in *y* and is determined with the pellet sign above plural strokes. I am unable to suggest an identification.

Line x+4

The tall vertical sign in the middle of this line is either an *s* or the man with hand to mouth determinative.

Commentary

No coherent context can be recovered from this fragment.

FRAGMENT 47 Plate 14

Dimensions: 2.3 × 11 cm. Colour: light brown.

Transliteration

1.].*ṯ*=*w ḫ*[
2.] *ꜥꜣ k*[*m*(?)
3.]...[
4. *S*]*bk* [..].[
5.]=*f sḥty* [
6.]. *Sbk nb* .[
7.]....=*f* [
8.]....[
9.]...[
10.]...[
11.]..[
12.] *st*(?)[

Translation

1.] of theirs *ḫ*[
2.] great bl[ack(?)

3.]...[
4. S]obek [..].[
5.] him illuminate [
6.]. Sobek lord of .[
7.].... him [
8.]....[
9.]...[
10.]...[
11.]..[
12.] *st*(?)[

Textual notes

Line 1

The partially preserved word at the beginning of the line is determined with the flesh sign. Perhaps restore something like [*ir*].*ṯ*=*w*, 'their [eye]s', or [*tr*].*ṯ*=*w*, 'their [hand]s'.

Line 2

(a) With the traces of *ꜥꜣ*, 'great', here, cf. the better preserved writing of that adjective in fragment 6, line 6. Compare also that of its feminine counterpart *ꜥꜣ.t* in fragment 1, line 8.

(b) The initial sign of the damaged word after *ꜥꜣ* is perhaps that of *km*, 'black'. Cf. better preserved examples of that adjective in fragment 3, lines 1, 6, and 9.

Line 4

With the traces of *Sbk* here, cf. the better preserved writing of that divine name in line 6 below.

Line 5

For the final *y* of *sḥty*, 'illuminate', here, cf. fragment 19, line x+7; *Glossar*, p. 450. For the sun determinative, see also writings of *sḥt*, 'shining, radiant', in fragment 3, lines 4 and 14.

Line 6

The trace after *nb* is presumably the beginning of a toponym identifying the city or locality of which Sobek is said to be lord. Given the provenance of our text, one might expect this to be Tebtunis. The name of that city can be written *Ḏbꜣ-tn* in demotic (cf. *Glossar*, p. 627), and the trace visible after *nb* here could well be the initial sign of such an orthography. Compare writings of *ḏbꜣ* in fragment 4, II, 15, and fragment 13, II, 15.

Line 8

The tall vertical sign immediately before the horizontal stroke at the end of the line could be the man with hand to mouth determinative.

Commentary

This fragment mentions the god Sobek twice, the only place where he figures in our cosmology. The context is obscure.

FRAGMENT 48 Plate 14

Dimensions: 2.1 × 3.7 cm. Colour: medium brown.

Transliteration

x+1.]....[
x+2.] *ḥry.t irm p3* .[
x+3.] *r pr* [
x+4.]. *p3 Nwn* [
x+5.] *p3 Nwn(?)* [

Translation

x+1.]....[
x+2.] upper, with the .[
x+3.] to go out [
x+4.]. the Primaeval Ocean [
x+5.] the Primaeval Ocean(?) [

Textual notes

Line x+1

The traces in this line might conceivably represent *r3*, 'mouth'. Cf. better preserved occurrences of that noun in fragments 5, II, 2, 18; 10, line x+5; and 19, line x+12.

Line x+2

For the writing of *ḥry.t* here, cf. e.g. fragment 1, line 9, and fragment 10, line x+10; also, P. Carlsberg 10029, line x+4. All three of these examples occur in the phrase *n t3 ly.t ḥry.t*, 'above', which perhaps should be restored here as well.

Line x+5

The three signs after *p3* appear to be *n*, *w*, and *n*, in that order. Cf. the better preserved writing of *Nwn* in the line immediately above. Here, however, there is also a diagonal stroke following the *p3*, which is difficult to reconcile with the initial consonant of that name.

Commentary

This fragment mentions the Primaeval Ocean in an obscure context.

PART III: SUMMARY

As reconstructed in the preceding pages, our cosmology presents a fairly coherent picture. The beginning of the text is now lost. Fragment 1 starts with a reference to a plant, concerning which it says: 'It will grow, after being cut off, up until today.' There follows an address by an unidentified group of beings, whether human or divine is unclear, to a god called *pꜣ-Šy pꜣ Nwn*, a name which could mean either 'Pshai the Primaeval Ocean' or 'Pshai (in) the Primaeval Ocean'.[600] The speakers make reference to various beneficial acts performed by this deity's son Ptah. Thereafter, the text changes from direct address to narrative, and begins to recount the story of how the cosmos came into existence.

According to this, *pꜣ-Šy pꜣ Nwn*, the initiator of creation, reaches a resting place of sand in the midst of the waters and stands up. Within the waters, there is also a plant, possibly the same one that was mentioned in the opening line of the fragment, which is said to grow into a large thicket. This is not motionless, but drifts through the water. Having originated in Elephantine, it floats northward until it comes to rest at a place which is perhaps to be identified as Hermopolis.[601] Upon coming to rest, the plant is given a new name, 'reed'. The god *pꜣ-Šy pꜣ Nwn* settles upon the newly stabilised reed thicket and scatters his seed.

A poorly preserved passage mentions some unnamed divinities and a serpent in a context which is obscure. Immediately afterwards, the god Ptah appears in the narrative for the first time. Since the address which begins the text identifies that deity as the son of *pꜣ-Šy pꜣ Nwn*, it is reasonable to suppose that the scattering of the latter god's seed was the catalyst for Ptah's coming into existence. The text goes on to describe how Ptah in turn settles upon the reed thicket. One may have here an account of the theophany of that god as the newly risen land, which emerges from the fluid element surrounding it at the point marked by the reeds.[602]

Fragment 2 begins with a reference to a town, the name of which is not preserved, and a further mention of the reed thicket. What follows is concerned with the creative activity of Ptah in his turn. This unfolds in three stages: first he takes thought, then he fashions some objects, the noun denoting which is unfortunately lost in a break, and finally he scatters his seed. The precise nature of the objects fashioned by Ptah is

[600] See commentary on fragment 1, line 2.
[601] See commentary on fragment 1, line 11.
[602] See commentary on fragment 1, lines 18-20.

impossible to determine. We learn later that they contain the members of the Hermopolitan Ogdoad. It is difficult, therefore, to avoid thinking of eggs or something similar.[603] The text states, somewhat cryptically, that the deity 'comes to be around' these enigmatic objects. This could mean, on the one hand, that he swallows them or otherwise ingests them into himself. Another possibility is that Ptah transforms himself into a liquid solution or medium in which the members of the Ogdoad have their latent existence.[604]

In due course, the eight Hermopolitan deities emerge from their place of concealment, the four males with the heads of frogs and the four females with the heads of serpents. Ptah names them with a play on words, calling the divinities 'our little images' (*nꜣy=n ẖm mny*), which gives rise to their designation as *nꜣ Ḫmnw*, 'the Ogdoad'. The names of the individual members of this corporation are then enumerated: Amun, Amaunet, Heh, Hauhet, Kek, Kauket, Nun, and Naunet.

After their initial appearance, these deities begin to merge together in a union 'according to name'. First the names of the four male members of the Ogdoad are said to meet, then those of the four female members. The end result of this process of merging is that all eight divinities are incorporated in a single one, Amun, who embodies the rest within himself. In fragment 3, this god is said to have the form of a black bull with eight hypostases or aspects, one for each of the Hermopolitan deities. The text extols his attributes in taurine form, sexual potency, might, ferocity, and so on.

Following this, in fragment 4, there is a reference to the four winds of the sky, which I have suggested are probably a further manifestation of Amun.[605] If my proposed reconstruction of the text is correct, these four winds merge into a single high wind, which brings the cosmos into existence by elevating the sky, thereby separating it from the earth, and forming or fertilising an egg from which the solar deity emerges.[606] This is none other than the primordial creator *pꜣ-Šy pꜣ Nwn* who was encountered at the very beginning of the narrative. Until this point, however, he has remained hidden within the depths of the Primaeval Ocean. But now he manifests himself in the created cosmos for the first time, an event which is marked by his assumption of a new name. Whereas before he has been called *pꜣ-Šy pꜣ Nwn*, 'Pshai the Primaeval Ocean' or 'Pshai (in) the Primaeval Ocean', now he is called simply Pshai.

This god, in turn, creates all the other divinities. A damaged passage in fragment 5 may state that these come forth from the sun god's mouth.[607] Particular attention is given to the creation of Thoth, after which event a conversation takes place, presumably between that deity and his maker, Pshai.

[603] See commentary on fragment 2, lines 5-6.

[604] See commentary on fragment 2, line 7.

[605] See commentary on fragment 4, Column II.

[606] See ibid.

[607] See commentary on fragment 5, I, 13.

At this point, only isolated words and phrases of the text can be made out. There are grounds for supposing, however, that our cosmology described the creation of Thoth as taking place in the middle of a rebellion against the sun god.[608] If so, then the ensuing narrative must have recounted how this uprising was opposed and ultimately suppressed. What is preserved of the relevant section of the manuscript (fragments 6-8) accords well with such an interpretation.[609] Deities who appear there include Thoth, *pꜣ Nwn*, 'the Primaeval Ocean', and the celestial cow goddess Mehetweret or Ahetweret, all of whom figure prominently in other accounts of rebellions against the sun god, as advisers or protectors. In some of those narratives, the solar deity is borne aloft into the sky at the conclusion of the revolt. With the sun thus raised to its proper position, the phase of the creation of the universe that is said to begin with the separation of the sky and earth comes to an end. Fragment 8 may have described this event as well.[610]

Fragments 9 and 10 recount how the eight Hermopolitan deities, having grown old, die and are buried in Thebes. It is Thoth who performs their burial rites, in his capacity as earthly deputy for the solar deity. Subsequently, in fragment 11, he too is described as entering the sky, presumably in his role as god of the moon. The ensuing text is badly preserved, but may have recounted how Ptah brought material sustenance into existence.[611]

Fragment 12 is concerned with the birth of Horus in the papyrus marshes of Khemmis in the delta and how he was concealed there by his mother Isis. Thoth, Sakhmet, Tefnut, and Nekhbet are among the other divinities who appear in this episode. The sun god is mentioned as well, but by a different name: Pre rather than Pshai. This could be the name which the god acquires after he has distanced himself from his creation and assumed his place in the sky.[612]

Subsequently, in fragment 13, the text moves from the past to the present, describing the activities of a deity whose name has not been preserved, but whom I have suggested is probably the sun god Pre.[613] He is said to ensure the well-being of all the other gods and rejuvenate them eternally. The text recounts how death came into existence, 'flying to every place which is under the heavens', and then uniting with life. Life is said to have emanated from the Primaeval Ocean, while death is an emanation of a being identified only as 'the serpent' (*syt*). The breath of life is what causes all things to live; the breath of death is what causes them to die. A third type of breath is then introduced: the breath of the Primaeval Ocean. This is what causes the Mediterranean Sea to come into existence, although the breath of death can quench it or dry it up.

[608] See commentary on fragment 5, II, 3-14.
[609] See commentaries on those fragments.
[610] See commentary ad loc.
[611] See commentary on fragment 11, I, 5-17.
[612] See commentary on fragment 12, II, x+10.
[613] See commentary on fragment 13, Column II.

The breath of the Primaeval Ocean is said to be responsible for renewing or regenerating hills and mountains, but it can also ruin them. It seems to be the case that whereas the breath of life animates or enlivens sentient beings, the breath of the Primaeval Ocean performs this service for non-sentient things like mountains or bodies of water. Moreover, unlike the breath of life, which is entirely beneficial, the power of the breath of the Primaeval Ocean is ambivalent - it can be productive or harmful. By contrast, the breath of death is always destructive, and exercises power over both sentient and non-sentient beings.

The text affirms that every living creature is subject to death, from which there is no return, including gods, human beings, animals, birds, fish, reptiles, and even plants. In this respect, it says, 'one is like another'. The powers of life and death are united in a single all-powerful deity, who is said to have taken possession of the earth. Although he is not named, this is probably the sun god, since it is he who is said to unite life and death in other Egyptian sources, the two being joined together in the unending daily cycle of rising and setting established by that deity after his ascent to the sky.[614]

This god is then said to enter the sound eye (= the full moon). In astronomical terms, full moon occurs when the sun and moon are in opposition. The Egyptians called this the union of the two bulls, an occasion when, according to them, the greater light of the sun entered and suffused the lunar disk. The god who enters the full moon is said to acquire all the powers and attributes of a lunar deity. As such he is given credit for bringing the Nile inundation up from the Primaeval Ocean, providing breath for all living things, causing animals to mate and reproduce, making seeds to germinate, and all plants to grow. A broken passage may allude to him regulating all the affairs of the underworld.[615]

Following this, in fragment 14, the text goes on to discuss stones and minerals. The stones of the mountains are said to belong to a pair of serpents. The identity of these is not made explicit. It is clear, nevertheless, that these are no ordinary snakes, but rather divine beings. Not only do the stones belong to the serpents, the bodies of those reptiles are themselves said to be made of stone, even though they are living things, and the minerals are said to be the emanation, literally the perspiration, of their flesh. The stones of the mountains grow and flourish because they are watered by the Nile inundation, which is itself said to be the emanation of the Primaeval Ocean. This recalls the passage in fragment 13, II, 8-10, where the breath of the Ocean is described as renewing or regenerating hills and mountains.

So much is preserved in the main sequence of fragments that make up our cosmology, those which I have identified as Group 1 in the Introduction. The fragments belonging to Group 2, each of which can be related to one or more of those in the first group, occasionally help to clarify or fill out this picture. Thus, as mentioned above, fragment 2 describes how Ptah creates some objects from which the Ogdoad emerge.

[614] See commentary on fragment 13, Column II.

[615] See commentary on fragment 13, II, 21.

The noun denoting these is lost in a break, but they are likely to have been eggs or egg sacs. The related fragment 15, which refers to eggs (*swḥy.w*) deposited by Ptah, helps to confirm this supposition. Similarly, fragments 16, 17 and 18, which are related to fragment 3, add further details to the description of Amun in the form of a black bull which is given in the latter. Fragment 21 may specify the form in which Pshai manifested himself when creating the other deities as having been that of a serpent, a fact omitted from, or not preserved in, the related fragment 5 which actually describes the gods' creation.[616] Additional instances of the sort of supplementary information which can be gleaned from the fragments in Group 2 are cited in the relevant commentaries.

The fragments of Group 3 have their own contribution to make, albeit of a somewhat different nature. Some of them recount events or episodes which are not mentioned in any of the fragments belonging to Group 1. However, because the fragments in the third group cannot be assigned a precise position within the overall sequence formed by those in the first, it is impossible to determine at what point in the longer narrative such events were said to have occurred. Examples of episodes in this category include the creation and naming of the goddess Sothis, recounted in fragment 30, and the fashioning of two disks in or belonging to the underworld by a male being whose name is said to be 'darkness' (*kky*), described in fragment 38. Other fragments in Group 3 mention deities or objects which occur nowhere else in our cosmology. Thus, the god Sobek appears uniquely in fragment 47; references to a birth brick (*msẖny*) occur in fragment 44. One can only speculate as to the part which these played in the narrative.

The problems and uncertainties that affect the reconstruction of our cosmology have been set out in some detail, both in the Introduction and in the commentaries on individual fragments. Even when two fragments can be placed in sequence, however, and the general order of the events recounted in them established, one faces further difficulties when seeking to ascertain precisely how these events unfold, owing to the gaps in the narrative between the points where one fragment ends and the next one begins. It is clear, for instance, that the text recounts how the eight members of the Ogdoad merge together and are incorporated in a single deity, Amun. The beginning of this process is described in fragment 2, its culmination in fragment 3. But what are the intermediate stages, after the initial unions involving, respectively, the four male Hermopolitan deities and their four female counterparts? How does Amun assume the taurine form attributed to him in fragment 3? Similarly, if I am correct in suggesting that the four winds mentioned at the beginning of fragment 4 are a further manifestation of Amun,[617] how does the god pass from his previous form into this one? Such questions cannot be answered with certainty.

[616] See commentary on fragment 21.

[617] See reference cited in note 605 above.

Many individual features of our cosmology have parallels in other ancient Egyptian sources. In the commentary, I have made frequent reference to these parallels, and without them large parts of the text would remain obscure. As might be expected, there are close affinities with other cosmogonical/cosmological works of the Graeco-Roman Period, for example, hieroglyphic inscriptions from the temples of Edfu, Esna, Dendera, Kom Ombo, and Karnak, as well as the account of the creation of the cosmos preserved in the demotic P. Berlin 13603. But there are also parallels with earlier sources, e.g. the Coffin Texts and the New Kingdom Book of the Heavenly Cow.

It is impossible to say when our cosmology was originally composed. I can see no compelling reason to think that it was much earlier than the date of the papyrus in which the text is preserved, i.e. the first half of the second century AD. Nevertheless, the account of the creation of the cosmos given in our papyrus is thoroughly Egyptian in character and conforms to the traditions established by earlier Egyptian works of the same genre. It is a synthetic or composite cosmogony according to the terminology employed by Sauneron and others, in the sense that it combines ideas and motifs drawn from a range of different 'local' traditions.[618] However, the diverse strands of material have been worked into a single organic whole, and the text would certainly have been seen as such by its readers in antiquity, as Sauneron rightly emphasises in connection with other ancient Egyptian cosmogonies.[619]

The papyrus inscribed with our cosmology originates from the Fayyum. One might well ask, therefore, whether the creation account preserved in it displays any distinctively Fayyumic characteristics. Two features are worthy of note in this regard. The first is the important role attributed to *pꜣ Nwn*, 'the Primaeval Ocean', in our text. The Ocean is mentioned by name thirty-five times in all,[620] far more frequently than any other divinity. Its waters were of particular significance in the Fayyum, since local tradition identified the region's most prominent natural feature, Lake Moeris, as a manifestation of them.[621] The Primaeval Ocean also plays an important part in the cosmogonical sections of the Book of the Fayyum, the chief repository of the area's indigenous mythological and theological lore.[622] Thus, the prominence accorded to the Ocean in our cosmology may be a reflection of the work's Fayyumic origin.

I have suggested above that some fragments of our text may preserve the remains of an account of a rebellion against the solar deity. If so, then this could be attributable

[618] See Sauneron, *Esna* 5, pp. 273-4; idem and Yoyotte, *La Naissance du monde*, pp. 32-4; Ryhiner, *L'Offrande du lotus*, pp. 212-13. On the validity of the concept 'local tradition', see commentary on fragment 2, lines 5-6, with particular reference to literature cited in note 89 there.

[619] See first reference cited in the preceding footnote.

[620] This figure includes only instances where the reading of the name is beyond doubt. For what may be additional occurrences, see textual note (a) on fragment 28, line x+9, and the textual note on fragment 45, line x+5.

[621] Beinlich, *Göttinger Miszellen* 100 (1987), pp. 15-18.

[622] Idem, *Das Buch vom Fayum*, pp. 287 and 312-14.

to its Fayyumic provenience as well. To be sure, such rebellion accounts have been recovered from many parts of Egypt, not just the Fayyum area. Nevertheless, this particular myth possessed a special significance there. Until the Graeco-Roman Period, the Fayyum was peripheral to the Nile Valley, both in terms of geography and of cosmogonical doctrine and theory. There was no tradition, for instance, that the sun god had been born in that region or that the primaeval mound arose there. In the local version of the myth of the rebellion against the solar deity, however, the Fayyum is the place of refuge to which the god withdraws in the face of his enemies. There too, he mounts upon the back of his mother, who ascends heavenward and transforms herself into the sky, thus completing the process whereby earth and heaven are separated.[623] In the Book of the Fayyum, the sun god's mother is said to be the female personification of Lake Moeris.[624] Therefore, heaven itself originates from that body of water. In the context of this myth, the Fayyum ceases to be peripheral and acquires an importance of the first rank. Accordingly, one finds frequent allusions to it in texts emanating from that region. Apart from the possible rebellion account in our text and the one in the Book of the Fayyum, there are several others preserved in demotic fragments of known Fayyumic provenience, most of them still unpublished.[625]

It will be remembered that Volten, in referring to the portion of our cosmology now in Copenhagen, described it as a scientific tract treating the primitive Ocean *Nwnw* and classified it in the category of 'hermetic literature'.[626] We are now in a position to consider how far he was justified in doing so. First of all, in what sense can our text be described as a scientific tract? Obviously, it is not scientific in the strict sense of the term, but I suspect that this is not really what Volten meant. Rather, he intended to indicate that the primary purpose of the text was didactic, to explain and provide detailed information about its particular subject. This may well be correct but, in my view, one cannot be absolutely certain. As we have seen, the greater part of the text is cast in the form of a narrative, with occasional passages of a descriptive nature, and this could have influenced Volten's perception of it. But it should be remembered that the Neith cosmogony from the temple of Esna is cast in the form of a narrative as well, yet this is explicitly identified as a ritual text, to be employed in the context of the feast

[623] For details, see Beinlich, *op. cit.*, pp. 314-19.

[624] Ibid., pp. 222-3.

[625] For a published specimen, see Smith in Frandsen and Ryholt (eds.), *The Carlsberg Papyri 3: A Miscellany of Demotic Texts and Studies*, pp. 95-112. Joachim Quack has drawn my attention to an unpublished fragment in the Beinecke Rare Book and Manuscript Library (P. CtYBR inv. 425), which describes Re hiding in the papyrus marshes from Apopis, just as Horus did when threatened by Seth. This is a further striking illustration of the parallelism which the Egyptians drew between the conflicts of Re and Horus with their respective enemies. See my remarks in the commentary on fragment 12.

[626] See second paragraph of the Introduction.

of the arrival of Neith-Mehetweret at her temple on the 13th of Epiphi.[627] Moreover, naming as a concomitant to the act of creation figures prominently in our cosmology.[628] As Finnestad has noted, calling things into being by uttering their names is chiefly characteristic of ritual contexts.[629] One cannot rule out the possibility, therefore, that our text had a ritual function, although this would not have lessened its value for didactic purposes.

A further point to bear in mind is that the work opens with a prayer or invocation to the deity responsible for initiating the process of creation.[630] In the commentary on fragment 1, I suggested the possibility that the ensuing narrative may have been intended for recitation by the same group who utter this initial prayer, glorifying the creator by means of an enumeration of his deeds. Comparable, in this respect, is the New Kingdom hymn to Amun-Re from Tura, which is also cast in the form of a narrative recounting how that god brought the world into being and, like our text, uses both the second and third persons in describing his activities.[631]

Both 'scientific' (in Volten's sense of the term) and ritual texts would have been kept in the same part of a temple: the *pr-ʿnḫ*, or 'house of life'. Fragment 34 of our cosmology actually mentions such an institution, unfortunately in a context which is obscure owing to damage. It is conceivable that this fragment once formed part of an introduction or colophon to our manuscript, specifying the location where it or the sources on which it was based were stored.

Volten identified the subject of our text as the Primaeval Ocean (*pꜣ Nwn*). His reasons for doing so are understandable. Attention has already been drawn to the important role played by the Ocean in the papyrus and the frequency with which it is mentioned there. Thus, one could well imagine that our text preserves the remains of an account of how the cosmos originated and continues to function which laid particular stress upon the part played by the Primaeval Ocean in those processes. The extant fragments of the manuscript are certain susceptible of being interpreted as a portion of such an account. Even in those fragments which make no explicit reference to the Ocean by name, a plausible connection with it is not far to seek. In fragment 3, for instance, which has as its subject the attributes and powers of the god Amun as a bull, this could have been the fact that he is sometimes equated with the Primaeval Ocean.[632] In the case of a fragment like no. 12, which deals with the birth of Horus in

[627] See Sauneron, *Esna* 5, pp. 247-9. The ritual aspect of this cosmogony is overlooked in the discussion of its grammar by Broze in *RdE* 44 (1993), pp. 3-9.

[628] See Introduction.

[629] Idem, *Image of the World and Symbol of the Creator*, pp. 68-74.

[630] The existence of the fragment which preserves this (no. 1) was unknown to Volten.

[631] See Bakir, *ASAE* 42 (1943), pp. 83-91; Assmann, *Re und Amun*, pp. 226-9.

[632] For this equation, see Zandee, *Hymnen aan Amon van Papyrus Leiden I 350*, pp. 101-2; van der Plas, *L'Hymne à la crue du Nil*, 1, p. 121; Assmann, *Re und Amun*, pp. 252 and 257; Leclant,

the papyrus marshes of Khemmis, it is less obvious what the connection could be. There, the Ocean might have played a part which is unrecorded in other accounts of that event.[633]

It is evident that, in our cosmology, the waters of the Primaeval Ocean are regarded as an active principle. What is more, the influence exercised by them on the development of the cosmos is not seen as having ceased after the initial stages of creation, but rather as continuing right up to the present. This is in marked contrast to the views expressed by some Egyptologists. Sauneron and Yoyotte, for instance, in their authoritative survey of ancient Egyptian cosmogonies, state that most such works ascribe to the Primaeval Ocean 'le rôle passif d'un berceau où s'élabora spontanément la première force active'. In other words, it is the place from which life emerged, but played no part in its creation.[634]

Our text, however, is not unique in presenting a quite different picture. A group of spells in the Coffin Texts portrays the Primaeval Ocean as working actively in conjunction with Atum to bring the cosmos into being.[635] In Spell 78, for instance, the Ocean is said to engender (*wtṯ*) the eight primordial chaos gods, while Atum is said to form them (*ir*).[636] A number of other sources likewise identify the Primaeval Ocean as the begetter or creator of the deities of the Ogdoad.[637] As a consequence, the Ocean may be described with some justice as the 'father of the gods',[638] and not merely by virtue of its 'antériorité chronologique'.[639] Coffin Text Spell 79 relates how the names of the chaos gods were created by the flesh of Atum *ḫft mdw Nww*, 'in accordance with the words of the Primaeval Ocean'.[640] Likewise, in Spell 76, those names are said to have been made by the former god *hrw pw mdw.n ꞽItm im=f ḥnꜥ Nww*, 'on that day when

Recherches sur les monuments thébains de la XXVe Dynastie dite Éthiopienne, pp. 240-2; Gabolde, *BIFAO* 95 (1995), p. 254. The names *ꞽImn-wr* and *Nwn-wr* occur together in an unclear context in the Book of the Fayyum. Cf. Beinlich, *Das Buch vom Fayum*, pp. 260-1 and 312-13. For Nun in the form of a bull, see Pécoil, *BSEG* 17 (1993), pp. 97-8; Leitz, *Tagewählerei*, p. 70.

[633] Cf. e.g. Chassinat, *Le Mammisi d'Edfou*, p. 55, line 9, where *Nwn* is mentioned obliquely in connection with Horus' birth. For that god's role elsewhere in the Horus-Seth-Osiris myth, cf. P. Salt 825, x+V, 4-5, which recounts the anger felt by the Primaeval Ocean when Seth murdered Horus' father Osiris, and how its waters rose up to conceal the latter's body from his enemy (Herbin, *BIFAO* 88 [1988], p. 103 and plate 6; Fischer-Elfert, *Göttinger Miszellen* 165 [1998], p. 106).

[634] See *La Naissance du monde*, p. 25.

[635] The spells in question are nos. 76-80 (de Buck, *The Egyptian Coffin Texts* 2, pp. 1-43). For discussion, see Pépin in Schoske (ed.), *Akten des Vierten Internationalen Ägyptologen Kongresses, München 1985*, vol. 3, pp. 339-40; Bickel, *La Cosmogonie égyptienne avant le Nouvel Empire*, pp. 46-9.

[636] De Buck, *The Egyptian Coffin Texts* 2, p. 20a.

[637] See references cited in footnote 99 of the commentary on fragment 2.

[638] For this title, see el-Banna, *BIFAO* 86 (1986), pp. 156-8.

[639] Thus Sauneron and Yoyotte, *La Naissance du monde*, p. 25.

[640] De Buck, *op. cit.*, pp. 23d-24a.

Atum spoke together with the Primaeval Ocean'.[641] Finally, in Spell 80, the Ocean acts as an adviser to Atum in the early stages of creation, counselling him about his children Maat and Shu.[642]

Closer in time to our cosmology, an inscription from the temple of Edfu speaks of the Primaeval Ocean as elevating or lifting up (*twꜣ*) the lotus in which the sun god Re is concealed, thereby allowing it to open and reveal the form of that deity.[643] An obscure passage in the Neith cosmogony at Esna may relate how the Ocean fertilised the solar egg from which Re ultimately emerged.[644] In Coffin Text Spell 648, the sun god credits its waters with having knit together (*ṯs*) his egg.[645] Thus, the active role ascribed to the Primaeval Ocean in our text is thoroughly in accordance with at least some earlier Egyptian traditions.

It is also maintained by some modern writers, again in contrast to our cosmology, that the Primaeval Ocean is only of importance at the very outset of creation. As Pépin has put it, its role 'cesse des l'apparition du démiurge'.[646] Thereafter, according to him, it is relegated to the status of 'deus otiosus'.[647]

Here too the Egyptian sources indicate otherwise. Since creation, in the Egyptian view, is cyclical and recurrent, not a single event that took place in the distant past,[648] the creative activity of the Primaeval Ocean must be the same. Thus, just as it is said to have elevated the lotus containing the sun god at the beginning of time in the Edfu text cited two paragraphs above, so too was it thought to raise him up into the sky each morning at the conclusion of his journey through the underworld.[649] Just as the Primaeval Ocean covered everything before the cosmos came into being, so too was the Nile inundation thought to flow forth from it each year and cover everything anew, thereby returning the world to its former state.[650]

In Egyptian mythology, the Primaeval Ocean certainly continues to play a wide variety of roles after the initial stages of creation. Reference has already been made in the commentary to the services rendered by it as adviser to Re in the myth of the rebellion against the sun god recounted in the Book of the Heavenly Cow.[651] Attention

[641] Ibid., p. 7e-f.

[642] Ibid., pp. 34g-35h.

[643] Chassinat, *Edfou* 6, p. 16, lines 5-7. Cf. Ryhiner, *L'Offrande du lotus*, pp. 146-7.

[644] Sauneron, *Esna* 3, p. 31 (206, 7-8). For the difficulties involved in the interpretation of this passage, see idem, *Esna* 5, pp. 262-4.

[645] De Buck, *The Egyptian Coffin Texts* 6, p. 270n.

[646] See second reference cited in note 635 above.

[647] Ibid., p. 344.

[648] See Finnestad, *Image of the World and Symbol of the Creator*, pp. 14 and 22-3.

[649] For representations of this act, see e.g. Hornung, *The Valley of the Kings*, pp. 85 (figure 52), 91, and 143 (figure 94).

[650] See commentary on fragment 13, II, 8-10.

[651] See commentary on fragment 5, II, 8.

has been drawn as well to another myth in which the Primaeval Ocean attempts to win the favour of a goddess with the aid of its crocodile familiars,[652] and to the prominent part which it has in various events narrated in the Book of the Fayyum.[653]

Further myths involving the Primaeval Ocean are recounted in the New Kingdom Calendar of Lucky and Unlucky Days. One describes how it had to be propitiated by a decree of Re-Atum so that it would emerge from its cavern in the form of an inundation of the correct height.[654] Another tells how the gods had to induce it from its underground lair with fire.[655] A third speaks of the going forth of the Ocean in order to set up the noble one (*šps*) in his place for the benefit of the gods who are in the latter's presence.[656] When Osiris was murdered by his brother Seth, the Primaeval Ocean protected his corpse, according to P. Salt 825.[657] A text of Ptolemaic date inscribed on the second pylon of the temple of Amun at Karnak narrates how, when the celestial cow was in distress after the birth of her child Re, the Primaeval Ocean summoned Amun in the form of the wind to support her.[658] The religious calendar preserved on the Ptolemaic gateway to the Mut precinct at Karnak connects the Ocean with the myth of the distant goddess, recounting how, on her journey back to Egypt, there was dancing and celebration, with *Nwn ḥr dwꜣ=s Šw ḥr wꜣš kꜣ=s*, 'the Primaeval Ocean extolling her and Shu paying homage to her ka'.[659] At Edfu, the Ocean is represented as a member of a corporation of deities called the *ḏꜣḏꜣ.t imy Nwn*, 'assembly which is in the Primaeval Ocean', one of whose functions is to bring forth the Nile inundation each year.[660]

Feasts were celebrated in honour of the Ocean[661] and hymns addressed to it.[662] Thus, it had a certain degree of importance in terms of cult. It could even be invoked in magical spells, where it was supposed to combat the harmful effects of poison.[663]

The evidence presented in the preceding paragraphs is far from being a complete dossier relative to the nature of the roles played by the Primaeval Ocean in the

652 See commentary on fragment 6, line 2.

653 See reference cited in note 622 above.

654 Leitz, *Tagewählerei*, pp. 147-9 and 406.

655 Ibid., pp. 137-9.

656 Ibid., pp. 89-90, where it is suggested that 'the noble one' could = the *ḏd*-pillar, i.e. Osiris.

657 See second reference cited in note 633 above.

658 Sethe, *Urkunden* 8, p. 105 (132k). Cf. idem, *Amun und die acht Urgötter von Hermopolis*, p. 93; Gutbub, *Textes fondamenteaux de la théologie de Kom Ombo*, p. 152, the latter citing a parallel from the temple of Esna (= Sauneron, *Esna* 2, p. 65 [25, 18]).

659 Sauneron, *La Porte ptolémaïque de l'enceinte de Mout à Karnak*, plate 9, line 26.

660 See Pécoil, *BSEG* 17 (1993), pp. 104-10.

661 See Leitz, *Tagewählerei*, pp. 107 and 436.

662 See Černý and Gardiner, *Hieratic Ostraca* 1, plate 93,1; Fischer-Elfert, *Literarische Ostraka der Ramessidenzeit in Übersetzung*, pp. 19-22.

663 Klasens, *A Magical Statue Base*, pp. 42-6, 61, and 101-9.

different spheres of Egyptian religious thought. It should suffice, however, to illustrate the point that it was in no sense regarded as a 'deus otiosus'. In this respect too, therefore, the picture of the Ocean presented in our cosmology accords well with what is found in other Egyptian texts.

Among the earlier sources surveyed above, the Coffin Texts lay particular stress upon the dynamic role played by the Primaeval Ocean in the process of creation, which is set in motion by the conversations which it initiates with the creator god Atum. As Bickel aptly notes: 'Le dieu Noun fut à l'origine de l'autogenèse d'Atoum par le fait qu'il lui adressa un discours qui permet à Atoum de prendre conscience et de se réaliser ...., le dieu Noun le stimule et l'incite à prendre conscience et par la même à inventer la forme d'existence de la vie.'[664]

There is a seeming contradiction here. How can one being initiate the autogenesis of another? Is it in fact autogenesis if one being comes into existence as a consequence of an act predicated of another? The answer would appear to be: only if the two beings are identical. This raises the possibility that, for the author(s) of the Coffin Text spells which record the dialogue between Atum and the Primaeval Ocean, the two, at least in the initial stage of creation, are simply different aspects of the same deity. If so, then the dialogue is an interior one, a colloquy with the self representing the initial act of a newly 'self'-conscious god. This would make the Primaeval Ocean not only the companion, confidante, and adviser of Atum, but his alter-ego as well.

Elsewhere in the Coffin Texts, there is evidence that the Egyptians were prepared to entertain such a view. In Spell 335, the primordial creator god, identified variously as Re or Atum, is made to say: *ink ꜥꜣ ḫpr ḏs=f*, 'I am the great one who came into existence by himself.' One version of the spell poses the question: 'Who is it, the great one who came into existence by himself?', and answers *mw.w Nww pw*, 'It is the waters of the Primaeval Ocean.'[665] As this passage reminds us, the creator of the universe is also the solar deity, and in fact the identification of that god and the Primaeval Ocean is attested as early as the Pyramid Texts, where Re is said to be the *ꜣgb wr*, 'great flood', in Spell 311.[666] The solar deity is similarly identified in the twelfth hour of the Amduat.[667] Among other New Kingdom underworld books, the Litany of Re calls the sun god the body (*ẖꜣ.t*) of the Primaeval Ocean,[668] while the Book of the Heavenly Cow states that he is the Ocean's ba.[669]

[664] Idem, *La Cosmogonie égyptienne avant le Nouvel Empire*, pp. 30-1.

[665] De Buck, *The Egyptian Coffin Texts* 4, p. 188b-c. Cf. Bickel, *op. cit.*, pp. 29 and 40-1.

[666] Sethe, *Die altägyptischen Pyramidentexte*, 1, p. 257 (499a). Cf. Volten, *MDAIK* 16 (1958), p. 364.

[667] Hornung, *Das Amduat* 1, p. 205; 2, p. 193; idem, *Texte zum Amduat* 3, p. 849.

[668] Hornung, *Das Buch der Anbetung des Re* 1, p. 25; 2, pp. 64 and 105, note 59. Cf. Roulin, *Le Livre de la Nuit*, p. 327.

[669] Hornung, *Der ägyptische Mythos von der Himmelskuh*, pp. 26 and 47.

In our cosmology, although great emphasis is placed upon the role of the Primaeval Ocean in the creation and maintenance of the universe, it is striking to note the important part attributed to the creator/solar deity as well. If my interpretation of the text is correct, this god is said to initiate the process of creation while still concealed in the Ocean's depths. At this stage, he is designated with the name *pꜣ-Šy pꜣ Nwn*.[670] Subsequently, he emerges from the watery abyss and manifests himself in the created cosmos for the first time, being revealed from within an egg. To mark this transition, he is given a new name, simply *pꜣ-Šy*, or 'Pshai'.[671] The newly emergent god then brings all the other divinities into existence.[672] A badly damaged section of the text may recount how a rebellion arose against Pshai, after the suppression of which he distanced himself from the world he had created by withdrawing to the sky.[673] It may be at this point that the solar deity acquires yet another name, *pꜣ-Rꜥ*, 'Pre', by which he is designated in a subsequent fragment.[674] When our text moves from narration of past events to a description of phenomena which occur in the present, it recounts vividly how the sun god exercises supreme dominion over the entire cosmos, uniting the powers of life and death in his person, and regulating the affairs of all that he created.[675]

Given the emphasis upon the solar deity in our cosmology, one might be tempted to amend Volten's statement that its subject is the Primaeval Ocean and say instead that it has two main subjects, the Ocean and the creator/sun god. Perhaps so, but it should be remembered that the name *pꜣ-Šy pꜣ Nwn* by which the latter is designated in the opening sections of our cosmology is ambiguous: it could be translated either 'Pshai the Primaeval Ocean' or 'Pshai (in) the Primaeval Ocean'.[676] If the former is correct, then for the author of our text, the creator god and the Ocean were one and the same at the initial stage of creation. As we have seen, this is a view which can be traced back as early as the Pyramid Texts, and is attested in Middle and New Kingdom sources as well. In the commentary on fragment 1, I cited a demotic magical spell on an ostracon now in Strasbourg which invokes Pshai and the Ocean as if they were a single being.[676] This illustrates that the conception of the creator and the waters which concealed him as one entity survived into the Graeco-Roman Period. Our cosmology may be a further attestation of it. If so, then Volten's statement does not require amending so much as it needs to be interpreted with an additional nuance: the chief subject of the text is the Primaeval Ocean in all its aspects, but one of the most important of these aspects is its

[670] See commentary on fragment 1, lines 2 and 6.
[671] See commentary on fragment 4, II, 22.
[672] See commentary on fragment 5, I, 13.
[673] See commentaries on fragments 6-8.
[674] See commentary on fragment 12, II, x+10.
[675] See commentary on fragment 13, Column II.
[676] See commentary on fragment 1, line 2.

identity with the creator god at the time when the universe originated. It is by virtue of this initial identity between Ocean and creator, one could argue, that the text devotes so much attention to the subsequent activities of the latter.

The only item in Volten's description of our cosmology that still remains to be considered is his characterisation of it as a specimen of hermetic literature. How accurate is this? Hermopolitan conceptions and deities certainly play an important part in the text. Chief among the latter are the members of the Ogdoad and Thoth. Other divinities who appear in our cosmology, like Amun and the celestial cow goddess, had important cults at Hermopolis during the Graeco-Roman Period, even though they did not originate there.[677] The manifestation of Amun in the form of a black bull, described in fragment 3, can be linked with the *km*, 'black bull', worshipped in that city.[678] If I am correct in suggesting that the egg which appears in fragment 4 is the solar egg, then this is a further case of Hermopolitan influence.[679] The city of Hermopolis itself is mentioned by name four, or possibly five, times in our cosmology.[680] Fragment 1 may have specified that city as the place where the first land emerged.[681] Fragment 6 speaks of the 'high sand' (= the primaeval mound) of Hermopolis, possibly in connection with the punishment of the sun god's enemies, who are said in some sources to have been overcome and executed there.[682]

Hermopolitan deities and conceptions play an important part in other cosmogonical/cosmological texts of the Graeco-Roman Period besides ours, irrespective of their provenience.[683] To some degree, all of these draw upon a stock of ideas associated with that city. At this remove, it is impossible to determine the extent to which authors of cosmogonies in, say, Esna or Thebes consciously borrowed from actual Hermopolitan sources, and that to which conceptions originating from such sources has been 'naturalised' in their own local traditions and consequently were available for them to exploit freely in whatever manner they wished.[684] It is probable that, for authors working in this genre, Hermopolis was an important *topos* both in a literal and in a literary sense.

[677] For the former, see Roeder, *Hermopolis 1929-1939*, pp. 65-72 and 177-8. For the latter, see references cited in notes 306-7 of the commentary on fragment 8.

[678] See Vernus, *Athribis*, pp. 355-6.

[679] See commentary on II, 19-22, of that fragment.

[680] In fragment 14, II, 22, the traces could represent either that toponym or the name of the Ogdoad. See textual note ad loc.

[681] See commentary on line 11 of that fragment.

[682] See commentary on the fragment in question.

[683] See, for instance, those cited in Sauneron and Yoyotte, *La Naissance du monde*, pp. 51-62.

[684] Cf. remarks of Assmann in *Re und Amun*, p. 222, on the 'naturalisation' of the Ogdoad at Thebes. Much the same point is made by Sauneron in *Esna* 5, pp. 274-5.

I have used the adjective 'Hermopolitan' deliberately to characterise the sorts of influences that I have been discussing, since what is involved is a purely Egyptian intellectual development.[685] Volten's term 'hermetic' is unsatisfactory, in my view, inasmuch as it implies a connection of some kind with the Greek hermetica. I have found no evidence for such a connection in the extant fragments of our cosmology, a point to which I shall return below. Accordingly, I think it better to avoid using this term in connection with it.

That being said, even if one does not agree with Volten's description of the text as hermetic, it still raises an interesting point. Our cosmology was inscribed in the first half of the second century AD. Does it display any affinities or parallels with contemporary Greek material of a cosmogonical/cosmological nature, either that known to have been composed and circulated in Egypt itself or that written elsewhere and alleged to have been influenced by Egyptian sources? Before such a question can be answered, it is necessary to define what is acceptable as a valid parallel in this context. As I see it, a true parallel, particularly if one intends to use it to argue direct influence in one direction or another, should fulfil five criteria:

1) The parallel must involve more than a mere general similarity of ideas or conceptions. Pépin and West, among others, have compared beliefs concerning the Primaeval Ocean expressed in Egyptian cosmogonical/cosmological works like our text with Orphic doctrines which affirm that water was the origin of everything.[686] However, the idea that water was the initial element is a widespread one, and can be found in a number of other ancient sources, both Greek and Near Eastern.[687] There is no compelling reason, therefore, to think that the Orphic texts have borrowed specifically from Egypt.

2) What is paralleled must be distinctive to a certain degree, i.e. not a ubiquitous or universal phenomenon. For instance, the breath of life (*ṯ3w n ʿnḫ*) plays an important part in fragment 13 of our text, where it is said to be what causes all things to live.[688] This is a conception which has a long history in Egypt.[689] In like manner, *pneuma*, 'breath' or 'spirit' is mentioned frequently in the Greek hermetica as a life-quickening agent.[690] Iversen has attempted to connect the two, seeing in the Egyptian conception the precursor of the hermetic idea.[691] In nature, however, it is a universal truth that

[685] It can be traced back at least as far as the New Kingdom. See references cited in note 683 above.

[686] See Pépin in Schoske (ed.), *Akten des Vierten Internationalen Ägyptologen Kongresses, München 1985*, vol. 3, p. 344; West, *The Orphic Poems*, pp. 183-90; Baumgarten, *The Phoenician History of Philo of Byblos*, p. 123.

[687] See West, *Early Greek Philosophy and the Orient*, pp. 205-13.

[688] See II, 5-6 and 10-13 of that fragment.

[689] See references cited in note 420 of the commentary on fragment 13.

[690] For references, see Copenhaver, *Hermetica*, pp. 100-101.

[691] Iversen, *Egyptian and Hermetic Doctrine*, pp. 19 and 35-6.

living things breathe while dead things do not. The connection between life and breath is easily made by any observant person. Thus, the authors of the Greek hermetica would hardly have needed to consult Egyptian sources to discover it.

3) The parallel must involve things which occur in similar contexts or appear in similar roles. Fragment 4 of our cosmology, if my suggested reconstruction of the text is correct, recounts how the sun god emerged from an egg which had been formed or fertilised by the wind.[692] The concept of the wind as the agent which quickens life within the egg is well-known from other Egyptian sources. Morenz, in discussing these, refers to a supposed Orphic doctrine concerning a wind-fertilised egg, which he would explain as having been developed under Egyptian influence.[693] However, the 'wind-egg' to which he makes reference is not one that has been fertilised by the wind, but rather one which is unfertilised, i.e. an egg produced by parthenogenesis.[694] Moreover, this doctrine, which appears in Aristophanes' *Birds*, is almost certainly only a parody of real Orphic beliefs.[695] Genuine Orphic cosmogonies do, in fact, sometimes refer to a cosmic egg from which a resplendent being resembling a solar deity emerges.[696] However, this being, variously named Protogonos or Phanes, is not the sun himself, but rather its creator.[697] What is more, instead of being generated by the wind, he is identified with it.[698] Thus, although wind and egg appear together in Orphic cosmogonies, there the latter gives rise to the former, and not the other way around as in Egyptian sources. For this reason, the Orphic texts are unlikely to be dependent upon Egyptian models, but may derive instead from an egg cosmogony originating elsewhere in the ancient Near East.[699]

4) What is paralleled must be expressed or formulated in similar language or terminology. Iversen, in his comparison of Egyptian and hermetic doctrines, assumes that the same ideas are expressed in two very different ways in native Egyptian sources and the Greek hermetica. The former employ what he calls concrete mythological terms; the latter the abstract terms of Greek philosophy.[700] For him, it is possible to discern the underlying Egyptian concepts which he believes are present in the hermetica by a process of 'translating' them from one mode of expression into the other. In my view, this amounts to granting oneself a license to interpret anything as being of Egyptian origin.

[692] See commentary on the second column of that fragment.
[693] Morenz, *Religion und Geschichte des alten Ägypten*, pp. 452-95.
[694] For this explanation of the Greek term ὑπηνέμιον ᾠόν, see West, *The Orphic Poems*, p. 201.
[695] West, *op. cit.*, pp. 111-12.
[696] Ibid., pp. 70 and 198-206.
[697] Ibid., pp. 206 and 210.
[698] Through his aspect as Eros/desire. Cf. West, *op. cit.*, pp. 200-1, 203 and 254.
[699] For a survey of such cosmogonies, see ibid., pp. 103-4.
[700] See reference cited in note 691 above.

5) The parallel must be based upon textual passages of which the meanings are clear and unambiguous. The cosmogony preserved in the *Phoenician History* of Philo of Byblos, for instance, mentions an egg shape and the creation of the sun in an obscure context.[701] Baumgarten emends the text to make it say that the sun emerged from an egg. This statement he then connects with Egyptian, specifically Hermopolitan, doctrines in which the sun's origin is explained in similar terms.[702] His reasons for doing so seem unconvincing to me, however. Having persuaded himself that Philo's text was influenced by Hermopolitan sources,[703] he then emends the text so that it conforms with this view. The circularity of such a procedure will be obvious.

Bearing the five criteria listed above in mind, I have searched for parallels with our cosmology in the following sources: the Corpus Hermeticum and related material, i.e. the *Perfect Discourse* in its Latin translation (*Asclepius*) and the Stobaean fragments,[704] the Orphic poems,[705] the cosmogony preserved in the *Phoenician History* of Philo of Byblos,[706] and the Greek magical papyri from Egypt.[707] Obviously, there are other works in addition to these which one might investigate, but it would be inappropriate to conduct an exhaustive search for parallels in the *editio princeps* of what is, after all, a very fragmentary text where so many uncertainties remain.

The only really striking parallels that I have noted are to be found in the Greek magical papyri. This is perhaps not surprising, since that corpus is known to preserve a great deal of authentic Egyptian material,[708] while in ancient Egypt there was no strict demarcation between the milieu of the priest who composed works like our

[701] See Baumgarten, *The Phoenician History of Philo of Byblos*, p. 12.

[702] See ibid., pp. 115-17. Other authors who accept the idea of Hermopolitan influence on this work include Marakhonova in *Ancient Egypt and Kush: in Memoriam Mikhail A. Korostovtsev*, pp. 277-89, and Carreira in *Sesto Congresso Internazionale di Egittologia: Atti* 2, pp. 69-76.

[703] See his comments in *The Phoenician History of Philo of Byblos*, pp. 108-10, 113, 120, 130, and 137-8.

[704] Text and translation: Nock and Festugière, *Corpus Hermeticum*, 4 vols. I have consulted in addition the following translations, commentaries, and discussions: Scott and Ferguson, *Hermetica*, 4 vols.; Copenhaver, *Hermetica*; Mahé, *Hermès en Haute-Égypte*, 2 vols; Fowden, *The Egyptian Hermes*; Iversen, *Egyptian and Hermetic Doctrine*; Derchain, *RHR* 161 (1962), pp. 175-98.

[705] For bibliography and discussion, see work cited in note 694 above.

[706] Text, translation, and commentary: Baumgarten, *The Phoenician History of Philo of Byblos*, pp. 12-13 and 94-139.

[707] Text: Preisendanz, *Papyri Graecae Magicae*, 2 vols. Translation: Betz (ed.), *The Greek Magical Papyri in Translation* (includes additional material not found in Preisendanz; cf. pp. xxvii-xxviii). For a general introduction to this corpus with annotated bibliography, see Brashear in Haase (ed.), *Aufstieg und Niedergang der römischen Welt* II, 18.5, pp. 3380-3684.

[708] See the textual notes of Ritner in Betz (ed.), *The Greek Magical Papyri in Translation*, passim; idem in Haase (ed.), *Aufstieg und Niedergang der römischen Welt* II, 18.5, pp. 3358-71.

cosmology and that of the magician.[709] In fact, Sauneron long ago drew attention to a parallel between the well-known *kosmopoiia* of Pap. Graec. Mag. (PGM) XIII (= P. Leiden I 395) and a passage in the Neith cosmogony at Esna.[710]

The closest parallel with our cosmology occurs in PGM IV, the so-called Great Paris Magical Papyrus (= Bib. Nat. Suppl. gr. 574), which dates to the fourth century AD and is of Theban origin.[711] If I have interpreted the second column of fragment 4 of our text correctly, it describes how the four winds of the sky coalesced into a single wind which either created or fertilised an egg from which the solar deity, identified specifically as Pshai, emerged.[712] PGM IV, 1596-1715, preserves an invocation to the sun god Helios in which he is called Ψοϊ φνουθι νινθηρ, a Greek transcription of the Egyptian words *pꜣ-Šy pꜣ ntr n nꜣ ntr.w*, 'Pshai, the god of the gods'.[713] Elsewhere in the same invocation, Helios/Pshai is addressed as ὁ ἀνατέλλων ἐκ τῶν τεσσάρων ἀνέμων, 'he who arose from the four winds'.[714] Some have understood this to mean that the sun god appears from the four directions represented by the winds,[715] but a more straightforward interpretation, in my view, is that he originates from the winds.[716] If so, then this is very like what one finds in our cosmology.

Similarly, in PGM III (= P. Louvre Mimaut 2391), also of fourth century date, Helios is invoked as Ψοει ω Ψοει ω π[ν]ουτε νεντηρ τηρ[ου], a Greek transcription of the Egyptian words *pꜣ-Šy ꜥꜣ pꜣ-Šy ꜥꜣ pꜣ ntr n nꜣ ntr.w tr=w*, 'Pshai the great (twice), the god of all the gods'.[717] Immediately following this, in lines 145-6, the magician tells the sun god: [ἐν]εύχομαί σοι κατὰ τοῦ [ᾠ]ου, 'I [ad]jure you by the [e]gg', doubtless the egg from which the solar deity was supposed to have emerged.[718]

[709] See Ritner, art. cit., pp. 3353-5; idem, *The Mechanics of Ancient Egyptian Magical Practice*, pp. 192-233.

[710] Sauneron, *Esna* 5, pp. 268-9; idem, *BSFE* 32 (1961), pp. 43-8. For further discussion, see Betz (ed.), *The Greek Magical Papyri in Translation*, pp. 175-8 and 184-6; West, *The Orphic Poems*, pp. 255-6; Fowden, *The Egyptian Hermes*, p. 63, note 65; Copenhaver, *Hermetica*, pp. 97-8. The lengthy treatment of the Leiden text by Merkelbach in *Abrasax* 3, passim, should be used with caution, in my view. I am particularly dubious of his interpretation (ibid., pp. 19-20) of the hissing and popping noises made by the creator god there as utterances of the name of the solar deity discussed in the paragraphs that follow.

[711] See Preisendanz, *Papyri Graecae Magicae* 1, p. 64.

[712] See commentary on the column in question.

[713] PGM IV, 1643 (Preisendanz, *op. cit.* 1, p. 124). For the transcription of the Greek, see Quaegebeur, *Le Dieu égyptien Shaï*, p. 169.

[714] PGM IV, 1605-7 (Preisendanz, *loc. cit.*).

[715] See e.g. Merkelbach and Totti, *Abrasax* 1, pp. 109, 116, and 140, who understand this as a reference to the sun god's 'Allherrschaft'.

[716] For this sense of ἀνατέλλω, see Liddell and Scott, *Greek-English Lexicon*, p. 123.

[717] PGM III, 144-5 (Preisendanz, *op. cit.* 1, p. 38). Cf. Quaegebeur, *Le Dieu égyptien Shaï*, p. 169.

[718] See remarks of Ritner in Betz (ed.), *The Greek Magical Papyri in Translation*, p. 22, note 37.

In PGM XII, the god Agathos Daimon is invoked as ὁ ἐκ τῶν δ' ἀν[έ]μων, 'the one (originating) from the four wi[n]ds'.[719] This text, which comes from Thebes and may be dated to the fourth century AD, is written on the verso of P. Leiden I 384, the recto of which preserves the famous demotic account of the myth of the solar eye.[720] The nearly contemporary PGM XIII preserves an identical invocation.[721] Agathos Daimon, as is well known, is the Greek translation of the Egyptian divine name Pshai.[722] Our text may have equated the latter deity with the Primaeval Ocean, at least in the initial stages of creation.[723] Accordingly, it is interesting to note that the spell just cited from PGM XIII goes on to address Agathos Daimon with the words σὺ εἶ ὁ ὠκεανός, 'You are the Ocean.'[724]

These parallels suggest to me that the description of the creation of the sun god in our cosmology and the statements concerning his origin in the above-named magical papyri may well derive from the same source. I have found no other parallels of sufficient importance to require mention here. Thus, the total of those identified is relatively small, although obviously one could increase their number by adopting a less rigorous definition of what constitutes a parallel. This paucity could indicate that there was only a limited degree of overlapping between our text and the Greek sources which I have selected to compare with it. On the other hand, it could reflect the manuscript's fragmentary state of preservation. Perhaps if more of it survived, then it would be possible to identify further parallels.

[719] PGM XII, 238 (Preisendanz, *op. cit.* 2, p. 74).

[720] See Preisendanz, *op. cit.* 2, p. 57; Spiegelberg, *Der ägyptische Mythus vom Sonnenauge*, p. 1.

[721] PGM XIII, 761 (Preisendanz, *op. cit.* 2, p. 122). On the date, see ibid., p. 86. This is the Leiden manuscript which preserves the parallel with the Esna cosmogony mentioned earlier. On its possible connection with the other magical papyri cited in the preceding paragraphs, see Fowden, *The Egyptian Hermes*, pp. 168-72.

[722] Quaegebeur, *Le Dieu égyptien Shaï*, pp. 170-6 and 263-4; Sternberg-El Hotabi, *Der Propylon des Month-Tempels in Karnak-Nord*, p. 92.

[723] See discussion earlier in this chapter.

[724] PGM XIII, 772-3 (Preisendanz, *op. cit.* 2, p. 122).

BIBLIOGRAPHY OF WORKS CITED

Allen, James P. *Genesis in Egypt.* The Philosophy of Ancient Egyptian Creation Accounts. Yale Egyptological Studies 2. New Haven: Yale Egyptological Seminar, 1988.

Alliot, Maurice. *Le Culte d'Horus à Edfou au temps des Ptolémées,* 2 vols. Bibliothèque d'Étude 20. Cairo: Institut Français d'Archéologie Orientale, 1949-54.

Altenmüller, Hartwig. *Die Apotropaia und die Götter Mittelägyptens.* Eine typologische und religionsgeschichtliche Untersuchung der sogenannten 'Zaubermesser' des Mittleren Reiches. München: no publisher, 1965.

— 'Chemmis', *Lexikon der Ägyptologie* 1. Wiesbaden: Otto Harrassowitz , 1975.

— 'Djed-Pfeiler', ibid.

Altenmüller-Kesting, Brigitte. *Reinigungsriten im ägyptischen Kult.* Hamburg: Lüdke bei der Uni, 1968.

Assmann, Jan. *Ägyptische Hymnen und Gebete.* Die Bibliothek der alten Welt. Reihe der alte Orient. Zürich and Munich: Artemis Verlag, 1975.

— *Der König als Sonnenpriester.* Ein kosmographischer Begleittext zur kultischen Sonnenhymnik in thebanischen Tempeln und Gräbern. Abhandlungen des Deutschen Archäologischen Instituts Kairo. Ägyptologische Reihe, Band 7. Glückstadt: J. J. Augustin, 1970.

— *Liturgische Lieder an den Sonnengott.* Untersuchungen zur altägyptischen Hymnik 1. Münchner ägyptologische Studien 19. Berlin: Verlag Bruno Hessling, 1969.

— *Re und Amun.* Die Krise des polytheistischen Weltbilds im Ägypten der 18.-20. Dynastie. Orbis Biblicus et Orientalis 51. Freiburg: Universitätsverlag/Göttingen: Vandenhoeck & Ruprecht, 1983.

— *Sonnenhymnen in Thebanischen Gräbern.* Theben 1. Mainz am Rhein: Philipp von Zabern, 1983.

— 'Die Verborgenheit des Mythos in Ägypten', *Göttinger Miszellen* 25 (1977), 7-43.

Aufrère, Sydney. 'Caractères principaux et origine divine des minéraux', *RdE* 34 (1982-83), 3-21.

— 'De l'influence des luminaires sur la croissance des végétaux. À propos d'une scène du papyrus funéraire de Nebhepet, Musée de Turin', *Memnonia* 6 (1995), 113-21.

— *L'Univers minéral dans la pensée égyptienne,* 2 vols. Bibliothèque d'Étude 105. Cairo: Institut Français d'Archéologie Orientale, 1991.

Baines, John. *Fecundity Figures.* Egyptian Personification and the Iconology of a Genre. Warminster: Aris and Phillips, 1985.

Baines, John. 'King, Temple, and Cosmos: an Earlier Model for Framing Columns in Temple Scenes of the Graeco-Roman Period', *Aspekte spätägyptischer Kultur.* Festschrift für Erich Winter zum 65. Geburtstag. Aegyptiaca Treverensia 7. Mainz: Verlag Philipp von Zabern, 1994.

Bakir, Abd el-Mohsen. 'A Hymn to Amon-Re at Tura', *ASAE* 42 (1943), 83-91.

el-Banna, Essam. 'À propos de la désignation "père des dieux" (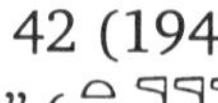, *it nṯrw*)', *BIFAO* 86 (1986), 151-70.

Barguet, Paul. 'Le Cycle lunaire d'après deux textes d'Edfou', *RdE* 29 (1977), 14-20.

— *La Stèle de la famine à Séhel.* Bibliothèque d'Étude 24. Cairo: Institut Français d'Archéologie Orientale, 1953.

Barucq, André. 'Les Textes cosmogoniques d'Edfou d'après les manuscrits laissés par Maurice Alliot', *BIFAO* 64 (1966), 125-67.

— and Daumas, François. *Hymnes et prieres de l'Égypte ancienne.* Littératures anciennes du proche Orient. Paris: Les Éditions du Cerf, 1980.

Baum, Nathalie. *Arbres et arbustes de l'Égypte ancienne.* La Liste de la tombe thébaine d'Ineni (no. 81). Orientalia Lovaniensia Analecta 31. Leuven: Departement Oriëntalistiek, 1988.

Baumgarten, Albert I. *The Phoenician History of Philo of Byblos: a Commentary.* Études préliminaires aux religions orientales dans l'empire romaine 89. Leiden: E. J. Brill, 1981.

Beinlich, Horst. 'Bemerkungen zum Schabaka-Stein', *Göttinger Miszellen* 122 (1991), 15-20.

— *Das Buch vom Fayum,* 2 vols. Zum religiösen Eigenverständnis einer ägyptischen Landschaft. Ägyptologische Abhandlungen 51. Wiesbaden: Otto Harrassowitz, 1991.

— 'Der Moeris-See nach Herodot', *Göttinger Miszellen* 100 (1987), 15-18.

Bell, Barbara. 'Climate and the History of Egypt: the Middle Kingdom', *American Journal of Archeology* 79 (1975), 223-69.

Bergman, Jan. *Ich bin Isis.* Studien zum memphitischen Hintergrund der griechischen Isisaretalogien. Acta Universitas Upsaliensis. Historia Religionum 3. Uppsala: Almqvist and Wiksell, 1968.

— *Isis-Seele und Osiris-Ei.* Zwei ägyptologische Studien zu Diodorus Siculus I 27, 4-5. Acta Universitas Upsaliensis. Historia Religionum 4. Uppsala: Upsaliensis S. Academiae, 1970.

Berlandini, Jocelyne. 'Ptah-démiurge et l'exaltation du ciel', *RdE* 46 (1995), 9-41.

Betz, Hans Dieter. *The Greek Magical Papyri in Translation, Including the Demotic Spells.* Volume 1: Texts (2nd edition). Chicago: University of Chicago Press, 1992.

Bickel, Susanne. *La Cosmogonie égyptienne avant le Nouvel Empire.* Orbis Biblicus et Orientalis 134. Freiburg: Éditions Universitaires Fribourg Suisse/Göttingen: Vandenhoeck & Ruprecht, 1994.

Blackman, Aylward M. *Middle-Egyptian Stories* 1. Bibliotheca Aegyptiaca 2. Brussels: Fondation Égyptologique Reine Élisabeth, 1932.

Bonneau, Danielle. *La Crue du Nil.* Études et commentaires 52. Paris: Librairie C. Klincksieck, 1964.

— 'La Divinité du Nil sous le principat en Égypte', *Aufstieg und Niedergang der römischen Welt,* Teil II, Band 18.5. Berlin and New York: Walter de Gruyter, 1995.

Borghouts, Joris F. 'A New Middle Kingdom Netherworld Guide', *Akten des Vierten Internationalen Ägyptologen Kongresses, München 1985,* vol. 3. Studien zur altägyptischen Kultur Beiheft 3. Hamburg: Helmut Buske Verlag, 1989.

Botti, Guiseppi. 'Quello che anche l'Egittologia deve a Carlo Anti', *Anthemon: Scritti di Archeologia e di Antichità classiche in onore di Carlo Anti.* Florence: G. C. Sansoni Editore, 1955.

Boylan, Patrick. *Thoth, the Hermes of Egypt.* A Study of Some Aspects of Theological Thought in Ancient Egypt. London: Oxford University Press, 1922.

Brashear, William. 'The Greek Magical Papyri: an Introduction and Survey; Annotated Bibliography (1928-1994)', *Aufstieg und Niedergang der römischen Welt,* Teil II, Band 18.5. Berlin and New York: Walter de Gruyter, 1995.

Breasted, James Henry. *The Edwin Smith Surgical Papyrus Published in Facsimile and Hieroglyphic Transliteration with Translation and Commentary in Two Volumes.* Oriental Institute Publications 3-4. Chicago: University of Chicago Press, 1930.

Broze, Michèle. 'La Creation du monde et l'opposition *sḏm.f - sḏm.n.f* dans le temple d'Esna', *RdE* 44 (1993), 3-10.

Brugsch, Heinrich. *Thesaurus Inscriptionum Aegyptiacarum,* 6 vols. Leipzig: J. C. Hinrichs, 1883-91.

Brunner-Traut, Emma. 'Krokodil', *Lexikon der Ägyptologie* 3. Wiesbaden: Otto Harrassowitz, 1980.

— 'Lotos', ibid.

Bucher, Paul. *Les Textes des tombes de Thoutmosis III et d'Amenophis II,* vol 1. Mémoires publiés par les membres de l'Institut Français d'Archéologie Orientale 60. Cairo: Institut Français d'Archéologie Orientale, 1932.

de Buck, Adriaan. *The Egyptian Coffin Texts,* 7 vols. Oriental Institute Publications 34, 49, 64, 67, 73, 81 and 87. Chicago: University of Chicago Press, 1935-61.

— *De Egyptische Voorstellingen betreffende den Oerheuvel.* Leiden: Eduard Idjo, 1922.

Budge, E. A. Wallis. *The Chapters of Coming Forth by Day or the Theban Recension of the Book of the Dead,* 3 vols. Books on Egypt and Chaldaea 28-30. London: Kegan, Paul, Trench, Trübner & Co., Ltd., 1910.

— *Hieroglyphic Texts from Stelae, &c., in the British Museum* 4. London: The Trustees of the British Museum, 1913.

Caminos, Ricardo A. 'Another Hieratic Manuscript from the Library of Pwerem son of Ḳiḳi (Pap. B.M. 10288)', *JEA* 58 (1972), 205-24.

Caminos, Ricardo A. *Late-Egyptian Miscellanies.* Brown Egyptological Studies 1. London: Oxford University Press, 1954.

Carreira, José Nunes. 'Hermopolitan Traditions in Philo Byblius' *Phoenician History*', *Sesto Congresso Internazionale di Egittologia: Atti 2.* Turin: Comitato Organizzativo del Congresso, 1993.

Cauville, Sylvie. 'La Chapelle de Thot-Ibis à Dendera édifiée sous Ptolémée Ier par Hor, scribe d'Amon-Rê', *BIFAO* 89 (1989), 43-66.

— *Dendara: les chapelles osiriennes*, 2 vols. Dendara 10. Cairo: Institut Français d'Archéologie Orientale, 1997.

— 'L'Hymne à Mehyt d'Edfou', *BIFAO* 82 (1982), 105-25.

de Cenival, Françoise. *Les Associations religieuses en Égypte d'après les documents démotiques.* Bibliothèque d'Étude 66. Cairo: Institut Français d'Archéologie Orientale, 1972.

— *Le Mythe de l'oeil du soleil.* Demotische Studien 9. Sommerhausen: Gisela Zauzich Verlag, 1988.

Černý, Jaroslav. *Late Ramesside Letters.* Bibliotheca Aegyptiaca 9. Brussels: Fondation Égyptologique Reine Élisabeth, 1939.

— 'A Note on the Recently Discovered Boat of Cheops', *JEA* 41 (1955), 75-9.

— and Gardiner, Alan H. *Hieratic Ostraca* 1. Oxford: Oxford University Press for the Griffith Institute, 1957.

Charpentier, Gérard. *Recueil de matériaux épigraphiques relatifs à la botanique de l'Égypte ancienne.* Paris: Trismégiste, 1981.

Chassinat, Émile. *Le Mammisi d'Edfou.* Mémoires publiés par les membres de l'Institut Français d'Archéologie Orientale du Caire 16. Institut Français d'Archéologie Orientale, 1939.

— *Le Mystère d'Osiris au mois de Khoiak*, 2 vols. Cairo: Institut Français d'Archéologie Orientale, 1966-68.

— and Daumas, François. *Le Temple de Dendara*, 9 vols. Publications de l'Institut Français d'Archéologie Orientale du Caire. Cairo: Institut Français d'Archéologie Orientale, 1934-87.

Chauveau, Michel. Review of *The Liturgy of Opening the Mouth for Breathing* by M. Smith. *CdE* 73 (1998), 75-7.

Copenhaver, Brian P. *Hermetica.* Cambridge: Press Syndicate of the University of Cambridge, 1992.

Cozi, Massimo. 'La Déesse Khefethernebes', *Göttingen Miszellen* 153 (1996), 17-31.

— 'La Nécropole de Khefethernebes', *Göttinger Miszellen* 151 (1996), 37-47.

Crum, W. E. *A Coptic Dictionary.* Oxford: The Clarendon Press, 1939.

— and Evelyn White, H. G. *The Monastery of Epiphanius at Thebes* 2. Publications of the Metropolitan Museum of Art Egyptian Expedition 4. New York: Metropolitan Museum of Art, 1926.

Cruz-Uribe, Eugene. 'The Khonsu Cosmogony', *JARCE* 31 (1994), 169-89.

Daressy, Georges. 'La Crue du Nil de l'an XXIX d'Amasis', *ASAE* 23 (1923), 47-8.

— 'Une inondation à Thèbes sous le règne d'Osorkon II', *Receuil de Traveaux* 18 (1896), 181-6.

Daumas, François. *Les Mammisis de Dendara.* Publications de l'Institut Français d'Archéologie Orientale du Caire. Cairo: Institut Français d'Archéologie Orientale, 1959.

— *Les Mammisis des temples égyptiens.* Annales de l'Université de Lyon. Troisième Série - Lettres. Fascicule 32. Paris: Société d'Édition 'Les Belles Lettres', 1958.

Davies, Norman de Garis. *The Temple of Hibis in El Khārgeh Oasis* 3. The Decoration. Publications of the Metropolitan Museum of Art Egyptian Expedition 17. New York: Metropolitan Museum of Art, 1953.

Dawson, Warren R. 'The Number "Seven" in Egyptian Texts', *Aegyptus* 8 (1927), 97-107.

Demarée, R. J. 'Recent Work on the Administrative Papyri in the Museo Egizio, Turin', *Sesto Congresso Internazionale di Egittologia: Atti* 2. Turin: Comitato Organizzativo del Congresso, 1993.

Derchain, Philippe. 'L'Authenticité de l'inspiration égyptienne dans le "Corpus Hermeticum"', *Revue de l'Histoire des Religions* 161 (1962), 175-98.

— 'Harkhebis, le Psylle-Astrologue', *CdE* 64 (1989), 74-89.

— 'La Mort ravisseuse', *CdE* 33 (1958), 29-32.

— 'Mythes et dieux lunaires en Égypte', *La Lune: mythes et rites.* Sources orientales 5. Paris: Éditions du Seuil, 1962.

— *Le Papyrus Salt 825 (B.M. 10051), rituel pour la conservation de la vie en Égypte.* Académie Royale de Belgique. Classe des Lettres. Mémoires. Collection in-8°. Deuxième série. Tome 58. Brussels: Palais des Académies, 1965.

— 'La Pêche de l'oeil et les mystères d'Osiris à Dendara', *RdE* 15 (1963), 11-25.

Derchain-Urtel, Maria Theresia. 'Die Schlange des "Schiffbrüchigen"', *SAK* 1 (1974), 83-104.

— *Thot à travers ses épithètes dans les scènes d'offrandes des temples d'époque gréco-romaine.* Rites égyptiens 3. Brussels: Fondation Égyptologique Reine Élisabeth, 1981.

Drioton, Étienne. 'Les Dédicaces de Ptolémée Évergète II sur la deuxième pylône de Karnak', *ASAE* 44 (1944), 111-62.

— 'Un Oudja à représentation hermopolitaine', *RdE* 1 (1933), 81-5.

— 'Les quatre Montou de Médamoud', *CdE* 6 (1931), 259-70.

— 'Trois documents pour l'étude de l'art copte', *BSAC* 10 (1944), 69-89.

Dunand, Françoise. 'Les Représentations de l'agathodémon à propos de quelque bas-reliefs du Musée d'Alexandrie', *BIFAO* 67 (1969), 9-48.

Edwards, I. E. S. *Hieroglyphic Texts from Egyptian Stelae, etc.,* vol. 8. London: The Trustees of the British Museum, 1939.

Egberts, Arno. *In Quest of Meaning.* A Study of the Ancient Egyptian Rites of Consecrating the *Meret*-Chests and Driving the Calves. 2 vols. Egyptologische Uitgaven 8. Leiden: Nederlands Instituut voor het Nabije Oosten, 1995.

Epigraphic Survey. *Medinet Habu* 1. Earlier Historical Records of Ramses III. Oriental Institute Publications 8. Chicago: University of Chicago Press, 1930.

— *Medinet Habu* 2. Later Historical Records of Ramses III. Oriental Institute Publications 9. Chicago: University of Chicago Press, 1932.

— *Ramses III's Temple within the Great Enclosure of Amon* 1. Reliefs and Inscriptions at Karnak 1. Oriental Institute Publications 25. Chicago: University of Chicago Press, 1936.

Erichsen, W. *Demotisches Glossar.* Copenhagen: Ejnar Munksgaard, 1954.

— *Papyrus Harris 1.* Hieroglyphische Transkription. Bibliotheca Aegyptiaca 5. Brussels: Fondation Égyptologique Reine Élisabeth, 1933.

— and Schott, Siegfried. *Fragmente memphitischer Theologie in demotischer Schrift (Pap. demot. Berlin 13603).* Akademie der Wissenschaften und der Literatur in Mainz. Abhandlungen der geistes- und sozialwissenschaftlichen Klasse. Jahrgang 1954. Nr. 7. Wiesbaden: Verlag der Akademie der Wissenschaften und der Literatur in Mainz in Kommission bei Franz Steiner Verlag GMBH, 1954.

Erman, Adolf and Grapow, Hermann. *Wörterbuch der ägyptischen Sprache,* 7 vols. plus 5 vols. Belegstellen. Leipzig: J. C. Hinrichs/Berlin: Akademie-Verlag, 1926-63.

Faulkner, Raymond O. *The Papyrus Bremner-Rhind (British Museum 10188).* Bibliotheca Aegyptiaca 3. Brussels: Fondation Égyptologique Reine Élisabeth, 1933.

Fauth, Wolfgang. *Helios Megistos.* Zur synkretistischen Theologie der Spätantike. Religions in the Graeco-Roman World 125. Leiden: E. J. Brill, 1995.

Finnestad, Ragnhild. *Image of the World and Symbol of the Creator.* Studies in Oriental Religions 10. Wiesbaden: Otto Harrassowitz, 1985.

Fischer, Henry George. 'Further Evidence for the Logic of Ancient Egyptian: Diminishing Progression', *JARCE* 10 (1973), 5-9.

Fischer-Elfert, Hans-Werner. 'Legenda Hieratika - I', *Göttinger Miszellen* 165 (1998), 105-12.

— *Lesefunde im literarischen Steinbruch von Deir el-Medineh.* Kleine ägyptische Texte 12. Wiesbaden: Harrassowitz Verlag, 1997.

— *Literarische Ostraka der Ramessidenzeit in Übersetzung.* Kleine ägyptische Texte. Wiesbaden: Harrassowitz Verlag, 1986.

— 'Uto', *Lexikon der Ägyptologie* 6. Wiesbaden: Otto Harrassowitz, 1986.

Fowden, Garth. *The Egyptian Hermes.* A Historical Approach to the Late Pagan Mind. Princeton: Princeton University Press, 1986.

Fox, Michael V. 'A Study of Antef', *Orientalia* 46 (1977), 393-423.

Gabolde, Marc. 'L'Inondation sous les pieds d'Amon', *BIFAO* 95 (1995), 235-58.

Gardiner, Alan H. 'Horus the Beḥdetite', *JEA* 30 (1944), 23-60.

— 'Hymns to Amon from a Leiden Papyrus', *ZÄS* 42 (1905), 12-42.

— *Late-Egyptian Miscellanies.* Bibliotheca Aegyptiaca 7. Brussels: Fondation Égyptologique Reine Élisabeth, 1937.

— *Late-Egyptian Stories.* Bibliotheca Aegyptiaca 1. Brussels: Fondation Égyptologique Reine Élisabeth, 1932.

Gasse, Annie. *Les Sarcophages de la Troisième Période Intermédiaire du Museo Gregoriano Egizio.* Museo Gregoriano Egizio Aegyptiaca Gregoriana vol. 3. Città del Vaticano: Monumenti, Musei e Gallerie Pontificie, 1996.

Gauthier, Henri. *Dictionnaire des noms géographiques contenus dans les textes hiéroglyphiques*, 7 vols. Cairo: Société Royale de Géographie d'Égypte, 1925-31.

Germer, Renate. *Flora des pharaonischen Ägypten.* Deutsches Archäologisches Institut, Abteilung Kairo. Sonderschrift 14. Mainz am Rhein: Philipp von Zabern, 1985.

— 'Tamariske', *Lexikon der Ägyptologie* 6. Wiesbaden: Otto Harrassowitz, 1986.

Gessler-Löhr, Beatrix. *Die Heiligen Seen ägyptischer Tempel.* Ein Beitrag zur Deutung sakraler Baukunst im alten Ägypten. Hildesheimer ägyptologischer Beiträge 21. Hildesheim: Gerstenberg Verlag, 1983.

Gilula, Mordechai. 'Peasant B 141-145', *JEA* 64 (1978), 129-30.

Glanville, S. R. K. *The Instructions of ʿOnchsheshonqy (British Museum Papyrus 10508).* Catalogue of Demotic Papyri in the British Museum 2. London: The Trustees of the British Museum, 1955.

Goyon, Georges. 'Les Travaux de Chou et les tribulations de Geb d'après le naos 2248 d'Ismaïlia', *Kêmi* 6 (1936), 1-42.

Goyon, Jean-Claude. 'Le Cérémonial de glorification d'Osiris du papyrus du Louvre I. 3079 (colonnes 110 à 112)', *BIFAO* 65 (1967), 89-156.

— 'Le Cérémonial pour faire sortir Sokaris. Papyrus Louvre I. 3079 (colonnes 112 à 114)', *RdE* 20 (1968), 63-96.

— *Les Dieux-gardiens et la genèse des temples d'après les textes égyptiens de l'époque gréco-romaine: les soixante d'Edfou et les soixante-dix-sept dieux de Pharbaethos*, 2 vols. Bibliothèque d'Étude 93. Cairo: Institut Français d'Archéologie Orientale, 1985.

— *Le Papyrus du Louvre N. 3279.* Bibliothèque d'Étude 42. Cairo: Institut Français d'Archéologie Orientale, 1966.

Graefe, Erhart. 'Phallus und Ei: Ptah als Urgott. Das Fragment einer magischen Stele der Bibliothèque Humaniste et Municipale de Sélestat', *Egyptian Religion: the last Thousand Years.* Studies Dedicated to the Memory of Jan Quaegebeur. 2 vols. Orientalia Lovaniensia Analecta 84-85. Leuven: Uitgeverij Peeters en Departement Oosterse Studies, 1998.

— *Untersuchungen zur Wortfamilie bjꜣ-.* Köln: no publisher, 1971.

Grapow, Hermann. *Anatomie und Physiologie.* Grundriss der Medizin der alten Ägypter 1. Berlin: Akademie-Verlag, 1954.

Grapow, Hermann. *Religiöse Urkunden. Ausgewählte Texte des Totenbuches.* Urkunden des Ägyptischen Altertums 5. Leipzig: J. C. Hinrichs, 1915-17.

— 'Die Welt vor der Schöpfung', *ZÄS* 67 (1931), 34-8.

Grieshammer, Reinhard. 'Flammeninsel', *Lexikon der Ägyptologie* 2. Wiesbaden: Otto Harrassowitz, 1977.

Griffith, F. L. and Thompson, Herbert. *The Demotic Magical Papyrus of London and Leiden*, 3 vols. London: H. Grevel and Company, 1904-9.

Griffiths, J. Gwyn. *Plutarch's De Iside et Osiride.* Cardiff: University of Wales Press, 1970.

Grimal, N.-C. *La Stèle triomphale de Pi(ʿankh)y au Musée du Caire, JE 48862 et 47086-47089.* Études sur la propagande royale égyptienne 1. Mémoires publiés par les membres de l'Institut Français d'Archéologie Orientale du Caire 105. Cairo: Institut Français d'Archéologie Orientale, 1981.

Guglielmi, Waltraud. 'Wortspiel', *Lexikon der Ägyptologie* 6. Wiesbaden: Otto Harrassowitz, 1986.

Gutbub, Adolphe. *Textes fondamentaux de la théologie de Kom Ombo*, 2 vols. Bibliothèque d'Étude 47. Cairo: Institut Français d'Archéologie Orientale, 1973.

— 'La Tortue animal cosmique bénéfique à l'époque ptolémaïque et romaine', *Hommages à la mémoire de Serge Sauneron* 1. Bibliothèque d'Étude 81. Cairo: Institut Français d'Archéologie Orientale, 1979.

Haikal, Fayza. 'Commentaire des textes du sarcophage no. 22954 du Musée Vatican', *Rivisti degli Studi Orientali* 63 (1989), 13-24.

— *Two Hieratic Funerary Papyri of Nesmin*, 2 vols. Bibliotheca Aegyptiaca 14 and 15. Brussels: Fondation Égyptologique Reine Élisabeth, 1970-72.

Harris, J. R. *Lexicographical Studies in Ancient Egyptian Minerals.* Deutsche Akademie der Wissenschaften zu Berlin. Institut für Orientforschung. Veröffentlichung 54. Berlin: Akademie-Verlag, 1961.

Heerma van Voss, M. S. H. G. 'Nechbet', *Lexikon der Ägyptologie* 4. Wiesbaden: Otto Harrassowitz, 1982.

Helck, Wolfgang. *Historisch-biographische Texte der 2. Zwischenzeit und neue Texte der 18. Dynastie.* Kleine ägyptische Texte. Wiesbaden: Otto Harrassowitz, 1975.

Herbin, François-René. 'Un hymne à la lune croissant', *BIFAO* 82 (1982), 237-82.

— *Le Livre de parcourir l'éternité.* Orientalia Lovaniensia Analecta 58. Leuven: Uitgeverij Peeters, 1994.

— 'Les Premières pages du Papyrus Salt 825', *BIFAO* 88 (1988), 95-112.

Hermsen, Edmund. *Lebensbaum Symbolik im alten Ägypten: eine Untersuchung.* Arbeitsmaterialen zur Religionsgeschichte 5. Köln: Religionswissenschaftliches Seminar der Universität Bonn in Kommission bei E. J. Brill, 1981.

— *Die zwei Wege des Jenseits*. Das altägyptische Zweiwegebuch und seine Topographie. Orbis Biblicus et Orientalis 112. Freiburg: Universitätsverlag Freiburg Schweiz/ Göttingen: Vandenhoeck & Ruprecht, 1991.

Hoenes, Sigrid-Eike. *Untersuchungen zu Wesen und Kult der Göttin Sachmet*. Habelts Dissertationsdrucke. Reihe: Ägyptologie. Heft 1. Bonn: Rudolf Habelt Verlag GMBH, 1976.

Hoffmann, Friedhelm. *Ägypter und Amazonen*. Mitteilungen aus der Papyrussammlung der Österreichischen Nationalbibliothek (Papyrus Erzherzog Rainer). Neue Serie 24. Wien: Verlag Brüder Hollinek, 1995.

— *Der Kampf um den Panzer des Inaros*. Mitteilungen aus der Papyrussammlung der Österreichischen Nationalbibliothek (Papyrus Erzherzog Rainer). Neue Serie 26. Wien: Brüder Hollinek, 1996.

— 'Das Wort *ḏfḏ*', *Göttinger Miszellen* 132 (1994), 37-8.

Hornung, Erik. *Der ägyptische Mythus von der Himmelskuh*. Eine Ätiologie des Unvollkommenen. Orbis Biblicus et Orientalis 46. Freiburg: Universitätsverlag/ Göttingen: Vandenhoeck & Ruprecht, 1982.

— *Das Amduat. Die Schrift des verborgenen Raumes*, 3 vols. Ägyptologische Abhandlungen 7 and 13. Wiesbaden: Otto Harrassowitz, 1963-7.

— *Das Buch der Anbetung des Re im Westen (Sonnenlitanei) nach den Versionen des Neuen Reiches*, 2 vols. Aegyptiaca Helvetica 2-3. Geneva: Ägyptologisches Seminar der Universität Basel and the Centre d'études orientales de l'Université de Genève, 1975-6.

— *Das Buch von den Pforten des Jenseits nach den Versionen des Neuen Reiches*, 2 vols. Aegyptiaca Helvetica 7-8. Geneva: Ägyptologisches Seminar der Universität Basel and the Centre d'études orientales de l'Université de Genève, 1979-80.

— 'Chaotische Bereiche in der geordneten Welt', *ZÄS* 81 (1956), 28-32.

— *Conceptions of God in Ancient Egypt: the One and the Many*. Ithaca: Cornell University Press, 1982.

— *Texte zum Amduat* 3. Aegyptiaca Helvetica 15. Geneva: Ägyptologisches Seminar der Universität Basel and the Centre d'études orientales de l'Université de Genève, 1994.

— *The Valley of the Kings*. New York: Timken Publishers Inc., 1990.

— *Zwei Ramessidische Königsgräber: Ramses IV. und Ramses VII*. Theben 11. Mainz am Rhein: Verlag Philipp von Zabern, 1990.

Hughes, George R. 'The Sixth Day of the Lunar Month and the Demotic Word for "Cult Guild"', *MDAIK* 16 (1958), 147-60.

Iversen, Erik. *Egyptian and Hermetic Doctrine*. Opuscula Graecolatina 27. Copenhagen: Museum Tusculanum Press, 1984.

Jackson Howard. 'Κόρη Κόσμου: Isis, Pupil of the Eye of the World', *CdE* 61 (1986), 116-35.

Janssen, J., van de Walle, B., and Vergote, J. 'Compléments au commentaire des *Hieroglyphica* d'Horapollon', *CdE* 22 (1947), 251-9.

Jasnow, Richard. Review of *Le Mythe de l'oeil du soleil* by Françoise de Cenival. *Enchoria* 18 (1991), 205-15.

— 'Three Notes on Demotic Lexicography', *Enchoria* 12 (1984), 7-13.

Jelínková-Reymond, E. *Les Inscriptions de la statue guérisseuse de Djed-Ḥer-Le-Sauveur.* Bibliothèque d'Étude 23. Cairo: Institut Français d'Archéologie Orientale, 1956.

Johnson, Janet H. 'Demotic Nominal Sentences', *Studies Presented to Hans Jakob Polotsky.* East Gloucester, Massachusetts: Pirtle and Polson, 1981.

— *The Demotic Verbal System.* Studies in Ancient Oriental Civilization 38. Chicago: The Oriental Institute, 1976.

Johnson, Ray. 'Aida, Opet, and the Great Flood of '94', *Chicago House Bulletin* 6, no. 1 (December 15, 1994), 1-4.

Junker, Hermann. *Der Auszug der Hathor-Tefnut aus Nubien.* Aus dem Anhang zu den Abhandlungen der Königl. Preuss. Akademie der Wissenschaften vom Jahre 1911. Berlin: Verlag der Königl. Akademie der Wissenschaften, 1911.

— *Der grosse Pylon des Tempels der Isis in Philä.* Österreichische Akademie der Wissenschaften. Philosophisch-historische Klasse. Denkschriften-Sonderband. Vienna: Rudolf M. Rohrer, 1958.

— 'Die sechs Teile des Horusauges und der "sechste Tag"', *ZÄS* 48 (1910), 101-6.

— and Winter, Erich. *Das Geburtshaus des Tempels der Isis in Philä.* Philä-Publikation 2. Österreichische Akademie der Wissenschaften. Philosophisch-historische Klasse. Denkschriften-Sonderband. Vienna: Hermann Böhlaus Nachfolger, 1965.

Kákosy, László. 'Ideas about the Fallen State of the World in Egyptian Religion: Decline of the Golden Age', *Studia Aegyptiaca* 7 (1981), 81-92.

— 'Ihet', *Lexikon der Ägyptologie* 3. Wiesbaden: Otto Harrassowitz, 1980.

— 'Mehet-weret', *Lexikon der Ägyptologie* 4. Wiesbaden: Otto Harrassowitz, 1982.

— 'The Ptah-Shu-Tefnut Triad and the Gods of the Winds on a Ptolemaic Sarcophagus', *Essays on Ancient Egypt in Honour of Herman te Velde.* Egyptological Memoirs 1. Groningen: Styx Publications, 1997.

— 'Sothis', *Lexikon der Ägyptologie* 5. Wiesbaden: Otto Harrassowitz, 1984.

Kaper, Olaf. 'The Astronomical Ceiling of Deir el-Haggar in the Dakhleh Oasis', *JEA* 81 (1995), 175-95.

Keimer, Ludwig. 'Flechtwerk aus Halfagras im alten und neuen Ägypten', *OLZ* 30 (1927), columns 76-85 and 145-54.

— *Die Gartenpflanzen im alten Ägypten* 2. Deutsches Archäologisches Institut, Abteilung Kairo. Sonderschrift 13. Mainz am Rhein: Verlag Philipp von Zabern, 1984.

Kitchen, K. A. *Ramesside Inscriptions, Historical and Biographical*, 8 vols. Oxford: B. H. Blackwell Ltd., 1975-89.

Klasens, Adolf. *A Magical Statue Base (Socle Behague) in the Museum of Antiquities at Leiden* (= *OMRO* 33 [1952]). Leiden: E. J. Brill, 1952.

Koefoed-Petersen, Otto. *Recueil des inscriptions hiéroglyphiques de la Glyptothèque Ny Carlsberg*. Bibliotheca Aegyptiaca 6. Brussels: Fondation Égyptologique Reine Élisabeth, 1936.

Koemoth, Pierre. *Osiris et les arbres*. Contribution à l'étude des arbres sacrés de l'Égypte ancienne. Aegyptiaca Leodiensia 3. Liège: Centre Informatique de Philosophie et Lettres, 1994.

— 'Osiris-Lune, l'horizon et l'oeil *oudjat*', *CdE* 71 (1996), 203-20.

— '*Snb*, le papyrus ou le cordon en papyrus de Pé', *Göttinger Miszellen* 130 (1992), 33-43.

Koenig, Yvan. *Le Papyrus Boulaq 6, transcription, traduction et commentaire*. Bibliothèque d'Étude 87, Cairo: Institut Français d'Archéologie Orientale, 1981.

el-Kordy, Zeinab. 'Deux études sur Harsomtous', *BIFAO* 82 (1982), 171-86.

Kurth, Dieter. *Die Dekoration der Säulen im Pronaos des Tempels von Edfu*. Göttinger Orientforschungen. Reihe 4: Ägypten. Band 11. Wiesbaden: Otto Harrassowitz, 1983.

— *Den Himmel Stützen*. Rites égyptiens 2. Brussels: Fondation Égyptologique Reine Élisabeth, 1973.

— 'Über Horus, Isis und Osiris', *Studia Aegyptiaca* 14 (1992), 373-83.

— 'Über den Ursprung des Tempels von Edfu', *Religion und Philosophie im alten Ägypten*. Festgabe für Philippe Derchain zu seinem 65. Geburtstag am 24. Juli 1991. Orientalia Lovaniensia Analecta 39. Leuven: Departement Oriëntalistiek and Uitgeverij Peeters, 1991.

— 'Wind', *Lexikon der Ägyptologie* 6. Wiesbaden: Otto Harrassowitz, 1986.

Labrique, Françoise. *Stylistique et théologie à Edfou*. Le Rituel de l'offrande de la campagne: étude de la composition. Orientalia Lovaniensia Analecta 51. Leuven: Uitgeverij Peeters en Departement Oriëntalistiek, 1992.

Lacau, Pierre. 'Liquides et matières en grains employés au pluriel', *BIFAO* 56 (1957), 161-72.

Lange, H. O. *Der magische Papyrus Harris*. Det Kgl. Danske Videnskabernes Selskab. Historisk-filologiske Meddelelser. Bind 14. No. 2. Copenhagen: A. F. Høst & Søn, 1927.

Leclant, Jean. 'La Chronique du Prince Osorkon', *Orientalia* 30 (1961), 407-15.

— *Recherches sur les monuments thébains de la XXVe Dynastie dite Éthiopienne*. Bibliothèque d'Étude 36. Cairo: Institut Français d'Archéologie Orientale, 1965.

Legrain, Georges. *Le Livre des transformations*. Paris: E. Leroux, 1890.

Leitz, Christian. *Tagewählerei*. Das Buch *ḥ3t nḥḥ pḥ.wy ḏt* und verwandte Texte. 2 vols. Ägyptologische Abhandlungen 55. Wiesbaden: Otto Harrassowitz, 1994.

Lepsius, R. *Über die Götter der vier Elemente bei den Ägyptern.* Abhandlungen der philosophisch-historischen Klasse der Königl. Akademie der Wissenschaften zu Berlin 1856. Berlin: Königl. Akademie der Wissenschaften, 1856.

Lexa, François. *Papyrus Insinger: Les Enseignements moraux d'un scribe égyptien du premier siècle après J.-C.* Paris: Librairie Orientaliste Paul Geuthner, 1926.

Liddell, Henry George and Scott, Robert. *Greek-English Lexicon* (9th Edition). Oxford: Clarendon Press, 1940. With 1968 Supplement.

Logan, Thomas K. 'Papyrus Harkness', *Studies in Honor of George R. Hughes.* Studies in Ancient Oriental Civilization 39. Chicago: University of Chicago Press, 1976.

Luft, Ulrich. *Beiträge zur Historisierung der Götterwelt und der Mythenschreibung* (= *Studia Aegyptiaca* 4 [1978]). Budapest: Les Chaires d'Histoire Ancienne de l'Université Loránd Eötvös de Budapest, 1978.

Macadam, M. F. Laming. *The Temples of Kawa* 1. The Inscriptions. Oxford University Excavations in Nubia. London: Oxford University Press on Behalf of the Griffith Institute, Ashmolean Museum, Oxford, 1949.

Mahé, Jean-Pierre. *Hermès en Haute-Égypte*, 2 vols. Bibliothèque Copte de Nag Hammadi. Section 'Textes'. Vols 3 and 7. Quebec: Les Presses de l'Université Laval, 1978 and 1982.

Marakhonova, Svetlana. 'A Version of the Origins of the World in Egyptian, Orphic and Gnostic Cosmogonies', *Ancient Egypt and Kush: in Memoriam Mikhail A. Korostovtsev.* Moscow: Nauka Oriental Literature Publishers, 1993.

Mariette, Auguste. *Dendérah.* Description générale du grand temple de cette ville. 5 vols. Paris: Librairie A. Franck, 1870-3 (plates)/F. Vieweg, Libraire-Éditeur, 1880 (text).

— *Les Papyrus égyptiens du Musée de Boulaq* 2. Paris: Librairie A. Franck, 1872.

Mathieu, Bernard. 'Les Hommes de larmes', *Hommages à François Daumas.* Montpellier: Publication de la Recherche-Université de Montpellier, 1986.

Meeks, Dimitri. 'Les "Quatre ka" du démiurge memphite', *RdE* 15 (1963), 35-47.

Merkelbach, Reinhold and Totti, Maria. *Abrasax.* Ausgewählte Papyri religiösen und magischen Inhalts. 4 vols. Abhandlungen der Rheinisch-Westfälischen Akademie der Wissenschaften. Sonderreihe Papyrologica Coloniensia 17.1-4. Opladen: Westdeutscher Verlag, 1990-6.

de Meulenaere, Herman. 'Premiers et seconds prophètes d'Amon au début de l'époque ptolémaïque', *Egyptian Religion: the Last Thousand Years.* Studies Dedicated to the Memory of Jan Quaegebeur. 2 vols. Orientalia Lovaniensia Analecta 84-5. Leuven: Uitgeverij Peeters en Departement Oosterse Studies, 1998.

Moftah, Ramses. *Die heiligen Bäume im alten Ägypten.* Beiträge zur Religionsgeschichte, Philologie, Archäologie und Botanik. Göttingen: no publisher, 1959.

Möller, Georg. *Die beiden Totenpapyrus Rhind des Museums zu Edinburg.* Demotische Studien 6. Leipzig: J. C. Hinrichs, 1913.

Morenz, Siegfried. *Religion und Geschichte des alten Ägypten.* Gesammelte Aufsätze. Weimar: Hermann Böhlaus Nachfolger, 1974.

Moret, Alexandre. *Le Rituel du culte divin journalier en Égypte d'après les papyrus de Berlin et les textes du temple de Séti Ier, à Abydos.* Annales du Musée Guimet. Bibliothèque d'Études 14. Paris: Ernest Leroux, 1902.

de Morgan, J. et al. *Kom Ombos*, 2 vols. Catalogue des monuments et inscriptions de l'Égypte antique. Première Série-Haute Égypte. Tome second et Tome troisième. Vienna: Adolphe Holzhausen, 1895-1909.

Münster, Maria. *Untersuchungen zur Göttin Isis vom Alten Reich bis zum Ende des Neuen Reiches.* Münchner ägyptologische Studien 11. Berlin: Verlag Bruno Hessling, 1968.

Nagel, Geo. 'Un papyrus funéraire de la fin du Nouvel Empire (Louvre 3292 Inv.)', *BIFAO* 29 (1929), 1-127.

Naville, Edouard. *Das ägyptische Todtenbuch der XVIII. bis XX. Dynastie aus verschiedenen Urkunden*, 2 vols. Berlin: A. Asher & Co., 1886.

Nock, A. D. and Festugière, A.-J. *Corpus Hermeticum*, 4 vols. Collection des Universités de France. Paris: Société d'Édition 'Les Belles Lettres', 1945-54.

Osing, Jürgen. *Hieratische Papyri aus Tebtunis* 1. The Carlsberg Papyri 2. Carsten Niebuhr Institute Publications 17. Copenhagen: Museum Tusculanum Press, 1998.

Otto, Eberhard. 'Geschichte einer religiösen Formel', *ZÄS* 87 (1962), 150-4.

— 'Das "Goldene Zeitalter" in einem ägyptischen Text', *Religions en Égypte hellénistique et romaine.* Colloque de Strasbourg 16-18 mai 1967. Bibliothèque des Centre d'Études Supérieures Spécialisés. Paris: Presses Universitaires de France, 1969.

— *Gott und Mensch nach den ägyptischen Tempelinschriften der griechisch-römischen Zeit.* Eine Untersuchung zur Phraseologie der Tempelinschriften. Abhandlungen der Heidelberger Akademie der Wissenschaften. Philosophisch-historische Klasse. Jahrgang 1964. Nr. 1. Heidelberg: Carl Winter/Universitätsverlag, 1964.

— *Topographie des thebanischen Gaues.* Untersuchungen zur Geschichte und Altertumskunde Ägyptens 16. Berlin: Akademie-Verlag/Leipzig: J. C. Hinrichs, 1952.

Parker, Richard A. and Lesko, Leonard H. 'The Khonsu Cosmogony', *Pyramid Studies and Other Essays Presented to I. E. S. Edwards.* Egypt Exploration Society Occasional Publications 7. London: The Egypt Exploration Society, 1988.

Pécoil, Jean-François. 'Les Sources mythiques du Nil et le cycle de la crue', *BSEG* 17 (1993), 97-110.

Pépin, Jean-François. 'Quelques aspects de Nouou dans les textes des pyramides et les textes des sarcophages', *Akten des Vierten Internationalen Ägyptologen Kongresses, München 1985*, vol 3. Studien zur altägyptischen Kultur Beiheft 3. Hamburg: Helmut Buske Verlag, 1989.

Piankoff, Alexandre. *Le Livre du Jour et de la Nuit.* Bibliothèque d'Étude 13. Cairo: Institut Français d'Archéologie Orientale, 1942.

Piehl, Karl. *Inscriptions hiéroglyphiques recueillies en Europe et en Égypte* 1. Stockholm and Leipzig: J. C. Hinrichs, 1886.

Pierret, Paul. *Études Égyptologiques.* Paris: Librairie A. Franck, 1873.

van der Plas, Dirk. *L'Hymne à la crue du Nil,* 2 vols. Leiden: Nederlands Instituut voor het Nabije Oosten, 1986.

Posener, Georges. *Catalogue des ostraca hiératiques littéraires de Deir el Médineh* 3. Documents de Fouilles 20. Cairo: Institut Français d'Archéologie Orientale, 1977.

Preisendanz, Karl. *Papyri Graecae Magicae. Die griechischen Zauberpapyri* (2nd edition). 2 vols. Sammlung wissenschaftlicher Commentare. Stuttgart: B. J. Treubner, 1973-4.

Quack, Joachim. 'Korrekturvorschläge zu einigen demotischen literarischen Texten', *Enchoria* 21 (1994), 63-72.

— 'Das Monumental-Demotisch', *Per Aspera ad Astra: Wolfgang Schenkel zum neunundfünfzigsten Geburtstag.* Kassel: Louise Gestermann, 1995.

Quaegebeur, Jan. *Le Dieu égyptien Shaï dans la religion et l'onomastique.* Orientalia Lovaniensia Analecta 2. Leuven: Leuven University Press, 1975.

— 'Mummy Labels: an Orientation', *Textes grecs, démotiques et bilingues.* Papyrologica Lugduno-Batava 19. Leiden: E. J. Brill, 1978.

— 'Les Quatre dieux Min', *Religion und Philosophie im alten Ägypten.* Festgabe für Philippe Derchain zu seinem 65. Geburtstag am 24. Juli 1991. Orientalia Lovaniensia Analecta 39. Leuven: Uitgeverij Peeters en Departement Oriëntalistiek, 1991.

Ray, J. D. *The Archive of Ḥor.* Texts from Excavations 2. London: The Egypt Exploration Society, 1976.

Revillout, Eugène. *Papyrus démotiques du Louvre.* Corpus Papyrorum Aegypti 1. Paris: Ernest Leroux, 1885-1902.

Reymond, E. A. E. *The Mythical Origin of the Egyptian Temple.* Manchester: Manchester University Press, 1969.

— 'The Primeval Djeba', *JEA* 48 (1962), 81-8.

Ricke, Herbert. 'Der "Hohe Sand in Heliopolis"', *ZÄS* 71 (1935), 107-11.

Ritner, Robert K. 'Egyptian Magical Practice under the Roman Empire: the Demotic Spells and their Religious Content', *Aufstieg und Niedergang der römischen Welt*, Teil II, Band 18.5. Berlin and New York: Walter de Gruyter, 1995.

— 'Horus on the Crocodiles: a Juncture of Religion and Magic in Late Dynastic Egypt', *Religion and Philosophy in Ancient Egypt.* Yale Egyptological Studies 3. New Haven: Yale Egyptological Seminar, 1989.

— *The Mechanics of Ancient Egyptian Magical Practice.* Studies in Ancient Oriental Civilization 54. Chicago: Oriental Institute of the University of Chicago, 1993.

de Rochemonteix, M. and Chassinat, Émile. *Le Temple d'Edfou*, 14 vols. Mémoires publiés par les membres de l'Institut Français d'Archéologie Orientale du Caire 10-11 and 20-31. Paris: Ernest Leroux/Cairo: Institut Français d'Archéologie Orientale, 1897-1934.

Roeder, Günther. *Hermopolis 1929-1939.* Ausgrabungen der Deutschen Hermopolis-Expedition in Hermopolis, Öber-Ägypten. Pelizaeus-Museum zu Hildesheim.

Wissenschaftliche Veröffentlichung 4. Hildesheim: Verlag Gebrüder Gerstenberg, 1959.

—'Zwei hieroglyphische Inschriften aus Hermopolis (Öber-Ägypten)', *ASAE* 52 (1954), 315-442.

Roulin, Gilles. *Le Livre de la Nuit.* Une composition égyptienne de l'au-delà. 2 vols. Orbis Biblicus et Orientalis 147. Freiburg: Éditions Universitaires Fribourg Suisse/ Göttingen: Vandenhoeck & Ruprecht, 1996.

Ryhiner, Marie-Louise. *L'Offrande du lotus dans les temples égyptiens de l'époque tardive.* Rites égyptiens 6. Brussels: Fondation Égyptologique Reine Élisabeth, 1986.

Sambin, Chantal. *L'Offrande de la soi-disant 'clepsydre'.* Le Symbole *šbt/wnšb/wtt* (= *Studia Aegyptiaca* 11 [1988]). Budapest: Chaire d'Égyptologie de l'Université Eötvös Loránd de Budapest, 1988.

Sandman, Maj. *Texts from the Time of Akhenaten.* Bibliotheca Aegyptiaca 8. Brussels: Fondation Égyptologique Reine Élisabeth, 1938.

Sandman Holmberg, Maj. *The God Ptah.* Lund: Gleerup, 1946.

Sauneron, Serge. 'La Conception égyptienne du bonheur à propos des "quatre ka"', *BIFAO* 57 (1958), 163-4.

—'Copte S. ⲕⲁⲗⲁϩⲏ', *Mélanges Maspero* 1. Orient Ancien. Quatrième fascicule. Mémoires publiés par les membres de l'Institut Français d'Archéologie Orientale du Caire 66. Cairo: Institut Français d'Archéologie Orientale, 1961.

— *Les Fêtes religieuses d'Esna aux derniers siècles du paganisme.* Esna 5. Cairo: Institut Français d'Archéologie Orientale, 1962.

— 'La Légende des sept propos de Methyer au temple d'Esna', *BSFE* 32 (1961), 43-51.

— *La Porte ptolémaïque de l'enceinte de Mout à Karnak.* Mémoires publiés par les membres de l'Institut Français d'Archéologie Orientale du Caire 107. Cairo: Institut Français d'Archéologie Orientale, 1983.

— 'Remarques de philologie et d'étymologie (en marge des textes d'Esna)', *Mélanges Mariette.* Bibliothèque d'Étude 32. Cairo: Institut Français d'Archéologie Orientale, 1961.

— 'Remarques de philologie et d'étymologie (nos. 19-25)' , *RdE* 15 (1963), 49-62.

— *Rituel de l'embaumement.* Cairo: Imprimerie Nationale, 1952.

— *Le Temple d'Esna.* Textes nos. 1-193. Esna 2. Cairo: Institut Français d'Archéologie Orientale, 1963.

— *Le Temple d'Esna.* Textes nos. 194-398. Esna 3. Cairo: Institut Français d'Archéologie Orientale, 1968.

— *Un Traité égyptien d'ophiologie.* Papyrus du Brooklyn Museum No. 47.218.48 et .85. Publications de l'Institut Français d'Archéologie Orientale. Bibliothèque Générale 11. Cairo: Institut Français d'Archéologie Orientale, 1989.

— 'Villes et légendes d'Égypte' (nos. VII-XI)', *BIFAO* 64 (1966), 185-91.

Sauneron, Serge and Yoyotte, Jean. 'La Naissance du monde selon l'Égypte ancienne', *La Naissance du monde.* Sources orientales 1. Paris: Éditions du Seuil, 1959.

el-Sayed, Ramadan. *La Déesse Neith de Saïs,* 2 vols. Bibliothèque d'Étude 86. Cairo: Institut Français d'Archéologie Orientale, 1982.

Schlögl, Hermann. *Der Gott Tatenen nach Texten und Bildern des Neuen Reiches.* Orbis Biblicus et Orientalis 29. Freiburg: Universitätsverlag Freiburg Schweiz /Göttingen: Vandenhoeck & Ruprecht, 1980.

— *Der Sonnengott auf der Blüte.* Eine ägyptische Kosmogonie des Neuen Reiches. Aegyptiaca Helvetica 5. Geneva: Ägyptologisches Seminar der Universität Basel and the Centre d'études orientales de l'Université de Genève, 1977.

Schott, Siegfried. *Die Reinigung Pharaos in einem memphitischen Tempel (Berlin P 13242).* Nachrichten der Akademie der Wissenschaften in Göttingen. I. Philologisch-historische Klasse. Jahrgang 1957. Nr. 3. Göttingen: Vandenhoeck & Ruprecht, 1957.

— *Zum Weltbild der Jenseitsführer des Neuen Reiches.* Nachrichten der Akademie der Wissenschaften in Göttingen. I. Philologisch-historische Klasse. Jahrgang 1965. Nr. 11. Göttingen: Vandenhoeck & Ruprecht, 1965.

Scott, Walter and Ferguson, A. S. *Hermetica.* The Ancient Greek and Latin Writings Which Contain Religious or Philosophic Teachings Ascribed to Hermes Trismegistus. 4 vols. Oxford: Clarendon Press, 1924-36.

Sethe, Kurt. *Die altägyptischen Pyramidentexte nach den Papierabdrücken und Photographien des Berliner Museums,* 4 vols. Leipzig: J. C. Hinrichs, 1908-22.

— *Amun und die acht Urgötter von Hermopolis.* Eine Untersuchung über Ursprung und Wesen des ägyptischen Götterkönigs. Abhandlungen der Preussischen Akademie der Wissenschaften. Philologisch-historische Klasse. Jahrgang 1929. Nr. 4. Berlin: Verlag der Akademie der Wissenschaften in Kommission bei Walter de Gruyter & Co., 1929.

— *Dramatische Texte zu altägyptischen Mysterienspielen.* Untersuchungen zur Geschichte und Altertumskunde Ägyptens 10. Leipzig: J. C. Hinrichs. 1928.

— and Firchow, Otto. *Thebanische Tempelinschriften aus griechisch-römischer Zeit.* Urkunden des ägyptischen Altertums 8. Berlin: Akademie-Verlag, 1957.

Simpson, R. S. *Demotic Grammar in the Ptolemaic Sacerdotal Decrees.* Griffith Institute Monographs. Oxford: The Griffith Institute, 1996.

Smith, H. S. and Tait, W. J. *Saqqâra Demotic Papyri* 1. Texts from Excavations 7. London: Egypt Exploration Society, 1983.

Smith, Mark. 'A Demotic Coffin Inscription: Berlin Äg. Inv. 7227', *Egyptian Religion: the Last Thousand Years.* Studies Dedicated to the Memory of Jan Quaegebeur. 2 vols. Orientalia Lovaniensia Analecta 84-5. Leuven: Uitgeverij Peeters en Departement Oosterse Studies, 1998.

— 'Did Psammetichus I Die Abroad?', *OLP* 22 (1981), 101-9.

— *The Liturgy of Opening the Mouth for Breathing.* Oxford: The Griffith Institute, 1993.

— *The Mortuary Texts of Papyrus BM 10507.* Catalogue of Demotic Papyri in the British Museum 3. London: British Museum Publications, 1987.

— 'New Middle Egyptian Texts in the Demotic Script', *Sesto Congresso Internazionale di Egittologia: Atti* 2. Turin: Comitato Organizzativo del Congresso, 1993.

— 'A New Version of a Well-known Egyptian Hymn', *Enchoria* 7 (1977) 115-49.

— 'P. Carlsberg 462: a Fragmentary Account of a Rebellion against the Sun God', *The Carlsberg Papyri* 3. A Miscellany of Demotic Texts and Studies. Carsten Niebuhr Institute Publications 22. Copenhagen: Museum Tusculanum Press, 2000.

— 'Remarks on the Orthography of Some Archaisms in Demotic Religious Texts', *Enchoria* 8 (1978), part 2, 17-27.

Spiegel, J. 'Ätiologie', *Lexikon der Ägyptologie* 1. Wiesbaden: Otto Harrassowitz, 1975.

Spiegelberg, Wilhelm. *Der ägyptische Mythus vom Sonnenauge (der Papyrus der Tierfabeln - 'Kufi') nach dem Leidener demotischen Papyrus I 384.* Strassburg: R. Schultz & Co., 1917.

— 'Amon als Gott der Luft oder des Windes (πνεῦμα)', *ZÄS* 49 (1911), 127-8.

— 'Aus der Strassburger Sammlung demotischer Ostraka', *ZÄS* 49 (1911), 34-41.

— *Demotische Inschriften und Papyri.* Die demotischen Denkmäler 3. Catalogue général des antiquités égyptiennes du Musée du Caire 50023-50165. Berlin: Reichsdruckerei, 1932.

— *Demotische Papyrus aus den königlichen Museen zu Berlin.* Leipzig and Berlin: Giesecke & Devrient, 1902.

— *Der demotische Text der Priesterdekrete von Kanopus und Memphis (Rosettana).* Heidelberg: by the author, 1922.

— *Die demotischen Inschriften.* Die demotischen Denkmäler 1. Catalogue général des antiquités égyptiennes du Musée du Caire 30601-31166. Leipzig: W. Drugulin, 1904.

— *Der Sagenkreis des Königs Petubastis nach dem Strassburger demotischen Papyrus sowie den Wiener und Pariser Bruchstücken.* Demotische Studien 3. Leipzig: J. C. Hinrichs, 1910.

Sternberg, Heike. *Mythische Motive und Mythenbildung in den ägyptischen Tempeln und Papyri der griechisch-römischen Zeit.* Göttinger Orientforschungen. Reihe 4: Ägypten. Band 14. Wiesbaden: Otto Harrassowitz, 1985.

Sternberg-El Hotabi, Heike. *Ein Hymnus an die Göttin Hathor und das Ritual 'Hathor das Trankopfer Darbringen' nach den Tempeltexten der griechisch-römischen Zeit.* Rites égyptiens 7. Brussels: Fondation Égyptologique Reine Élisabeth, 1992.

— *Der Propylon des Month-Tempels in Karnak-Nord.* Göttinger Orientforschungen. Reihe 4: Ägypten. Band 25. Wiesbaden: Otto Harrassowitz, 1993.

Störk, Lothar. 'Schlange', *Lexikon der Ägyptologie* 5. Wiesbaden: Otto Harrassowitz, 1984.

Stricker, B. H. *De Overstroming van de Nijl.* Mededelingen en Verhandelingen van het Vooraziatisch-Egyptisch Genootschaap 'Ex Oriente Lux' 11. Leiden: E. J. Brill, 1956.

Tait, W. J. 'Annexe', *The Carlsberg Papyri* 1. Demotic Texts from the Collection. Carsten Niebuhr Institute Publications 15. Copenhagen: Museum Tusculanum Press, 1991.

— 'The Fable of Sight and Hearing in the Demotic *Kufi* Text', *Acta Orientalia* 37 (1976), 27-44.

— 'P. Carlsberg 207: Two Columns of a Setna-text', *The Carlsberg Papyri* 1. Demotic Texts from the Collection. Carsten Niebuhr Institute Publications 15. Copenhagen: Museum Tusculanum Press, 1991.

— 'P. Carlsberg 230: Eleven Fragments from a Demotic Herbal', *The Carlsberg Papyri* 1. Demotic Texts from the Collection. Carsten Niebuhr Institute Publications 15. Copenhagen: Museum Tusculanum Press, 1991.

— *Papyri from Tebtunis in Egyptian and in Greek.* Texts from Excavations 3. London: Egypt Exploration Society, 1977.

Thissen, Heinz Josef. 'Ägyptologische Beiträge zu den griechischen magischen Papyri', *Religion und Philosophie im alten Ägypten.* Festgabe für Philippe Derchain zu seinem 65. Geburtstag am 24. Juli 1991. Orientalia Lovaniensia Analecta 39. Leuven: Departement Oriëntalistiek and Uitgeverij Peeters, 1991.

— *Die demotischen Graffiti von Medinet Habu.* Demotische Studien 10. Sommerhausen: Gisela Zauzich Verlag, 1989.

— *Die Lehre des Anchscheschonqi (P. BM 10508).* Papyrologische Texte und Abhandlungen 32. Bonn: Dr Rudolf Habelt GMBH, 1984.

Thomas, Elizabeth. 'Cairo Ostracon J. 72460', *Studies in Honor of George R. Hughes.* Studies in Ancient Oriental Civilization 39. Chicago: The Oriental Institute, 1976.

— 'Solar Barks Prow to Prow', *JEA* 42 (1956), 65-79.

Traunecker, Claude. 'Les Rites de l'eau à Karnak d'après les textes de la rampe de Taharqa', *BIFAO* 72 (1972), 195-236.

—, Le Saout, Françoise, and Masson, Olivier. *La Chapelle d'Achôris à Karnak*, 2 vols. Mémoires du Centre Franco-Égyptien d'Étude des Temples de Karnak 2. Recherche sur les grandes civilisations. Synthèse no. 5. Paris: Éditions A. D. P. F., 1981.

von Turajeff, Boris. 'Zwei Hymnen an Thoth', *ZÄS* 33 (1895), 120-5.

Vandersleyen, Claude. 'Deux nouveaux fragments de la stèle d'Amosis relatant une tempête', *RdE* 20 (1968), 127-34.

— 'Une tempête sous le règne d'Amosis', *RdE* 19 (1967), 123-59.

Vandier, Jacques. *La Famine dans l'Égypte ancienne.* Recherches d'Archéologie, Philologie et d'Histoire 7. Cairo: Institut Français d'Archéologie Orientale, 1936.

— *Le Papyrus Jumilhac.* Paris: Centre National de la Recherche Scientifique, 1961.

te Velde, Herman. 'The Theme of the Separation of Heaven and Earth in Egyptian Mythology', *Studia Aegyptiaca* 3 (1977), 161-7.

Vernus, Pascal. *Athribis.* Textes et documents relatifs à la géographie, aux cultes, et à la histoire d'une ville du delta égyptien à l'époque pharaonique. Bibliothèque d'Étude 74. Cairo: Institut Français d'Archéologie Orientale, 1978.

Vittmann, Günter. 'Ein neuer religiöser demotischer Text', *ZÄS* 117 (1990), 79-88.

Volten, Aksel. *Demotische Traumdeutung (Pap. Carlsberg XIII und XIV verso).* Analecta Aegyptiaca 3. Copenhagen: Einar Munksgaard, 1942.

— 'Das Harpunierergestirn', *MDAIK* 16 (1958), 346-66.

— 'The Papyrus-Collection of the Egyptological Institute of Copenhagen', *ArOr* 19 (1951), 70-4.

Vos, R. L. *The Apis Embalming Ritual: P. Vindob. 3873.* Orientalia Lovaniensia Analecta 50. Leuven: Uitgeverij Peeters en Departement Oriëntalistiek, 1993.

Waitkus, Wolfgang. *Die Texte in den unteren Krypten des Hathortempels von Dendera. Ihre Aussagen zur Funktion und Bedeutung dieser Räume.* Münchner ägyptologische Studien 47. Mainz: Verlag Philipp von Zabern, 1997.

Weber, Manfred. 'Lebenshaus', *Lexikon der Ägyptologie* 3. Wiesbaden: Otto Harrassowitz, 1980.

West, M. L. *Early Greek Philosophy and the Orient.* Oxford: Clarendon Press, 1971.

— *The Orphic Poems.* Oxford: Clarendon Press, 1983.

Westendorf, Wolfhart. *Altägyptische Darstellungen des Sonnenlaufes auf der abschüssigen Himmelsbahn.* Münchner ägyptologische Studien 10. Berlin: Verlag Bruno Hessling, 1966.

— 'Schepsi', *Lexikon der Ägyptologie* 5. Wiesbaden: Otto Harrassowitz, 1984.

Wild, Henri. 'Statue de Hor-néfer au Musée des Beaux-Arts de Lausanne', *BIFAO* 54 (1954), 173-222.

Willems, Harco. *The Coffin of Heqata (Cairo JdE 36418).* A Case Study of Egyptian Funerary Culture of the Early Middle Kingdom. Orientalia Lovaniensia Analecta 70. Leuven: Uitgeverij Peeters en Departement Oriëntalistiek, 1996.

Wilson, Penelope. *A Ptolemaic Lexikon.* A Lexicographical Study of the Texts in the Temple of Edfu. Orientalia Lovaniensia Analecta 78. Leuven: Uitgeverij Peeters en Departement Oosterse Studies, 1997.

Winlock, H. E. and Crum, W. E. *The Monastery of Epiphanius at Thebes* 1. Publications of the Metropolitan Museum of Art Egyptian Expedition 3. New York: Metropolitan Museum of Art, 1926.

de Wit, Constant. *Les Inscriptions du temple d'Opet à Karnak*, 3 vols. Bibliotheca Aegyptiaca 11-13. Brussels: Fondation Égyptologique Reine Élisabeth, 1958-68.

Wolf, Walther. 'Der Berliner Ptah-Hymnus (P 3048, II-XII)', *ZÄS* 64 (1929), 17-44.

Wreszinski, Walter. *Der Papyrus Ebers* 1. Leipzig: J. C. Hinrichs, 1913.

Yoyotte, Jean. 'Apopis et la montagne rouge', *RdE* 30 (1978), 147-50.

Zandee, Jan. *Der Amunhymnus des Papyrus Leiden I 344, verso.* Leiden: Rijksmuseum van Oudheden, 1992.

Zandee, Jan. *Death as an Enemy According to Ancient Egyptian Conceptions.* Studies in the History of Religions (Supplements to *Numen*) 5. Leiden: E. J. Brill, 1960.

— *De Hymnen aan Amon van Papyrus Leiden I 350* (= *OMRO* 28 [1947]). Leiden: Rijksmuseum van Oudheden te Leiden, 1947.

— 'Sargtexte, Spruch 80 (*Coffin Texts* II 27d-43)', *ZÄS* 101 (1974), 62-79.

— 'Das Schöpferwort im alten Ägypten', *Verbum: Essays on Some Aspects of the Religious Function of Words Dedicated to Dr H. W. Obbink.* Studia Theologica Rheno-Traiectina 6. Utrecht: V/H Kemink en Zoon N. V., 1964.

Zauzich, Karl-Theodor. 'Einleitung', *The Carlsberg Papyri* 1. Demotic Texts from the Collection. Carsten Niebuhr Institute Publications 15. Copenhagen: Museum Tusculanum Press, 1991.

— 'Paläographische Herausforderungen I', *Enchoria* 19/20 (1992/93), 165-79.

Zivie-Coche, Christiane. 'Fragments pour une théologie', *Hommages à Jean Leclant* 4. Bibliothèque d'Étude 106/4. Cairo: Institut Français d'Archéologie Orientale, 1994.

GLOSSARY

This glossary lists every occurrence of each word in the constituent fragments of the cosmology edited here, with the exception of the definite articles, the genitival adjective, the dative, the prepositions *n* and *r*, and the suffix pronouns *w* and *f*. It is divided into three main parts: general vocabulary, divine names, and toponyms. Words of uncertain reading and numbers are listed at the end of the first part.

As is to be expected in a fragmentary text like this one, there are numerous instances of damaged or imperfectly preserved words. Where these can be identified with certainty, they are listed in the appropriate place in the glossary, followed by the symbol [d] (= damaged). If enough is preserved of a word to permit a tentative identification, but this cannot be established with certainty, it is entered in the appropriate place with a query, again followed by the symbol [d]. Where more than one such reading can be suggested, the word is listed, with a query, under each relevant entry and these are suitably cross-referenced.

In the context of this glossary, any word which is not completely preserved is considered to be damaged and labelled as such, even if only a single sign or part of one has been lost. In the transliteration and translation, by contrast, damage is indicated in a slightly different way. There, only signs of which no trace remains have been placed within square brackets; those of which a part survives have been left unbracketed or, more rarely, placed within half-brackets. It is possible, therefore, that a word for which no indication of damage is given in the transliteration or translation will be marked with the symbol [d] in the glossary. This is because the intention in the former sections of the book is to show as precisely as possible which individual signs can actually be read on the papyrus and which cannot, whereas here, for obvious practical reasons, it is only the state of preservation of a word as a whole that is taken into account.

Some words in the text are so poorly preserved that, although it is possible to read one or more of their constituent signs, not even a tentative identification can be suggested for the whole. These are recorded in a separate section following the first three named above: traces of unidentifiable words. The signs in question have been listed in order of their occurrence in the manuscript, starting with those in fragment 1 and ending with those in fragment 47 (there are none in fragment 48). Each entry in this section of the glossary consists of two main parts: a report of the identifiable signs or traces, e.g. '*y* + sun determinative', and the fragment and line number in which they appear. Occasionally, I have added a brief comment to an entry as well, for instance a reference to a textual note in which the sign or signs in question are discussed.

Unless stated otherwise, it should be assumed that all signs reported in an entry of this type belong to the same word. Where these are a determinative or a phonetic sign followed by a determinative, it is obvious that one has to do with that word's end. With other phonetic signs, it is not always so simple to ascertain their original positions in the words of which they are now the sole remnants. In a few cases, it is clear that a particular sign was word-initial. These I have labelled as such. Where no certainty is possible, however, I have refrained from speculation about where a sign may have stood. Tentative identifications of signs are marked with a query. In this section, traces for which not even a tentative identification can be suggested are left out of consideration.

A. GENERAL VOCABULARY

ꜣ

ꜣꜥ.t	'back'. Frag 33, line x+1 [d]. For the writing, see textual note (b) ad loc. See also *ꜣt*.
ꜣll	'shout, cry'. Frag 1, line 2 [d] (or *šll* or *tll* ?). See textual note (a) ad loc.
ꜣḥ	noun, meaning unknown. Frag 3, line 11. See textual note (b) ad loc.
ꜣsy	'reed'. Frag 1, lines 12 and 20 [d]; Frag 2, lines 2 and 3 [d].
ꜣt	'back'. Frag 14, II, 18 [d]. See also *ꜣꜥ.t*.
ḥr-ꜣṯ	'upon'. Frag 1, line 10; Frag 4, II, 16 [d]; Frag 7, line x+5 [d]. All with following suffix pronoun.
ꜣty.w	'cows' (lit. vulvas). Frag 13, II, 19. See textual note (a) ad loc.

i

i-ir	see under *ir*.
iꜣw.wt	'cattle'. Frag 1, line 4.
iy	'come'. Frag 4, II, 13.
iꜥḥ	'moon'. Frag 44, line x+11 [d]. In the expression *r rꜥ iꜥḥ*, 'forever and ever'. See textual note ad loc.
iw=	marker of present I. Forms: *iw=k* (2nd person singular masculine pronoun subject). Frag 1, line 7*.

iw=f (3rd person singular masculine pronoun subject). Frag 1, line 10 [d]; Frag 7, line x+8* [d]; Frag 12, II, x+8* [d]; Frag 13, II, 1 (twice); Frag 21, line x+3* [d]; Frag 23, II, 8* [d]; Frag 24, line x+11* [d]; Frag 27, line x+3* [d].

iw=s (3rd person singular feminine pronoun subject). Frag 6, line 8(?)* [d].

Examples marked with * could also be circumstantial present.

iw	circumstantial converter. Frag 24, line x+8 [d]; Frag 25, line x+3(?) [d]; Frag 44, line x+7(?) [d]. See also *r*.
iw=	marker of circumstantial present. Forms:

iw=f (3rd person singular masculine pronoun subject). Frag 1, line 10(?) [d]; Frag 2, line 3; Frag 3, line 6; Frag 5, II, 1; Frag 11, I, 2, 9; Frag 12, I, 13; Frag 13, II, 2 (twice), 16 [d]; Frag 14, II, 9 [d]; Frag 15, line x+5 [d]; Frag 18, line x+5; Frag 38, line x+15.

iw=s (3rd person singular feminine pronoun subject). Frag 30, line 6 (twice, once [d]).

iw=w (3rd person plural pronoun subject). Frag 1, line 4 [d]; Frag 4, II, 22 [d]; Frag 14, I, 11, II, 13 [d].

For possible additional examples, see entry for marker of the present I above.

iw= (r)	marker of the future. Forms:

iw=y (1st person singular pronoun subject). Frag 5, II, 3.

iw=f r (3rd person singular masculine pronoun subject). Frag 1, line 7; Frag 39, line x+4 [d] (or conditional?).

iw=n r (1st person plural pronoun subject). Frag 6, line 4.

r-iw= r	marker of circumstantial future. Form: *r-iw=f r* (3rd person singular masculine pronoun subject). Frag 4, II, 11. No other forms attested.
iw=	marker of conditional. Form: *iw=f* (3rd person singular masculine pronoun subject). Frag 39, line x+4 (or future?).
iw	'come'. Frag 1, line 10; Frag 5, II, 6; Frag 11, I, 12 [d]; Frag 22, line 9 [d] (or *irm*?).
iwf.w	'flesh'. Frag 14, II, 2 [d], 3 [d], 6 [d], 7; Frag 26, line x+5 [d].
iwt	'among, between'. Frag 3, line 2; Frag 4, II, 23; Frag 5, I, 11.
ipt.w	'birds'. Frag 13, II, 11 [d].
imn	'be hidden'(?). Frag 3, line 6 [d]. See textual note (b) ad loc.

imnt — 'west'. Frag 42, line x+3 [d]. For *imnt* in the combination *pr-imnt*, see under *pr*. Perhaps this combination should be restored here.

in-sny — 'pass out, emanate'. Frag 5, II, 1 [d].

iny — 'stone'. Frag 2, line 11 [d]; Frag 10, line x+11 [d]; Frag 14, I, 12, 21 [d], II, 6; Frag 16, line x+3(?) [d]; Frag 28, line x+10 [d]. Plural form *iny.w*. Frag 14, II, 1 [d], 4 (twice, both [d]).

intm — 'phallus'(?). Frag 3, line 13. See textual note (b) ad loc.

ir — 'make, do'. Frag 1, line 4; Frag 2, line 5 (in expression *ir mꜣwy*, 'take thought'), line 9 [d] (or *ḏ*?); Frag 4, II, 18; Frag 5, I, 8 [d] (or *ḏ*?), II, 8, 11; Frag 6, line 3; Frag 10, lines x+1, x+7 [d]; Frag 11, I, 2 [d], 9(?) [d], 13 [d]; Frag 12, I, 9 [d] (or *ḏ*?); Frag 19, line x+10 (twice, both [d], or is the second one *ḏ*?); Frag 22, lines 4, 6; Frag 27, line x+2 [d]; Frag 34, line x+4; Frag 35, line 4 (?)[d].

Special form *ir=f* (with 3rd person singular masculine pronoun subject). Frag 2, line 6; Frag 4, II, 15 [d], 16.

For *ir* in *b-ir* and *b-ir-tw*, see under *b*.

i-ir — past participle. Frag 1, lines 3, 4; Frag 3, line 16*; Frag 4, II, 6*; Frag 5, II, 4; Frag 7, line x+6(?)[d]; Frag 10, line x+9; Frag 13, II, 13 [d], 19 [d]; Frag 14, I, 9 [d]; Frag 19, line x+4; Frag 25, line x+4* [d]; Frag 28, line x+5* [d]; Frag 36, line x+3 [d]; Frag 41, line x+6 [d].

Examples marked with * occur in broken contexts and could also be the second tense marker.

i-ir — second tense marker. With noun subject. Frag 2, line 15 [d]; Frag 6, line 5 [d]; Frag 8, line x+8; Frag 12, II, x+9 [d]; Frag 13, II, 15 [d]; Frag 14, II, 8, 15 [d]; Frag 38, line x+1 [d] (or *i-ir=s*?), line x+13(?) [d]. Other forms: *i-ir=f* (3rd person singular masculine pronoun subject). Frag 1, lines 11, 22 [d]; Frag 2, lines 1 [d], 10 [d]; Frag 4, II, 3; Frag 14, I, 8 [d]; Frag 38, line x+4 [d]; Frag 41, line x+5. *i-ir=s* (3rd person singular feminine pronoun subject). Frag 38, line x+1 [d] (or *i-ir* + *nꜣ* or *tꜣ*?). *i-ir=w* (3rd person plural pronoun subject). Frag 5, II, 9; Frag 19, line x+16; Frag 24, line x+2 [d]; Frag 28, line x+6(?) [d].

For possible additional examples, see the preceding entry.

ir.t — 'eye'. Frag 3, lines 12 [d], 18 [d]; Frag 5, II, 20(?) [d]; Frag 17, line x+2 [d]; Frag 18, line x+4 [d]; Frag 30, lines 2 (written *ir.ṱ* + suffix pronoun), 3 [d]; Frag 32, line x+4 (*ir.ṱ* + suffix); Frag 37, line x+1 [d].

ir.t	'companion'. Frag 14, II, 11.
irm	'and, together with'. Frag 1, line 5 [d]; Frag 4, II, 18; Frag 5, II, 15 [d]; Frag 11, I, 3 [d], 11; Frag 12, I, 14, II, x+11; Frag 13, II, 2, 5, 7, 11, 15; Frag 19, lines x+12 [d], x+15, x+19 [d]; Frag 20, line 3 [d]; Frag 22, line 9 [d] (or *iw*?); Frag 24, line x+10; Frag 38, lines x+6, x+7; Frag 42, line x+2 [d]; Frag 48, line x+2.
ihy	noun, meaning unknown. Frag 45, line 1. See textual note ad loc.
iḥ.(t)	'cow'. Frag 8, line x+7. Plural form *iḥ.w(t)*. Frag. 3, line 17(?) [d]. See textual note (a) ad loc.
iḫy.t	'horizon'. Frag 25, line x+2(?) [d]. See textual note ad loc.
ist	'ground, platform'. Frag 1, line 19 [d]; Frag 2, line 4; Frag 10, line x+8; Frag 14, I, 4; Frag 24, line x+7 [d].
itn	'disk'. Frag 13, II, 16; Frag 38, lines x+4 [d], x+6 [d].
iṯ	'father'. Frag 2, lines 9 (with initial stroke omitted through haplography), 17 [d]; Frag 5, I, 9 [d]; Frag 11, I, 1 [d].

y

ym	'sea'. Frag 13, II, 6 [d]. Plural form *ym.w*. Frag 14, II, 11 [d].

ʿ

ʿꜣ	'great'. Frag 6, line 6; Frag 11, I, 2 [d]; Frag 47, line 2 [d]. Feminine form *ʿꜣ.t*. Frag 1, line 8 [d].
ʿꜣ.t	'stone'. Frag 14, I, 19 [d]; Frag 26, line x+2.
ʿʿe.w	'limbs'. Frag 3, line 2 [d]; Frag 4, II, 12; Frag 19, line x+17.
ʿwy	'house'. Frag 6, line 10 [d].
ʿb.w	'horns'. Frag 3, lines 8 [d], 10, 12 [d], 22 [d]; Frag 18, line x+5 [d].
ʿn	'again, also, as well, still'. Frag 2, line 14; Frag 5, I, 17 [d]; Frag 14, I, 16 [d], 21, II, 6, 7, 8; Frag 24, lines x+2, x+4 [d]; Frag 28, lines x+4 [d], x+7; Frag 38, line x+7; Frag 39, line x+4 [d].
ʿnw	'be fair, beautiful'. Frag 11, I, 9 [d].
nꜣ.w-ʿn	meaning as last. Frag 3, line 13 [d].
ʿnḫ	verb 'live'. Frag 1, line 5(?) [d] (see textual note (d) ad loc.); Frag 13, II, 6 [d], 12 [d].
ʿnḫ	noun 'life'. Frag 13, II, 3 [d], 5, 6, 12 [d], 15.
	For *ʿnḫ* in the compound *pr-ʿnḫ*, see under *pr*.

ʿnḫ	'living'. Frag 11, I, 7 [d].
ʿḥʿ	'stand, stand up'. Frag 1, line 6; Frag 2, line 11(?) [d] (in the compound *ti ʿḥʿ*, 'set up, erect'?); Frag 12, II, x+5 [d].
ʿšm	'quench'. Frag 13, II, 8.
ʿq	'enter'. Frag 11, I, 2 [d], 4; Frag 13, II, 16; Frag 38, line x+10 [d].

w

w	3rd person plural suffix pronoun. Passim.
w	stative ending. Third person singular masculine subject. Frag 5, II, 1. Third person plural subject. Frag 14, I, 11. *w* occurs in broken contexts in Frag 2, line 5; Frag 4, II, 21; Frag 5, I, 4; Frag 7, line x+3; and Frag 38, line x+9, where it could conceivably be a stative ending but might equally well be a suffix pronoun or plural marker.
wꜣy	'be distant'. Frag 5, II, 14 [d].
wꜣḥ	(a) 'lay, set down'. Frag 19, line x+6 [d]; Frag 31, line x+3 [d]. (b) 'come to rest, stop'. Frag 1, line 12(?) [d]. See textual note (b) ad loc.
wꜣḥ	marker of the perfect. Frag 19, line x+9.
wyn	'light'. Frag 19, line x+8 [d].
wʿ	noun 'one'. Frag 13, II, 13 (twice, once [d]).
wʿ	indefinite article. Frag 13, II, 4. Feminine form *wʿ.t*. Frag 1, lines 6, 8, 15(?) [d]; Frag 4, II, 19 [d] (or *w*?). See textual note (b) ad loc.
wpy	'open'. Frag 1, line 9 [d]; Frag 5, II, 18 [d], 19(?) [d]; Frag 8, line x+6 [d].
wn	'open'. Frag 30, line 3; Frag 31, lines x+2 [d], x+4 [d].
wn-mtw	'have, possess'. Frag 4, II, 19 [d]; Frag 5, II, 13 [d].
(*r*)-*wn-nꜣ.w*	imperfect participle. Frag 8, line x+9 [d] (*nꜣ.w wn-nꜣ.w* written for *nꜣ r-wn-nꜣ.w*); Frag 41, line x+4 [d].
wnw.t	'hour'. Frag 3, line 7 [d].
wnmy	noun, meaning unknown. Frag 28, lines x+4(?) [d], x+5 [d].
wnḥ	'appear'. Frag 1, line 19(?) [d]. See textual note (c) ad loc.
wr	'great, great one'. Frag 8, line x+5 [d]; Frag 43, line 2(?) [d].
wḫꜣ	verb 'desire'. Frag 38, line x+9 [d].
wsm	'ruin'. Frag 13, II, 10.

wsẖ	'hall'. Frag 33, line x+2 [d].
wtḥ	verb 'smelt'. Frag 26, line x+4 [d].
wtḥ	'vessel'. Frag 26, line x+3(?) [d]. See textual note (a) ad loc.
wḏꜣ	'be well'. Frag 13, II, 1.
wḏy	'sound eye'. Frag 13, II, 16 [d].

b

b-ir	marker of negative habitude. Frag 5, II, 14; Frag 13, II, 9 [d], 14; Frag 15, line x+2; Frag 22, line 3 [d].
b-ir-tw	marker of negative perfect. Frag 1, line 19 [d]; Frag 11, I, 10 [d].
bꜣ	'thicket'. Frag 1, lines 8 [d], 18 [d], 20 [d]; Frag 2, line 2 [d].
m-bꜣḥ	'before'. Frag 1, line 2.
bw	'exhalation'(?). Frag 13, II, 3 [d]. See textual note (a) ad loc.
bn-pw	marker of negative past. Frag 11, I, 4 [d], 5.
bnpy	'iron'. Frag 43, line 2(?) [d]. See textual note (a) ad loc.
bnr	'outside, exterior'. Frag 5, II, 14; Frag 36, line 3 [d]. In both passages, what stood before *bnr* has been lost. Perhaps restore as one of the following.
n-bnr	'except'. Frag 11, I, 4.
r-bnr	adverb 'out'. Frag 2, line 8; Frag 5, I, 13 [d], II, 1, 17, 19 [d]; Frag 11, I, 1(?) [d]; Frag 14, II, 7; Frag 19, line x+13.
bl	'blind one'. Frag 8, line x+4(?) [d]. See textual note ad loc.
bš	'shed'. Frag 4, II, 14.

p

p.t	'sky'. Frag 2, line 15 [d]; Frag 4, II, 4, 5 [d], 18 [d]; Frag 5, I, 3, II, 3 [d]; Frag 11, I, 2(?) [d], 3 [d], 4 [d], 5(?) [d]; Frag 12, I, 17(?) [d]; Frag 13, II, 2, 21; Frag 22, line 6 [d]; Frag 34, line x+5 [d]; Frag 38, lines x+14 [d], x+15 [d]; Frag 44, line x+12(?) [d]; Frag 45, line 2 [d].
pꜣ	masculine singular definite article. Passim.
pꜣy	masculine singular demonstrative article. Frag 20, line 2.
pꜣy	copula. Frag 2, line 1 (twice, once [d], or is the first example *f*?), line 11; Frag 4, II, 13 [d]; Frag 12, I, 11; Frag 13, II, 3, 4, 14; Frag 19, line x+6(?); Frag 30, line 5; Frag 41, line x+5.

pꜣy	'leap up'. Frag 10, line x+6 [d].
pꜣy	'ejaculate'. Frag 13, II, 18.
pꜣy=	formative element of possessive article used before masc. sing. nouns. Forms: *pꜣy=k* (2nd person singular masculine possessor). Frag 1, line 3 [d]. *pꜣy=f* (3rd person singular masculine possessor). Frag 6, line 7 (once, possibly twice, both [d]); Frag 13, II, 21(?) [d]; Frag 15, line x+4(?) [d]. *pꜣy=w* (3rd person plural possessor). Frag 2, lines 7 (twice), 9, 12 [d], 16; Frag 11, I, 1 [d]; Frag 22, line 2 [d]; Frag 24, line x+8 [d]; Frag 44, line x+11 [d].
pr	verb 'come out or forth, emerge'. Frag 1, line 11; Frag 4, II, 21 [d]; Frag 5, I, 6, II, 4 [d]; Frag 7, lines x+2, x+6 [d]; Frag 11, I, 16 [d]; Frag 13, II, 3 [d]; Frag 14, II, 9, 13 [d]; Frag 19, lines x+2, x+19 [d]; Frag 26, line x+3 [d]; Frag 30, line 2 [d]; Frag 34, line x+4 [d]; Frag 36, line 2 [d]; Frag 48, line x+3 [d].
pr	noun 'emergence'. Frag 14, I, 18 [d]; Frag 19, line x+15 [d].
pr-imnt	'the west'. Frag 10, line x+12 [d]; Frag 38, line x+5 [d]; Frag 42, line x+3(?) [d]. See also *imnt*.
pr-ꜥnḫ	'house of life'. Frag 34, line x+3 [d].
pr.t	'seed'. Frag 13, II, 19 [d].
plḫ	'spread out'. Frag 13, II, 18.
plg	'detach, free'. Frag 1, line 5 [d] (or *glg* or *tlg*?). See textual note (c) ad loc.
pḥ	'reach, arrive at'. Frag 1, line 6; Frag 4, II, 21; Frag 14, II, 8; Frag 29, line x+2.
pḥ.ṱ	'hindquarters'. Frag 3, lines 3 [d], 5; Frag 31, line x+5.

f

f	3rd person singular masculine suffix pronoun. Passim.
fy	'carry, rise, fly up'. Frag 4, II, 1 [d]; Frag 5, I, 20 [d]; Frag 36, line 1 [d].
ftyw	'efflux'. Frag 14, II, 5 [d], 7 [d], 8 [d], 9 [d]. For reading and meaning, see textual note (a) on Frag 14, II, 5.

m

m	preposition. For *m* in the compounds *m-bꜣḥ*, *m-ḥ.ṯ*, *m-sꜣ*, *m-šs*, and *m-qty*, see under the second elements.
mꜣ	'place'. Frag 1, line 7 [d]; Frag 13, II, 2 [d]; Frag 23, II, 3 [d].
mꜣꜣ	'see'. Frag 3, line 17.
mꜣwy	'thought'. Frag 2, line 5.
mw	'water'. Frag 1, line 6; Frag 5, I, 15, 22; Frag 7, line x+3 [d]; Frag 11, I, 7(?) [d], 8 [d], 10; Frag 13, II, 17 [d]; Frag 14, II, 8 (twice, once [d]); Frag 21, line x+1 [d]; Frag 27, line x+2; Frag 29, line x+2; Frag 32, line x+2 [d]; Frag 39, line x+1. Plural form *mw.w*. Frag 19, line x+18; Frag 27, line x+5(?) [d].
mw	'like, as'. Frag 3, line 15(?) [d]. See textual note (a) ad loc.
mwt	'death'. Frag 13, II, 2, 3 [d], 5, 6 [d], 8 [d], 12, 15 (twice).
mny	'image'. Frag 2, lines 9 [d], 10 [d], 11 [d].
mny	'daily'. Frag 13, II, 10.
mḥ	'burn'. Frag 30, line 6 [d].
mḥ	formative element in ordinal numbers, q.v.
mḥt	'north'. Frag 1, line 12; Frag 33, line x+1 [d]; Frag 38, line x+2 [d].
mḥt.t	adjective 'northern'. Frag 38, line x+14 [d].
mḥṯ	'seize, take possession of'. Frag 31, II, 13 [d].
ms	'bear, give birth, fashion'. Frag 9, line x+5 [d]; Frag 12, I, 12; Frag 13, II, 8 (written *ms.ṯ* + suffix pronoun); Frag 27, line x+6 [d]; Frag 38, line x+4.
msḥ	'crocodile'. Frag 23, I, unnumbered line(?) [d]. See textual note ad loc. Plural form *msḥ.w*. Frag 6, line 2 [d]; Frag 7, line x+4.
msḫny	'birth brick'. Frag 44, lines x+1 [d], x+10 [d]. See textual note on the former.
mt	verb 'speak'. Frag 5, II, 2; Frag 10, line x+15 [d] (or *ḫpr* or *ḫrw*?). See textual note ad loc.
mt	noun 'word, speech, matter, thing'. Frag 1, line 18(?) [d]; Frag 12, II, x+6 [d]; Frag 41, line x+3.
mtw	marker of conjunctive. With noun subject. Frag 4, II, 20 [d]; Frag 5, I, 15; Frag 11, I, 3 (twice, or both written for *nt-iw*?); Frag 13, II, 8. Other forms: *mtw=f* (3rd person singular masculine pronoun subject). Frag 2, line 11; Frag 5, II, 11 [d]; Frag 15, line x+1 [d]; Frag 36, line 2 [d]. *mtw=s* (3rd person singular feminine pronoun

	subject). Frag 44, line x+12. *mtw=w* (3rd person plural pronoun subject). Frag 14, I, 19 [d], II, 1; Frag 29, line x+1 [d].
mtw=	formative element of independent pronoun. Forms: *mtw=k* (2nd person singular masculine). Frag 5, II, 2(?) [d], 5. *mtw=f* (3rd person singular masculine). Frag 13, II, 8 [d], 14 (twice, once [d]), 16, 17 [d], 18 [d], 19 (twice, both [d]), 20 [d], 21; Frag 36, line 3 [d]. *mtw=w* (3rd person plural). Frag 13, II, 4 (twice).
mtw	'belonging to'. Frag 2, line 3. For *mtw* in the combination *wn-mtw*, see under *wn*.
mtw	writing of relative converter *nt-iw*, q.v. Frag 5, I, 14 (or *nt ṯ*?); Frag 11, I, 3 (twice, or both conjunctive?); Frag 12, I, 11; Frag 13, II, 5 [d], 9, 21 [d]; Frag 14, II, 4, 6 (twice); Frag 22, line 2.
mtwy	'seed, semen'. Frag 1, line 13 [d]; Frag 2, line 6; Frag 11, I, 6 [d].
mtre	'meet'. Frag 2, lines 14 [d], 15 [d].
mḏ	'phallus'. Frag 3, line 4.

n

n	genitival adjective. Passim.
n	dative 'to, for'. Passim. Forms with following suffix pronoun: *n=f*. Frag 1, line 12; Frag 2, line 11(?) [d]; Frag 4, II, 3; Frag 5, II, 6, 11; Frag 9, line x+5 [d]; Frag 14, I, 2; Frag 19, line x+10; Frag 21, line x+1; Frag 28, line x+6. *n=s*. Frag 12, I, 9, II, x+8(?) [d]; Frag 30, line 4. *n=n*. Frag 1, line 3. *n=w*. Frag 2, line 9; Frag 4, II, 23.
n	preposition (< earlier *m*). Passim. Forms with following suffix pronoun: *n-im=k*. Frag 5, II, 5. *n-im=f*. Frag 2, line 12; Frag 4, II, 13, 21; Frag 13, II, 13; Frag 36, line 2 [d]. *n-im=s*. Frag 1, line 5. *n-im=w*. Frag 13, II, 13. For possible additional examples of *n-im=w*, see the following entry.

n-im=w — adverb 'there'. Frag 5, II, 9; Frag 11, I, 10 [d]; Frag 28, line x+1 [d]. All three examples occur in broken contexts and could also be the preposition *n* (< earlier *m*) + suffix pronoun *w*.

n3 — plural definite article. Passim.

n3y — plural demonstrative pronoun. Frag 2, line 5 [d]; Frag 6, line 5; Frag 9, line x+4.

n3y= — formative element of possessive article used before plural nouns. Forms:

> *n3y=f* (3rd person singular masculine possessor). Frag 3, lines 2, 12; Frag 4, II, 12, 21; Frag 17, line x+2 [d]; Frag 18, line x+4(?) [d].
>
> *n3y=n* (1st person plural possessor). Frag 2, lines 9 [d], 10.
>
> *n3y=w* (3rd person plural possessor). Frag 2, line 13 [d]; Frag 35, line 2.

n3.w — plural demonstrative pronoun 'those'. Frag 38, line x+7.

n3.w — plural copula. Frag 2, lines 10, 13 [d], 14; Frag 5, I, 19; Frag 9, line x+6; Frag 13, II, 5 [d]; Frag 14, I, 10 [d], II, 3 [d], 5, 6 (twice, once [d]), 12; Frag 19, line x+14(?) [d].

n3.w — for *n3* + prothetic *r* of participle *r-wn-n3.w*. Frag 8, line x+9.

n3.w- — formative element in adjective verbs. For *n3.w-* in the adjective verbs *n3.w-ʿn*, *n3.w-sbk*, *n3.w-sḥt*, and *n3.w-q3*, see under the second elements. The verb following *n3.w-* in this construction at the end of Frag 3, line 5 has been lost.

nw — 'see'. Frag 1, line 12 [d]; Frag 5, I, 22 [d]; Frag 7, line x+3; Frag 22, line 7 [d]; Frag 38, line x+5 [d].

nb — 'lord'. Frag 14, II, 6 [d]; Frag 33, line x+3 [d] (or *nb*, 'every'?); Frag 34, line x+7 [d] (or *nb*, 'every'?); Frag 47, line 6 [d]. For the writing of *nb* in Frag 14, II, 6, see textual note (c) ad loc.

nb — 'every'. Frag 2, line 3; Frag 4, II, 2; Frag 5, II, 4, 12; Frag 10, line x+13 [d] (or *r*?); Frag 13, II, 2, 6, 19 [d], 20; Frag 14, I, 3 [d], II, 13; Frag 15, line x+5; Frag 33, line x+3 [d] (or *nb*, 'lord'?); Frag 34, line x+7 [d] (or *nb*, 'lord'?). For the writing of *nb* in Frag 14, I, 3, see textual note ad loc.

nm — 'tamarisk' or 'pole'. Frag 19, line x+10. For the possible meanings of the word, see textual note (a) ad loc.

nhy — 'sycomore'. Frag 12, II, x+4.

nhm — 'rejoice'. Frag 1, lines 3 [d], 4.

nhsy — 'rise up, awaken'. Frag 10, line x+10 [d]; Frag 13, II, 7 (written *nhsy.ṯ* + suffix pronoun).

nḥḥ — 'eternity'. Frag 19, line x+14; Frag 24, line x+8.

nkt — 'property'. Frag 14, II, 5 [d].

nt — relative converter. Frag 2, line 2; Frag 4, II, 2, 23 (or *ṯ*?); Frag 5, I, 14 (or *mtw* instead of *nt ṯ*?), II, 4, 16; Frag 10, lines x+4 [d] (or *r-ḏbꜣ*?), x+6, x+11, x+12 [d]; Frag 12, II, x+5 [d]; Frag 13, II, 1, 2, 4 (twice), 5, 6 (3 times), 7, 8 (twice), 9, 10, 12, 14 (twice), 16, 17, 18 (twice), 20; Frag 14, I, 10, II, 2, 9, 10(?) [d]; Frag 19, line x+5; Frag 20, line 3; Frag 26, lines x+2 [d], x+3; Frag 27, line x+5; Frag 30, line 2 [d]; Frag 32, line x+3. Other forms: *nt-iw* (with noun subject different to antecedent). Frag 3, line 17. See also *mtw*. *nt-iw=f* (with 3rd person singular masculine pronoun subject different to antedecent). Frag 2, line 12.

ntr — 'god'. Frag 1, line 17(?) [d]; Frag 2, lines 8 [d], 14; Frag 3, lines 7 [d], 9 [d], 13, 15; Frag 5, II, 12; Frag 8, line x+8; Frag 11, I, 2 [d]; Frag 12, II, x+12 [d]; Frag 19, lines x+6, x+19; Frag 22, line 2 [d]; Frag 23, II, 8 [d]; Frag 42, line x+4(?) [d]. Plural form *ntr.w*. Frag 1, lines 12 [d], 15 [d]; Frag 5, I, 1, II, 9 [d]; Frag 6, line 3 [d]; Frag 7, line x+7 [d]; Frag 11, I, 3 [d]; Frag 13, II, 1 [d], 10 [d].

r

r — preposition. Passim. Forms with following suffix pronoun:

- *r-r=s*. Frag 1, lines 13, 16 [d]; Frag 14, II, 8.
- *r-r=w*. Frag 5, I, 9; Frag 8, line x+3(?) [d]; Frag 11, I, 1 [d], 5; Frag 12, I, 16; Frag 14, II, 2 [d]; Frag 22, line 7; Frag 23, II, 4 [d]; Frag 29, line x+2 [d]; Frag 36, line 4 [d].

r — circumstantial converter. Frag 1, line 19; Frag 6, line 5; Frag 10, line x+5; Frag 11, I, 10 [d]; Frag 13, II, 1; Frag 14, II, 11; Frag 22, line 3 [d]; Frag 30, line 3; Frag 31, line x+4. See also *iw* and *r-iw= r*.

r — marker of past relative. Frag 2, line 10; Frag 10, line x+7; Frag 11, I, 13; Frag 15, line x+6 [d]; Frag 44, line x+4.

r — in imperfect participle *r-wn-nꜣ.w*. See under *wn-nꜣ.w*.

rꜣ — 'mouth'. Frag 5, I, 13 [d], II, 2, 17 [d], 18 [d]; Frag 7, line x+7 [d]; Frag 10, lines x+5, x+6 [d]; Frag 18, line x+2(?) [d]; Frag

	19, lines x+12 [d], x+13; Frag 21, lines x+2, x+4 [d], x+5 [d]; Frag 48, line x+1(?) [d].
rꜣ	'door'. Frag 2, lines 14, 15 [d].
rꜥ	'sun'. Frag 44, line x+11 [d]. In the expression *r rꜥ iꜥḥ*, 'forever and ever'. See textual note ad loc.
rpy.w	'temples'. Frag 14, I, 20 [d].
rmt	'man'. Frag 2, line 11. Plural form *rmt.w*. Frag 14, II, 2.
rn	'name'. Frag 1, lines 2, 11, 12 (twice, once [d]), 14 [d], 22(?) [d], 23 [d]; Frag 2, lines 10, 11, 12, 13 [d], 14, 15 [d], 16 (twice, once [d], or is first example *ḏ*?); Frag 4, II, 6 [d], 22; Frag 5, II, 6 [d], 11 [d], 18; Frag 8, line x+9; Frag 11, I, 9 [d], 12, 17 [d], 18 [d]; Frag 12, II, x+2, x+3 [d], x+7 [d]; Frag 22, line 2 [d]; Frag 23, II, 7; Frag 28, lines x+6 [d], x+8 [d]; Frag 30, line 4 [d]; Frag 38, lines x+3, x+8 [d].
rnp	'be renewed, rejuvenated'. Frag 13, II, 1, 9 [d].
rḫ	'know, be able' (to do something). Frag 1, line 11; Frag 11, I, 12 [d]; Frag 13, II, 15; Frag 22, line 3.
rt	'grow'. Frag 1, lines 1, 7 [d], 8 [d]; Frag 4, I, x+6 [d]; Frag 7, line x+8 [d]; Frag 10, line x+14 [d]; Frag 14, I, 11 [d].
rt	'feet'. Frag 2, line 4 [d] (written *rṯ* + suffix pronoun); Frag 3, line 20 [d]; Frag 36, line 1 [d] (*rṯ* + suffix pronoun).

l

ly.t	'side'. Frag 1, line 9 [d]; Frag 10, line x+10 [d]; Frag 11, I, 8 [d]; Frag 23, II, 5(?) [d]. The first three examples occur in the phrase *n tꜣ ly.t ḥry.t*, 'above'.
lym.w	'fish'. Frag 13, II, 11 [d].
lwš	'care, concern'. Frag 38, line x+12.
lhny	'resting place'. Frag 1, lines 6, 14(?) [d]; Frag 2, line 5 [d]; Frag 4, II, 11. For the doubtful example of *lhny* in Frag 1, line 14, see textual note (b) ad loc.

h

hyny.w	plural indefinite article. Frag 2, line 6 [d]; Frag 14, II, 3 [d], 4 [d].
hb	'ibis'. Frag 5, I, 20 [d], II, 13 [d]; Frag 20, lines 2 [d] (or *hb*, 'send'?), 4 [d].

hb	'send'. Frag 5, II, 3 [d]; Frag 20, line 2 (twice, both [d], or is second one *hb*, 'ibis'?).
r-hn-r	'up to, until'. Frag 1, line 1 [d]; Frag 4, II, 14-15 [d]; Frag 5, II, 15-16 [d]; Frag 10, line x+2 [d]; Frag 12, I, 1 [d]; Frag 13, II, 5 [d]; Frag 19, line x+15 [d]; Frag 32, line x+3 [d].
šʿ-hn-r	'up to, until'. Frag 2, line 17 [d].
hrw	'day'. Frag 1, line 1; Frag 2, line 17; Frag 4, II, 15; Frag 5, II, 16 [d]; Frag 13, II, 5; Frag 19, line x+15. All in the expression *pꜣ hrw*, 'today'. Plural form *hrw.w*. Frag 19, line x+6 [d].

ḥ

ḥ.t	'front'. Frag 44, line x+9 [d]. The context is broken; perhaps restore *m-ḥ.t* or *n-ḥ.t* as below. The trace immediately after the *t* could represent the *ṯ* added to writings of *ḥ.t* when a suffix pronoun follows.
m-ḥ.ṯ	'before, in front of'. Frag 3, line 14 [d] (+ suffix pronoun).
(*n*)-*ḥ.ṯ*	meaning as last. Frag 16, line x+4 (+ suffix pronoun).
ḥꜣt	'heart'. Frag 7, line x+5 [d] (written *ḥꜣṯ* + suffix pronoun); Frag, 23, II, 6 [d].
ḥwy	'deposit'. Frag 15, lines x+1 [d], x+6 [d]. For the meaning, see textual note on the former line.
ḥwt	noun 'male'. Frag 2, lines 8 [d], 14 [d].
ḥbʿ	verb 'play'. Frag 3, line 11.
ḥbs	'jubilee'. Frag 17, line x+3 [d].
ḥp	'hide'. Frag 9, line x+3; Frag 10, line x+7.
ḥf	'serpent'. Frag 1, lines 16 [d], 23 [d]; Frag 14, I, 10 [d], II, 5; Frag 21, line x+4 [d]; Frag 38, line x+5 [d]. Plural form *ḥf.w*. Frag 14, I, 1 [d], 16 [d]; Frag 20, line 2(?) [d]; Frag 22, line 10 [d]; Frag 26, line x+5 [d]; Frag 30, lines 3 [d], 7 [d]; Frag 31, lines x+1 [d], x+2 [d].
ḥf.t	'female serpent'. Frag 38, lines x+2 [d], x+3.
ḥmsy	'sit, sit down'. Frag 1, line 20 [d]; Frag 4, II, 16 [d]; Frag 8, line x+9 [d].
ḥr	'face, head'. Frag 2, line 8; Frag 5, I, 12; Frag 19, line x+16 [d].
ḥr	preposition 'upon, concerning, because of'. Frag 1, lines 18, 20; Frag 2, lines 1, 2 [d], 4; Frag 4, II, 2(?) [d]; Frag 5, I, 1 [d], 21;

Frag 10, line x+11 [d]; Frag 13, II, 5, 18 [d]; Frag 14, I, 10, 11 [d]; Frag 19, line x+14; Frag 23, II, 6; Frag 33, line x+1; Frag 34, line x+8; Frag 39, lines x+3 [d], x+5. For *ḥr* in the compound *ḥr-ꜣṯ*, see under *ꜣṯ*. For *ḥr* in the compound *ḥr-ib*, see the following entry.

ḥr-ib ‘within’. Frag 3, line 9. The element *ḥr* in this compound is written with the flesh determinative, as if it were *ḥr*, ‘face’.

ḥry ‘upper part’. Frag 30, line 1. In the expression *r pꜣ ḥry*, ‘upwards’.

r-ḥry adverb ‘up’. Frag 10, line x+10 [d]; Frag 13, II, 17; Frag 44, line x+7 [d].

ḥry.t adjective ‘upper’. Frag 1, line 9 [d]; Frag 10, line x+10 [d]; Frag 11, I, 8 [d]; Frag 48, line x+2.

ḥl ‘fly’. Frag 13, II, 2.

ḥlly ‘flower’. Frag 13, II, 20 [d].

ḥlḥ ‘protect’. Frag 11, I, 3(?) [d].

ḥq verb ‘rule’. Frag 13, II, 15.

ḥty meaning unknown. Frag 28, line x+5. See textual note (a) ad loc.

ḥtp verb (or noun?) ‘rest’. Frag 5, I, 2. The word occurs in a damaged context and it is impossible to be certain what part of speech it is.

ḫ

ḫꜣy (a) ‘rise up’. Frag 1, line 9.

(b) ‘be long’. Frag 3, lines 8 [d], 10, 14, 16 [d]; Frag 4, II, 8(?) [d].

ḫꜥ verb ‘appear’. Frag 2, line 7; Frag 6, line 8; Frag 12, I, 8 [d]; Frag 13, II, 16 [d]; Frag 26, line x+2 [d].

ḫꜥ noun ‘appearance’. Frag 12, I, 17 [d].

ḫpr (a) ‘come into existence, be, occur, happen’. Frag 1, lines 3, 18 [d]; Frag 2, lines 2 [d], 3, 7 (twice, once [d]), 11(?) [d]; Frag 4, II, 8; Frag 5, I, 8; Frag 6, line 5; Frag 9, line x+6; Frag 10, lines x+9, x+15 [d] (or *mt* or *ḫrw*?); Frag 11, I, 8(?) [d], 13 [d], 16 [d]; Frag 12, II, x+6 [d], x+9 [d]; Frag 13, II, 2, 14, 20 [d]; Frag 14, II, 1, 3, 7 [d]; Frag 30, line 6 [d] (or *sṯ*?); Frag 38, lines x+7, x+8, x+13; Frag 40, line 3; Frag 41, line x+3(?) [d]; Frag 42, line x+4(?) [d].

(b) ‘begin’ (to do something), followed by circumstantial clause. Frag 11, I, 2.

ḫpr	noun 'fact'. Frag 11, I, 5; Frag 13, II, 15 [d] (in the phrase *r-ḏbꜣ ḫpr*).
ḫpr	'because'. Frag 1, line 7; Frag 14, II, 8 [d].
ḫm	'little'. Frag 2, lines 9 [d], 10.
ḫm-ẖl.w	'children'. Frag 2, line 10 [d].
ḫnṱ	'go upstream'. Frag 19, line x+3(?) [d].
ḫrw	'voice'. Frag 10, line x+15 [d] (or *mt* or *ḫpr*?). See textual note ad loc.
ḫt	'tree'. Frag 12, II, x+2 [d] (or *šn*?).

ẖ

ẖ.ṱ	'body'. Frag 19, line x+14 (+suffix pronoun).
ẖn	'in, from'. Frag 1, lines 6 [d], 22 [d]; Frag 2, lines 2, 8; Frag 4, II, 10; Frag 5, I, 4; Frag 6, line 1 [d]; Frag 7, line x+3 [d]; Frag 9, line x+3; Frag 11, I, 10 [d]; Frag 12, I, 14 [d], II, x+5; Frag 13, II, 1 [d], 16 (twice); Frag 14, I, 4 [d], 17, II, 1 [d]; Frag 20, line 3(?) [d]; Frag 22, line 1 [d] (written *ẖn n*); Frag 27, line x+5 [d]; Frag 32, line x+3; Frag 41, line x+4.
ẖny	'skin'. Frag 4, II, 14 [d]; Frag 10, line x+10; Frag 14, II, 6 [d].
ẖr	'under, bearing, filled'. Frag 13, II, 2 [d], 20 [d]; Frag 38, line x+12(?) [d]. Form with following suffix pronoun: *ẖr-r=w*. Frag 24, line x+3 [d].
ẖrt	'child'. Frag 3, line 17.
ẖl.w	in the compound *ḫm-ẖl.w*. See under *ḫm*.
ẖlꜣ	'sustenance'. Frag 1, line 3.
ẖly	'lower part'. Frag 22, line 9(?) [d]; Frag 24, line x+7. Both examples occur in broken contexts; perhaps restore as the following.
r-ẖly	adverb 'down'. Frag 1, line 10 [d].
ẖty	'go downstream'. Frag 19, line x+3(?) [d].

s

s	3rd person singular feminine suffix pronoun. Frag 1, lines 4 [d], 14, 22; Frag 2, line 1; Frag 4, II, 15 [d], 16 [d]; Frag 12, I, 8, II, x+7 [d]; Frag 13, II, 1; Frag 14, II, 7; Frag 30, line 6; Frag 32, line x+5 [d] (or *nꜣ* or *tꜣ*?); Frag 38, line x+1 [d] (or *nꜣ* or *tꜣ*?); Frag 41, line x+5; Frag 44, line x+5.

s	3rd person singular dependent pronoun. Frag 1, lines 5, 10, 11, 13; Frag 38, line x+9 [d].
s-ḥm.t	noun 'female, woman'. Frag 2, lines 8, 15; Frag 15, line x+2. Plural form *s-ḥm.wt*. Frag 12, I, 2 [d].
m-s3	'after'. Frag 1, line 1 [d]; Frag 2, line 4 [d]; Frag 4, II, 7; Frag 9, line x+4 [d]; Frag 10, line x+13 [d]; Frag 11, I, 11.
sy	'be satisfied'. Frag 1, line 4 [d].
syw.w	'stars'. Frag 5, I, 5 [d]; Frag 19, line x+5 [d]. See also *sw*.
sym	'plant'. Frag 11, I, 9 [d]. Plural form *sym.w*. Frag 13, II, 12, 20 [d].
syt	'serpent'. Frag 13, II, 3.
sw	'star'. Frag 44, line x+8. See also *syw.w*.
swḥy.t	'egg'. Frag 4, II, 19 [d]; Frag 41, line x+4(?) [d]. Plural form *swḥy.w*. Frag 15, line x+6 [d].
swsw	'lift'. Frag 19, line x+4 [d].
swty	'withdraw'(?). Frag 44, line x+8 [d]. See textual note (b) ad loc. See also *sṯ*.
n3.w-sbk	'be short'. Frag 3, lines 1 [d], 8 [d], 12 [d].
sbt.t	'hill'. Frag 27, line x+4 [d] (or plural?). Plural form *sbt.wt*. Frag 13, II, 8 [d]; Frag 14, I, 12 [d], II, 12 [d].
sp	'time, occasion'. Frag 13, II, 4 [d].
sfy	'sword'. Frag 43, line 2(?) [d]. See textual note (a) ad loc.
smne	'fix, establish'. Frag 10, lines x+5 [d], x+12 [d]; Frag 11, I, 11; Frag 22, line 5 [d]; Frag 24, lines x+6 [d], x+9 [d].
smt	'manner, likeness'. Frag 13, II, 10 [d]; Frag 14, II, 3 (twice, once [d]).
sn	'brother'. Frag 34, line x+3(?) [d].
sny	in *in-sny*, 'pass out, emanate'. See under *in*.
srpt	'lotus'. Frag 12, I, 6 [d].
sḥn	'ordain'. Frag 13, II, 18.
sḥt	'be shining, radiant'. Frag 3, lines 3(?) [d], 4 [d], 14 [d]; Frag 18, line x+1 [d]. For the doubtful example of *sḥt* in Frag 3, line 3, see textual note (a) ad loc.
n3.w-sḥt	meaning as last. Frag 3, line 21 [d].
sḥty	'illuminate'. Frag 19, line x+7 [d]; Frag 47, line 5 [d].
sẖ.t	'field'. Frag 11, II, x+1(?) [d]; Frag 13, II, 19. For the doubtful example of *sẖ.t* in Frag 11, II, x+1, see textual note ad loc.
sẖf	meaning unknown. Frag 3, line 3 [d]. See textual note (b) ad loc.

sẖny	'event' or 'constellation'. Frag 13, II, 20. For the possible meanings of the word, see commentary ad loc.
sẖny	'settle, alight'. Frag 1, line 13 [d]. See textual note (a) ad loc.
sẖ	verb 'write'. Frag 2, line 18 [d] (or noun 'writing'?).
sẖ	noun 'writing'. Frag 1, line 17 [d]. See also preceding entry.
sšm.w	'*sšm*-beasts'. Frag 13, II, 11. See textual note (a) ad loc.
sšn	'lotus'. Frag 12, I, 4 [d].
sšt	noun 'form'. Frag 15, line x+4 [d].
sk	'collection, assemblage'. Frag 24, line x+8 [d]; Frag 44, line x+11.
st	'tail'. Frag 3, line 14 [d]; Frag 18, line x+1 [d].
sty	'scatter' (seed). Frag 1, lines 3, 21 [d]; Frag 2, line 6 [d]; Frag 11, I, 6 [d]; Frag 15, line x+3 [d].
sty	'engendering'. Frag 13, II, 17. In the expression *kꜣ sty*, 'engendering bull'.
sty	'flame'. Frag 5, II, 4 [d]; Frag 22, line 4 [d]; Frag 30, line 5 [d]; Frag 31, line x+3 [d]; Frag 44, line x+2 [d].
sṯ	(a) 'lead, conduct'. Frag 5, I, 10 [d]; Frag 30, line 6 [d] (or *ẖpr*?).
	(b) 'do again, repeat'. Frag 13, II, 14. In the construction *b-ir=f sṯ=f* followed by *r* + infinitive. See textual note (b) ad loc.
	See also *swty*.
sḏm	verb 'hear' or noun 'hearing'. Frag 14, II, 10 [d].
sḏr	'lie down, sleep'. Frag 39, line x+3 [d].

š

šy	'nose'. Frag 16, line x+5 [d].
šy.ṯ	'two feathers'. Frag 3, line 4 [d]. For the writing, see textual note ad loc.
šꜥ	verb 'begin, initiate'. Frag 1, line 4.
šꜥ	preposition 'until'. Frag 11, I, 1; Frag 13, II, 1; Frag 14, I, 18 [d], 22, II, 2; Frag 22, line 7 [d]; Frag 27, line x+1 [d]. For *šꜥ* in *šꜥ-hn-r*, see under *hn*.
šꜥꜣ	'sand'. Frag 1, lines 6 [d], 7 [d]; Frag 6, line 6; Frag 11, I, 10; Frag 14, I, 3 [d], 9 [d]; Frag 28, line x+6 [d].
šꜥṯ	'cut off'. Frag 1, line 1 [d].
šb	'change'. Frag 13, II, 9 [d]; Frag 14, II, 9 [d] (or *qty*?)

šbs	noun 'noble one'. Frag 41, line x+6.
šbs	adjective 'noble'. Frag 19, line x+19. See also *šps*.
šp	'receive'. Frag 13, II, 19 [d]; Frag 44, line x+5. In the first passage, *šp* has the extended sense 'receive' (semen, seed), i.e. be impregnated. See textual note (a) ad loc.
šps	'noble'. Frag 4, II, 23(?) [d]. See also *šbs*.
špš.w	'arms'. Frag 5, II, 7 [d]; Frag 16, line x+6; Frag 19, line x+20 [d]. *špš.w* in Frag 16, line x+6, is used with reference to the forelegs of a bull. See commentary ad loc.
šm	'go'. Frag 4, I, x+4(?) [d], II, 13.
šn	'tree'. Frag 12, II, x+2 [d] (or *ẖt*?). Plural form *šn.w*. Frag 13, II, 20.
šnb	'join, unite'. Frag 12, I, 16 [d]; Frag 13, II, 2, 4 [d].
šr	'son'. Frag 1, line 3; Frag 19, line x+11(?).
šll	'pray'. Frag 1, line 2 [d] (or *ꜣll* or *tll*?). See textual (a) note ad loc.
m-šs	'very much, greatly'. Frag 1, lines 5 [d], 9 [d].

q

nꜣ.w-qꜣ	'be exalted, high'. Frag 3, line 5 [d].
qmꜣ	'create'. Frag 38, line x+7 [d].
qmꜣ	'rush, reed'. Frag 1, lines 5, 18 [d]; Frag 11, I, 17. For the meaning(s) of the word, see commentary on Frag 1, lines 5 and 12.
qty	'go around, circulate'. Frag 14, II, 9 [d] (or *šb*?).
qty	'surroundings, vicinity'. Frag 2, line 7 [d]; Frag 7, line x+4 [d]. In the expression (*n*) *pꜣ qty*, 'around'.
m-qty	'like'. Frag 2, line 16 [d]; Frag 4, II, 14 [d]; Frag 5, I, 14 [d]; Frag 13, II, 13; Frag 30, line 5 [d]; Frag 41, line x+7.

k

kꜣ	'bull'. Frag 3, lines 1 [d], 6 [d], 9; Frag 13, II, 17 [d]. Plural form *kꜣ.w*. Frag 40, line 2 [d].
kꜣ	'another'. Frag 11, I, 9 [d]; Frag 42, line x+2.
km	'black'. Frag 3, lines 1 [d], 6, 9 [d]; Frag 16, lines x+2, x+3 [d]; Frag 47, line 2(?) [d].
kky	'darkness'. Frag 4, II, 10 [d]; Frag 38, line x+3 [d].

g

gꜣy	'way, manner'. Frag 5, I, 6 [d]; Frag 22, line 3 [d]; Frag 40, line 3 [d]; Frag 44, line x+9.
gm	'find'. Frag 1, line 10 [d]; Frag 11, I, 5 [d]; Frag 29, line x+1 (written *gm.ṯ* + suffix pronoun).
gmꜥ	'damage'. Frag 14, II, 10 [d].
glg	'hunt'. Frag 1, line 5 [d] (or *plg* or *tlg*?). See textual note (c) ad loc.
gtg	'hasten'. Frag 14, II, 14 [d].

t

tꜣ	feminine singular definite article. Passim.
tꜣ	'land'. Frag 2, lines 12 [d], 16(?) [d]; Frag 4, I, x+7 [d], II, 7(?) [d], 18 [d]; Frag 5, I, 1 [d], 10, II, 6; Frag 10, lines x+4 [d], x+11 [d]; Frag 11, I, 11 [d]; Frag 13, II, 5, 6 [d], 13; Frag 14, I, 3 [d], 10, 11; Frag 19, lines x+7 [d], x+8, x+16 [d]; Frag 28, line x+2 [d]; Frag 34, line x+5 [d].
tꜣy	feminine singular demonstrative pronoun. Frag 25, line x+3 (or copula?).
tꜣy	copula. Frag 14, I, 7 [d], II, 8, 11 [d]. See also preceding entry.
tꜣy=	formative element of possessive article used before fem. sing. nouns. Forms: *tꜣy=f* (3rd person singular masculine possessor). Frag 1, line 13 [d]; Frag 2, line 6; Frag 3, line 18 [d]; Frag 4, II, 14; Frag 11, I, 6; Frag 13, II, 7; Frag 33, line x+1. *tꜣy=s* (3rd person singular feminine possessor). Frag 14, II, 11 [d]. *tꜣy=w* (3rd person plural possessor). Frag 14, II, 6.
tꜣwy	'dawn'. Frag 5, I, 2 [d].
ti	'cause, give' (infinitive). Frag 1, line 3; Frag 2, line 11 [d] (*ti ꜥḥꜥ*?); Frag 6, line 8; Frag 13, II, 1 (twice), 4, 6, 9, 12 [d], 14 (twice, once [d]), 17 [d], 18 [d], 19 [d], 20 [d]; Frag 14, II, 9; Frag 38, line x+15. See also *tw*.
ty	adverb 'here'. Frag 2, line 7(?) [d].
tyꜣ	meaning unknown. Frag 23, II, 1 [d].
tyly	'raise, rise'. Frag 5, II, 7; Frag 22, line 8(?) [d].

tw ‘cause, give’ (*sḏm=f*/past relative). Frag 38, line x+11 [d]; Frag 40, line 5(?) [d]; Frag 44, line x+4 [d]. See also *ti*.

tw in negative perfect *b-ir-tw*. See under *b*.

tw ‘mountain’. Frag 10, line x+12; Frag 24, line x+9. Plural form *tww.w*. Frag 13, II, 9; Frag 14, II, 4 [d].

tw=s ‘behold, here is’. Frag 2, line 13; Frag 4, II, 13(?) [d]; Frag 23, II, 2.

twꜣ.t Frag 4, II, 18 [d]; Frag 13, II, 21; Frag 19, line x+10 [d]; Frag 20, line 3 [d]; Frag 35, line 3; Frag 38, line x+6.

twn ‘arise’. Frag 3, line 11.

twt ‘image, statue’. Frag 2, line 11; Frag 11, I, 7 [d].

twt ‘union, assemblage’. Frag 2, lines 9(?) [d], 12 [d].

tmy ‘town’. Frag 2, lines 1 [d], 3.

tr= ‘all, entire’. Frag 4, II, 12; Frag 7, line x+7 [d]; Frag 10, lines x+4 [d], x+11 [d]; Frag 11, I, 11 [d]; Frag 13, II, 5 [d]; Frag 14, II, 12; Frag 19, lines x+7 [d], x+16; Frag 20, line 1 [d].

tr.ṱ ‘hand’. Frag 6, line 4 [d] (+ suffix pronoun).

n-tr.t ‘by, through’. Frag 1, line 3 [d].

trš ‘red’. Frag 41, line x+7.

tll ‘rejoice’. Frag 1, line 2 [d] (or *ꜣll* or *šll*?). See textual note (a) ad loc.

tlg ‘uproot’. Frag 1, line 5 [d] (or *plg* or *glg*?). See textual note (c) ad loc.

ṱ

ṱꜣ ‘time’. Frag 2, line 7 [d].

ṯ

ṯ ‘take, take away’. Frag 5, I, 14 (or *mtw* instead of *nt ṯ*?); Frag 13, II, 6 (written *ṯ.ṱ* + suffix pronoun), 14, 15 (*ṯ.ṱ* + suffix); Frag 39, line x+4 (*ṯ.ṱ* + suffix).

ṯꜣw ‘wind, breath’. Frag 4, I, x+8(?) [d], II, 5, 8; Frag 13, II, 3, 5 [d], 6 [d], 7, 8, 9 [d], 10, 12 (twice), 18 [d]; Frag 14, I, 15(?) [d].

ṯsy ‘lift, elevate’. Frag 4, II, 17 [d]; Frag 44, lines x+7 [d], x+12 [d].

ṯsy ‘muster’ (a generation). Frag 19, line x+11 [d].

ḏ

ḏ — 'call, say'. Frag 1, lines 12, 18(?) [d]; Frag 2, lines 1 (written *ḏ.ṱ* + suffix pronoun), 9 (twice, once [d], or is 1st example *ir*?), 10 [d], 12 [d], 16 (twice, both [d], or is 1st one *rn*?); Frag 4, II, 3, 22 [d], 23; Frag 5, I, 8 [d] (or *ir*?), II, 3, 5 [d], 6 [d], 9 (twice), 10; Frag 11, I, 15, 18(?) [d]; Frag 12, I, 9 [d] (or *ir*?), II, x+1; Frag 14, I, 2 [d], 8 [d]; Frag 19, line x+10 [d] (or *ir*?); Frag 20, line 2; Frag 23, II, 8 [d]; Frag 28, line x+6; Frag 30, line 4 [d]; Frag 35, line 5; Frag 36, line 3 [d]; Frag 38, line x+8 [d]; Frag 41, line x+5 (*ḏ.ṱ* + suffix); Frag 43, line 1 [d].

ḏ — 'saying' (introducing direct speech). Frag 1, line 7; Frag 2, line 9; Frag 5, II, 2, 10, 16; Frag 28, line x+8 [d]; Frag 38, line x+13.

ḏ — 'that' (introducing indirect statement). Frag 1, line 11.

ḏ — 'namely'. Frag 2, line 12.

ḏ — 'because'. Frag 13, II, 3 [d].

ḏwf — 'papyrus'. Frag 12, I, 9 [d], 11 [d], 13 [d], 14.

r-ḏbꜣ — 'because, on account of'. Frag 1, line 4 [d] (written *r-ḏbꜣ.ṱ* + suffix pronoun); Frag 2, lines 4 [d], 11 [d]; Frag 4, II, 15 [d] (*r-ḏbꜣ.ṱ* + suffix pronoun); Frag 10, lines x+3 [d], x+4 [d] (or *nt*?); Frag 11, I, 17 (*r-ḏbꜣ.ṱ* + suffix); Frag 12, I, 15; Frag 13, II, 15 (in the phrase *r-ḏbꜣ ḫpr*); Frag 14, II, 7 (*r-ḏbꜣ.ṱ* + suffix); Frag 19, line x+21(?) [d]; Frag 22, line 11 [d]; Frag 32, lines x+3 [d], x+6 [d]; Frag 41, line x+5(?) [d].

ḏf — 'pupil' (of eye). Frag 3, line 10.

ḏm — 'generation'. Frag 19, line x+11 [d].

ḏt — 'eternity'. Frag 2, line 16 [d]; Frag 13, II, 1, 17, 21 [d]; Frag 19, line x+14; Frag 24, line x+8 [d].

ḏtf.w — 'serpents, worms'. Frag 13, II, 11 [d]. See textual note (c) ad loc.

Words of Uncertain Reading

inꜣ.t/mnꜣ.t(?) — feminine noun with flesh determinative. Frag 14, II, 12 [d].

ꜥꜥ.. — 'herds'(?). Frag 13, II, 11 [d]. See textual note (b) ad loc.

nm.. — verb. Frag 19, line x+9 [d]. See textual note ad loc.

sꜥ.. — masculine noun with house determinative. Frag 14, I, 9 [d].

 — masculine noun. Frag 13, II, 13. See textual note (a) ad loc.

 — (?). Frag 14, II, 13 [d].

Numbers

2 — 'two'. Frag 13, II, 4; Frag 14, I, 10, II, 5; Frag 21, line x+5 [d] (or *s* at start of new word?); Frag 24, line x+9 [d]; Frag 38, lines x+4, x+6.

4 — 'four'. Frag 2, lines 8 [d], 14 (twice, once [d]), 15 [d], 16 [d]; Frag 4, II, 5; Frag 19, line x+6 (written above line); Frag 30, line 7 [d]. Feminine form *4.t*. Frag 2, lines 8 [d], 15.

7 — 'seven'. Frag 6, line 7 [d]; Frag 22, line 2.

8 — 'eight'. Frag 2, lines 8, 14; Frag 3, line 13.

mḥ-4 — 'fourth'. Frag 3, line 7 [d].

mḥ-5 — 'fifth'. Frag 3, line 9 [d].

mḥ-6 — 'sixth'. Frag 3, line 11 [d].

mḥ-7 — 'seventh'. Frag 3, line 13 [d].

mḥ-8 — 'eighth'. Frag 3, line 15.

B. DIVINE NAMES

ꜣh.t-wly — 'Ahet-weret'. Frag 8, line x+7 [d] (or *Mḥ.t-wly*?). See textual note and commentary ad loc.

ꜣst — 'Isis'. Frag 25, line x+5 [d].

ỉmn — 'Amun'. Frag 2, line 13; Frag 3, lines 7, 9, 11, 13 [d], 15 [d].

ỉmn.t — 'Amaunet'. Frag 2, line 13.

Ptḥ — 'Ptah'. Frag 1, lines 3, 18 [d]; Frag 2, lines 5, 9 [d], 15; Frag 11, I, 5(?) [d], 14 [d]; Frag 14, II, 23 [d]; Frag 15, lines x+3 [d], x+8 [d].

Mḥ.t-wly — 'Mehet-weret'. Frag 8, line x+7 [d] (or *ꜣh.t-wly*?). See textual note and commentary ad loc.

Nwn — 'Nun', member of the Ogdoad. Frag 2, line 13.

Nwn.t — 'Naunet'. Frag 3, line 15 [d]. See textual note (b) ad loc.

pꜣ Nwn — 'the Primaeval Ocean'. Frag 1, lines 1 [d], 2 (*pꜣ-Šy pꜣ Nwn*), 20 [d], 21 [d]; Frag 2, line 16 [d]; Frag 4, II, 9 [d], 20 [d] (*pꜣ-Šy pꜣ Nwn*); Frag 5, I, 7 [d], II, 8 [d], 15 [d], 20 [d]; Frag 6, line 1 [d]; Frag 11, I, 1 [d], 11 [d]; Frag 13, II, 3 [d], 4, 7, 9, 10 [d], 17 [d]; Frag 14, I, 2 [d], II, 9 [d], 13 [d], 15 [d], 19 [d]; Frag 22, line 1 [d]; Frag 24, lines x+1 [d], x+2 [d]; Frag 28, lines x+2 [d], x+3 [d], x+4 [d], x+7 [d], x+9 (twice, both [d], or is 1st one the end of *p.t*, 'sky'?); Frag 39, line x+2 [d]; Frag 44, line x+12 [d] (or end of *p.t*, 'sky'?); Frag 48, lines x+4 [d], x+5(?) [d]. For the name *pꜣ-Šy pꜣ Nwn* which occurs in Frag 1, line 2, and Frag 4, II, 20, see commentary on the former.

Nḫbꜣ.t	'Nekhbet'. Frag 12, I, 15 [d].
pꜣ-Rꜥ	'Pre'. Frag 12, II, x+10.
Ḥr	'Horus'. Frag 12, I, 12.
Ḥḥ	'Heh'. Frag 2, line 13 [d].
Ḥḥ.t	'Hauhet'. Frag 2, line 13 [d].
nꜣ Ḫmnw	'the Ogdoad'. Frag 2, line 12; Frag 9, lines x+2 [d], x+3 [d]; Frag 14, II, 22 [d] (or *Ḫmnw*, 'Hermopolis'?).
Sbk	'Sobek'. Frag 47, lines x+4 [d], x+6 [d].
Spt.t	'Sothis'. Frag 30, line 4 [d].
Sḫmy	'Sakhmet'. Frag 12, I, 10 [d].
Šy	'Shai'. Frag 4, II, 22. This name appears as *pꜣ-Šy* elsewhere in the text. See the following entry.
pꜣ-Šy	'Pshai'. Frag 1, line 2 (*pꜣ-Šy pꜣ Nwn*); Frag 4, II, 20 [d] (*pꜣ-Šy pꜣ Nwn*); Frag 5, I, 21 [d], II, 5, 8 [d], 10 [d], 17 [d]; Frag 7, line x+6 [d]; Frag 10, line x+9 [d]; Frag 20, line 1(?) [d]; Frag 40, line 1(?) [d]. For the name *pꜣ-Šy pꜣ Nwn* which occurs in Frag 1, line 2, and Frag 4, II, 20, see commentary on the former.
Šw	'Shu'. Frag 3, line 9. See textual note (b) ad loc.
Kk	'Kek'. Frag 2, line 13 [d].
Kky.t	'Kauket'. Frag 2, line 13.
Tfny	'Tefnut'. Frag 12, I, 10.
Ḏḥwty	'Thoth'. Frag 5, I, 8, II, 6, 9 [d], 11; Frag 6, line 3 [d]; Frag 8, line x+8 [d]; Frag 10, line x+7 [d]; Frag 11, I, 2, 12 [d]; Frag 12, II, x+1(?) [d], x+12 [d]; Frag 13, II, 13 [d].

C. TOPONYMS

Yb	'Elephantine'. Frag 1, line 11 [d].
Nw.t	'Thebes'. Frag 9, line x+4 [d].
Ḫmnw	'Hermopolis'. Frag 1, line 11 [d]; Frag 6, line 6 [d]; Frag 8, line x+5 [d]; Frag 14, II, 22 [d] (or [*nꜣ*] *Ḫmnw*, 'Ogdoad'?); Frag 24, line x+10 [d].
Ḫl	'Syria'. Frag 13, II, 7.

Kmy	'Egypt'. Frag 11, I, 15 [d], 18 [d].
Ḏbꜣ-tn	'Tebtunis'. Frag 47, line 6(?) [d]. See textual note ad loc.

D. TRACES OF UNIDENTIFIABLE WORDS

m + plant determinative	Frag 1, line 8. Not *sym*, see textual note (a) ad loc.
y + walking legs determinative	Frag 1, line 14.
house determinative	Ibid.
seated child determinative(?)	Frag 1, line 15.
y + seated child determinative	Ibid.
word-initial *m*	Frag 1, line 19.
plant determinative	Frag 2, line 2.
seated man determinative	Frag 2, line 12.
sun determinative	Frag 2, line 13.
word-initial *s*, or man with hand to mouth determinative	Frag 3, line 3. See textual note (b) ad loc.
brazier determinative	Frag 3, line 10.
word-initial *s*	Frag 4, II, 4. Preceded by article *pꜣ*.
word-initial *m* or *š* + *y*	Frag 4, II, 12.
word-initial *m* preceding break followed by *n* or *y* + man with hand to mouth determinative	Frag 4, II, 19. See textual note (a) ad loc.
word-initial *n*	Frag 4, II, 21.
š + man with hand to mouth determinative	Frag 5, I, 3.
walking legs determinative	Frag 5, I, 4.
y	Frag 5, I, 18.
word-final *h*	Frag 6, line 3.
flesh determinative	Frag 6, line 7.
ẖ	Frag 7, line x+1.
flesh determinative + *ṯ*	Frag 7, line x+5.
m(?)	Frag 8, line x+3.
feminine *t* ending	Frag 9, line x+6. Part of noun preceded by article *tꜣ*.
strong arm determinative	Frag 10, line x+3.

phallus determinative	Frag 10, line x+12.
pellet determinative above horizontal plural strokes	Frag 11, I, 2.
word-initial *ḥ* + *q* + vertical stroke + *ḏ* + vertical stroke	Frag 11, I, 4. Possibly parts of more than one word.
n(?)	Ibid.
determinative found in words like *twt*, 'statue', and *mny*, 'image'	Frag 11, I, 16.
word initial *nn*(?)	Frag 11, II, x+1. See textual note ad loc.
y + jar determinative	Frag 12, I, 3.
plant determinative	Frag 12, I, 11.
man with hand to mouth determinative	Frag 12, I, 15.
word-initial *b*	Frag 13, II, 1.
divine determinative	Frag 13, II, 21.
word-initial *m*	Ibid.
pellet determinative above horizontal plural strokes	Frag 14, I, 7.
s or man with hand to mouth determinative	Frag 14, I, 8.
stone determinative	Frag 14, I, 13.
stone determinative + plural stroke	Frag 14, I, 23. Not *iny.w*.
ḥ(?)	Frag 14, II, 3.
word-initial *w*(?)	Frag 14, II, 4.
man with hand to mouth determinative + suffix *w* or plural stroke	Frag 14, II, 5. Not *tr=w*. See textual note (b) ad loc.
man with hand to mouth determinative	Frag 14, II, 10.
man with hand to mouth determinative	Frag 14, II, 14.
word-initial *m* + *ẖ*(?)	Frag 14, II, 17.
s or man with hand to mouth determinative	Frag 14, II, 20.
divine determinative	Frag 15, line x+7.
e or *y* + flesh determinative	Frag 16, line x+2.
flesh determinative	Frag 17, line x+1.

flesh determinative	Frag 17, line x+4.
l + *ẖ*	Frag 18, line x+3.
m(?)	Frag 19, line x+2.
bookroll determinative above horizontal plural strokes	Frag 19, line x+13.
y + flesh determinative	Frag 19, line x+17.
ḏ + *y*	Frag 19, line x+18.
word-initial *w*	Frag 20, line 1.
divine + serpent determinatives	Frag 21, line x+3.
flesh determinative	Frag 21, line x+5.
wood determinative	Frag 24, line x+3.
serpent + divine determinatives	Frag 24, line x+4.
word-initial *s*	Frag 24, line x+10. Possibly a toponym. See textual note ad loc.
s or man with hand to mouth determinative	Frag 25, line x+6.
word-initial *m*	Frag 26, line x+4. Preceded by article *nꜣ*.
man with hand to mouth determinative	Frag 27, line x+3.
y + flesh determinative	Frag 28, line x+5.
y	Frag 30, line 8.
e or *y* + man with hand to mouth determinative	Frag 31, line x+1.
pellet determinative above horizontal plural strokes	Frag 31, line x+4.
b	Frag 32, line x+4.
š + *n* or *t*	Ibid.
word-initial *b* or *i*, strong arm determinative	Frag 32, line x+5.
m	Frag 33, line x+2.
ḏ + *w*	Frag 34, line x+6.
y + determinative used in words like *ẖꜣy*, 'high', and *qꜣ*, 'be exalted'	Frag 35, line 1.
word-initial *ẖ*	Ibid. Possibly preceded by article *pꜣ*.
walking legs determinative	Frag 35, line 2.

flesh determinative	Frag 35, line 4.
y + *w*	Frag 37, line x+2.
strong arm determinative	Frag 38, line x+16.
serpent determinative	Frag 39, line x+2.
word-initial *ẖ*(?)	Frag 39, line x+5. Preceded by article *nꜣ*.
word-initial *m*	Frag 40, line 2.
n + *w* + man with hand to mouth determinative	Frag 40, line 4.
strong arm determinative(?)	Frag 40, line 5. See textual note ad loc.
word-initial *s*	Frag 41, line x+6.
e or *y* + wood determinative	Frag 41, line x+7.
sun + divine determinatives	Frag 44, line x+4.
word-initial *m*	Frag 44, line x+6. Preceded by article *pꜣ*.
walking legs determinative	Ibid.
ṱ(?) + *wn*(?) + *n*(?)	Frag 44, line x+8. Possibly parts of more than one word.
w + *g* + evil determinative	Frag 46, line x+2. Not *swg*.
y + pellet determinative above horizontal plural strokes	Frag 46, line x+3. Preceded by article *nꜣ*.
s or man with hand to mouth determinative.	Frag 46, line x+4.
flesh determinative + *ṱ*	Frag 47, line 1.
word-initial *ḫ*	Ibid.
man with hand to mouth determinative.	Frag 47, line 8.
word-initial *s* + *t*(?)	Frag 47, line 12.

THE PLATES

Sample of Greek hand on front of papyrus

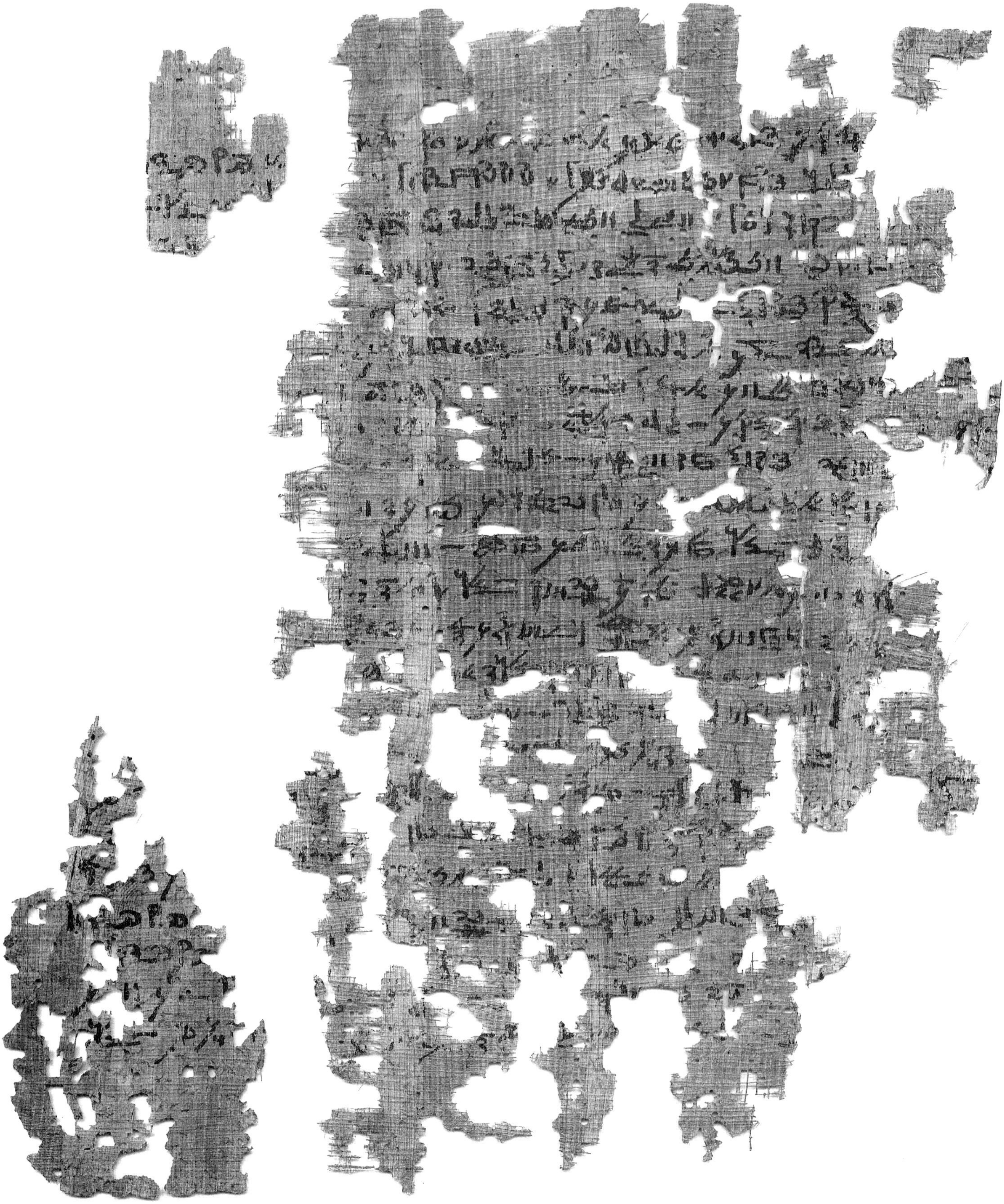

Fragment 1

Fragment 2

PLATE 4

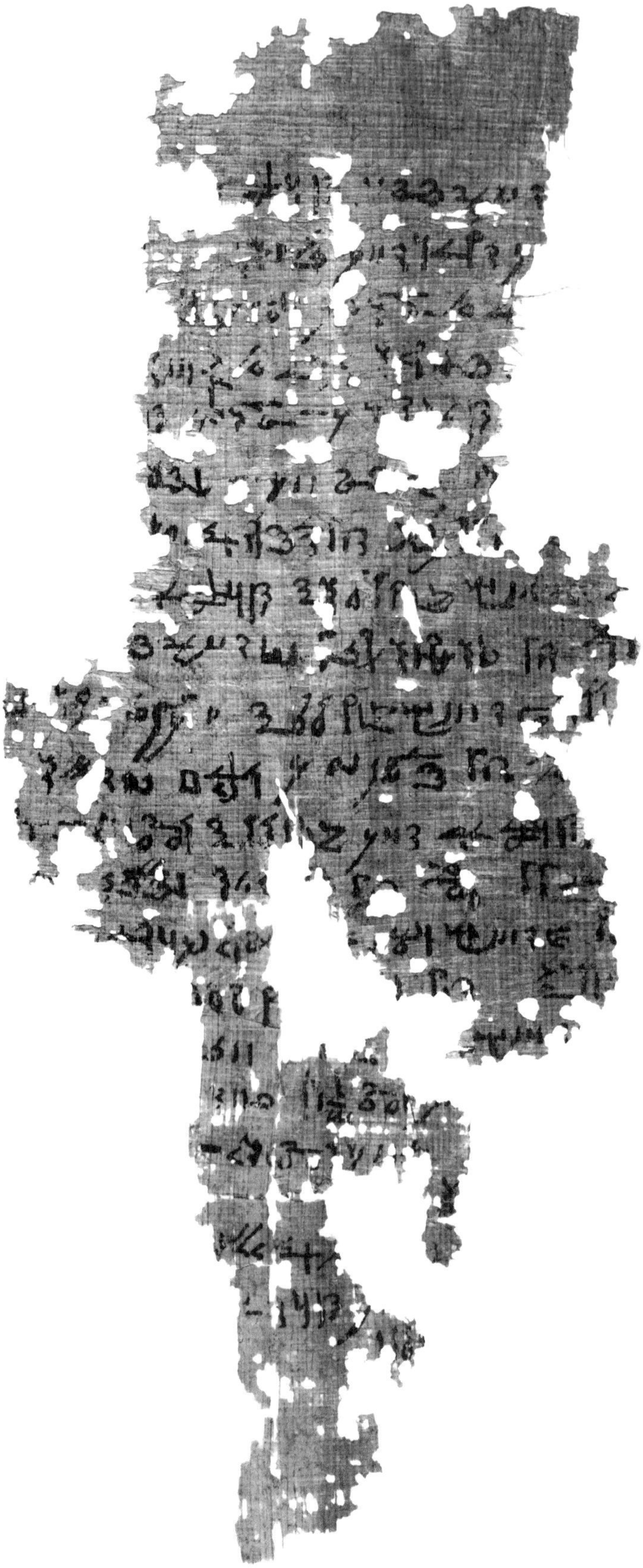

Fragment 3

Fragment 4

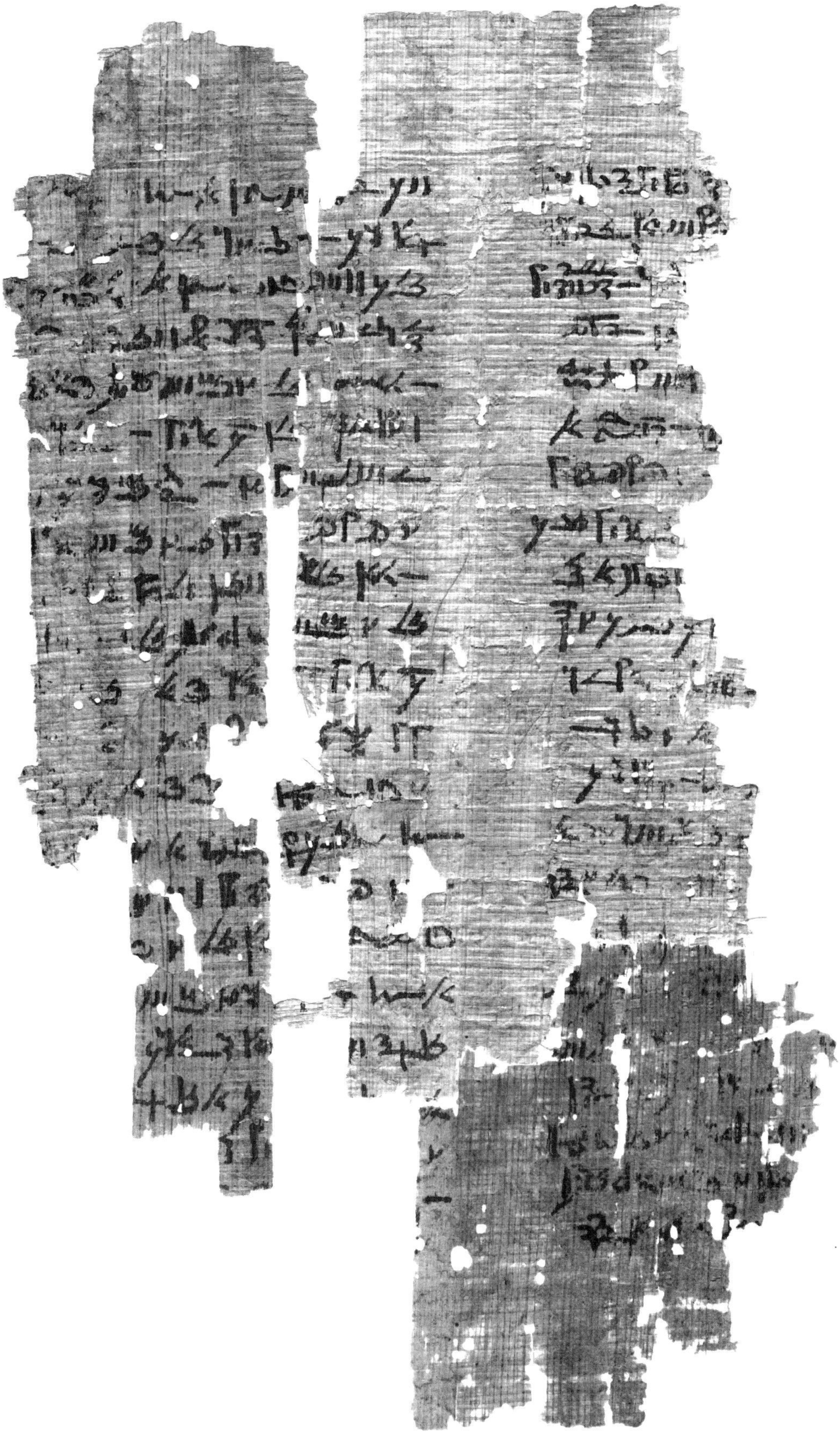

Fragment 5

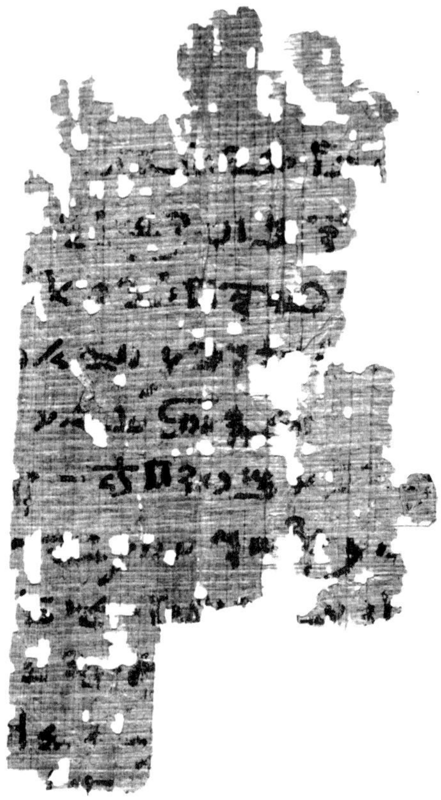

Fragment 6

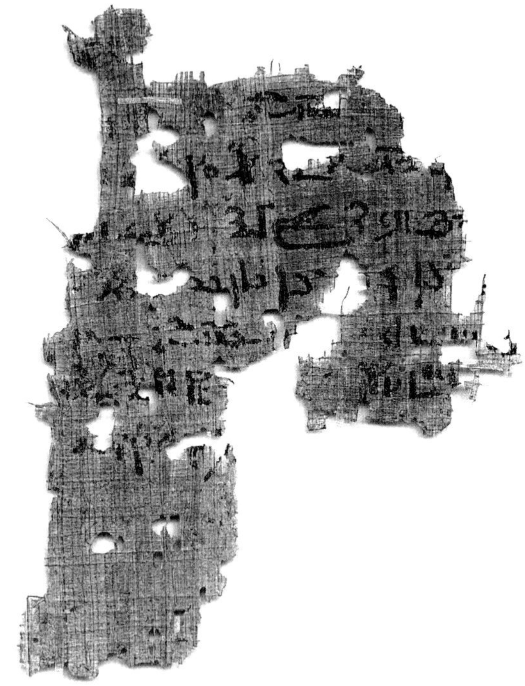

Fragment 7

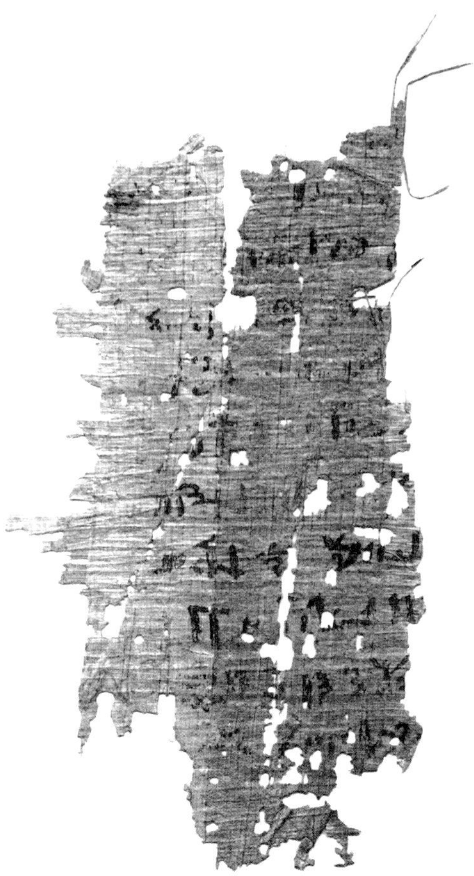

Fragment 8

Fragment 9

Fragment 10

Fragment 11

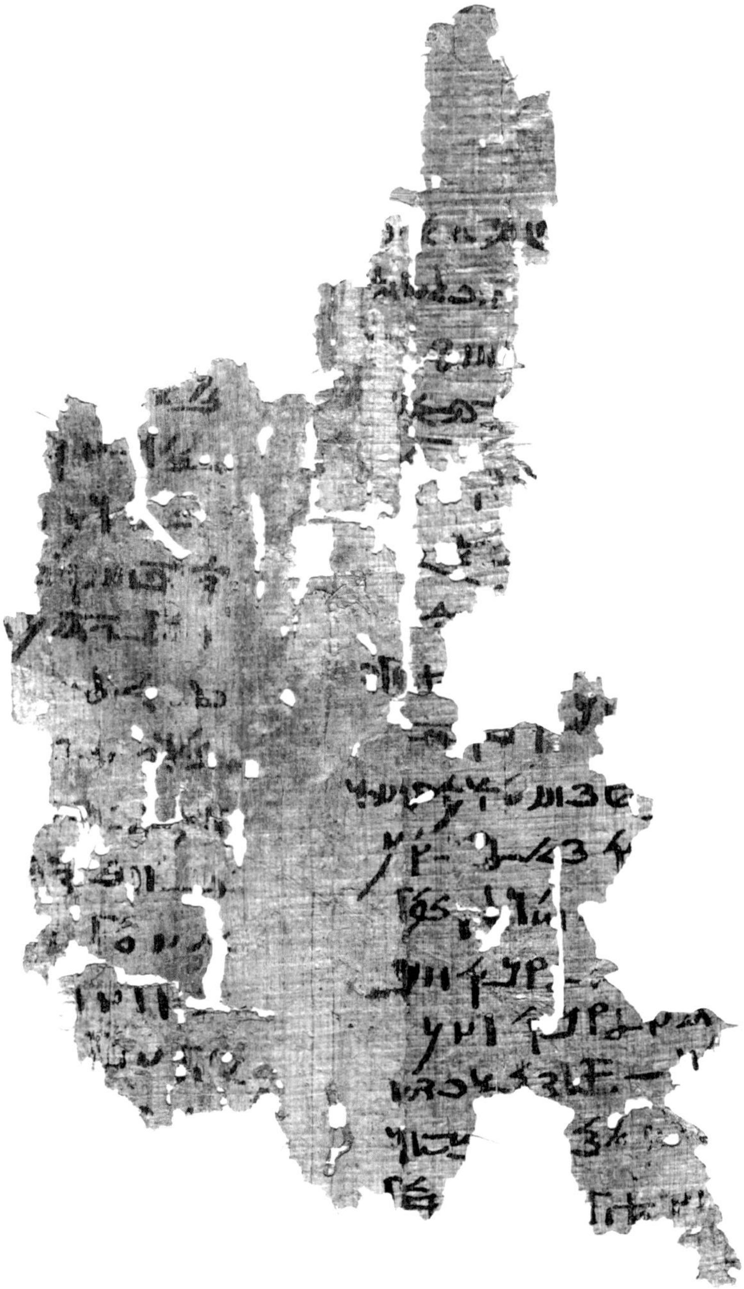

Fragment 12

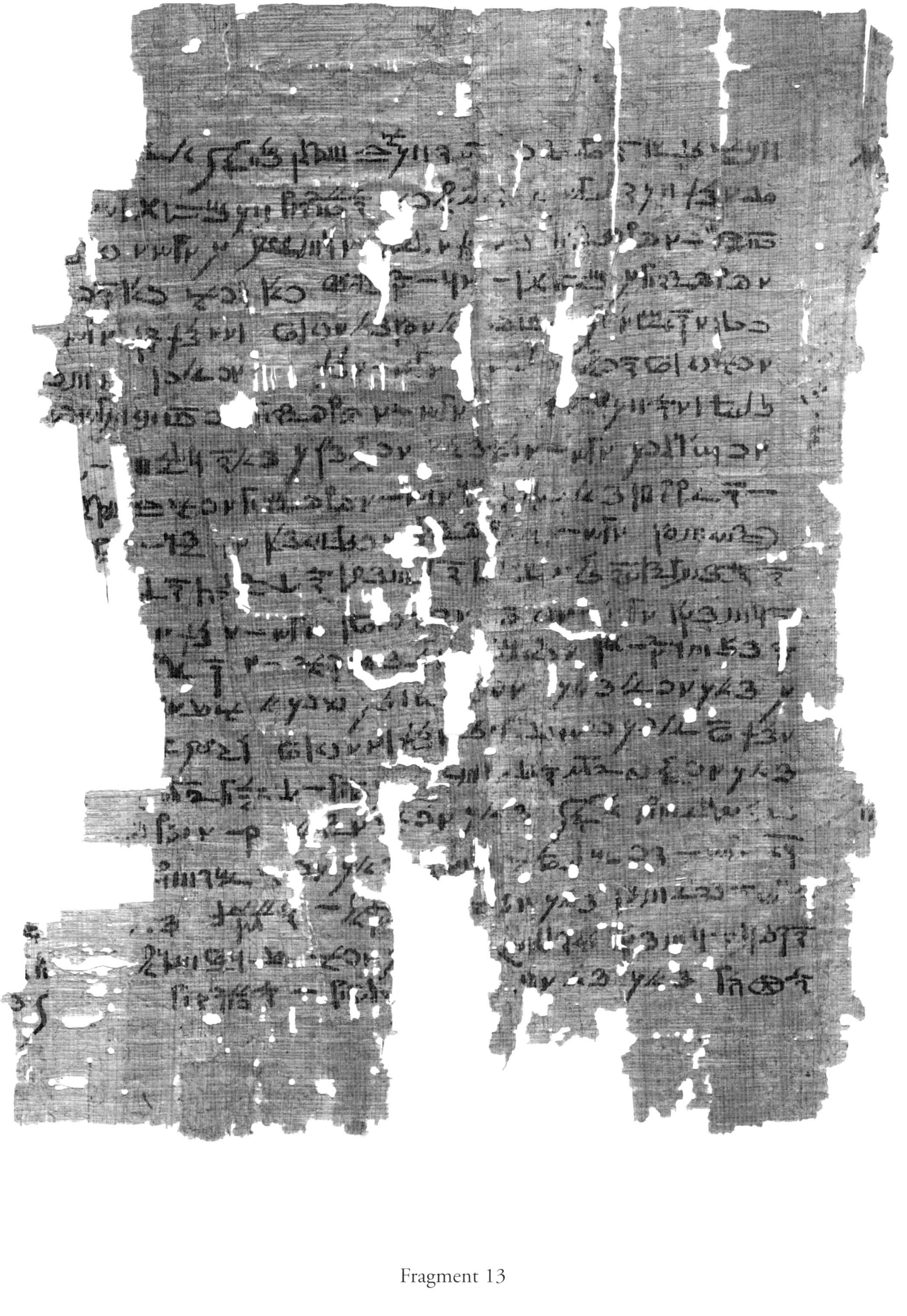

Fragment 13

PLATE 10

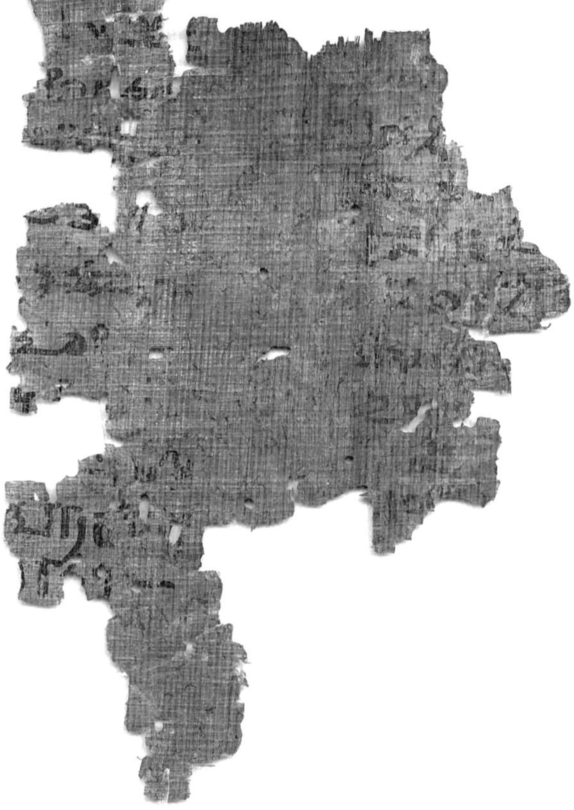

Fragment 15

Fragment 14

PLATE 11

Fragment 16

Fragment 17

Fragment 18

Fragment 19

Fragment 20

Fragment 21

Fragment 22

PLATE 12

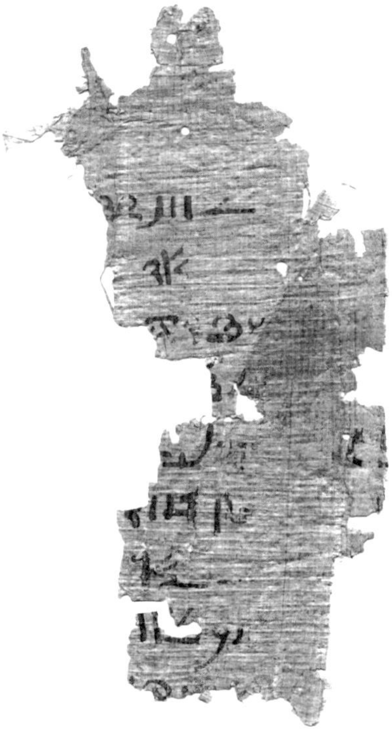

Fragment 23

Fragment 24

Fragment 25

Fragment 26

Fragment 27

Fragment 28

Fragment 29

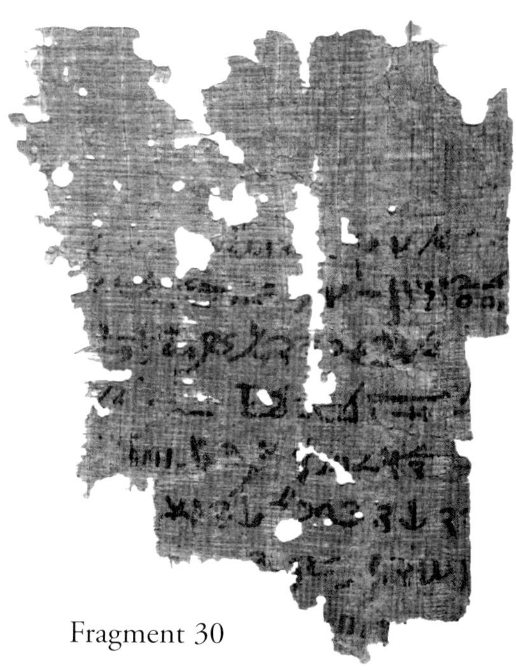

Fragment 30

PLATE 13

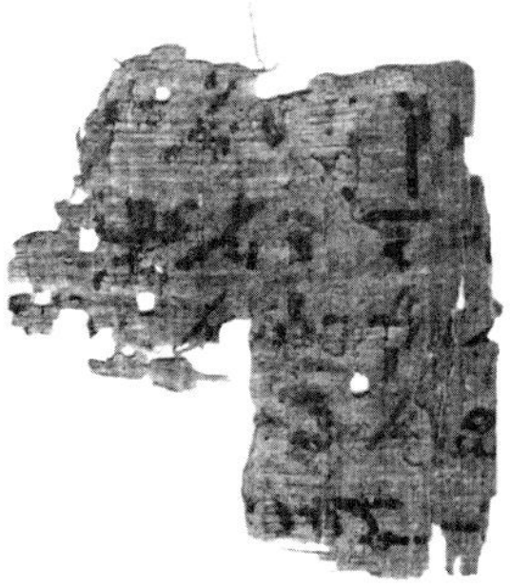
Fragment 31

Fragment 32

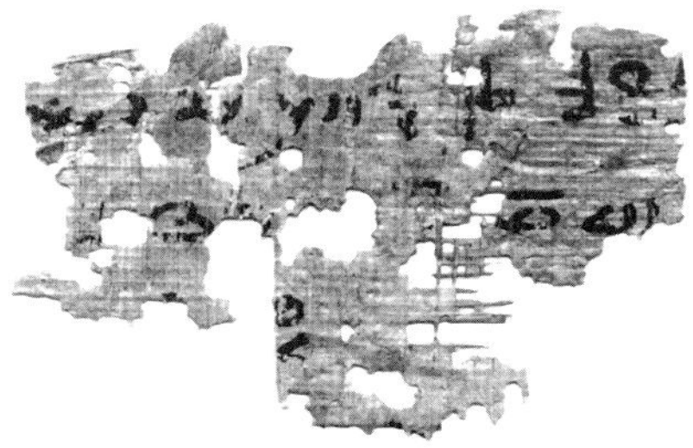
Fragment 33

Fragment 34

Fragment 35

Fragment 36

Fragment 37

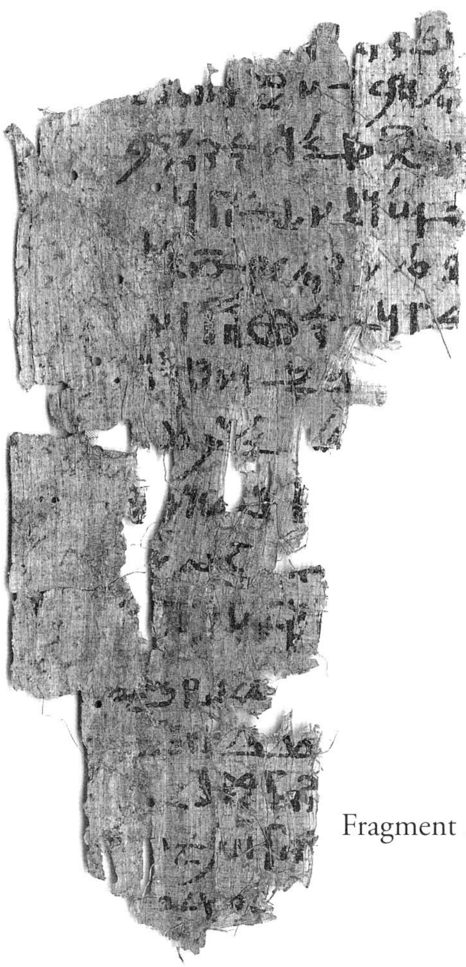
Fragment 38

Fragment 39

PLATE 14

Fragment 40

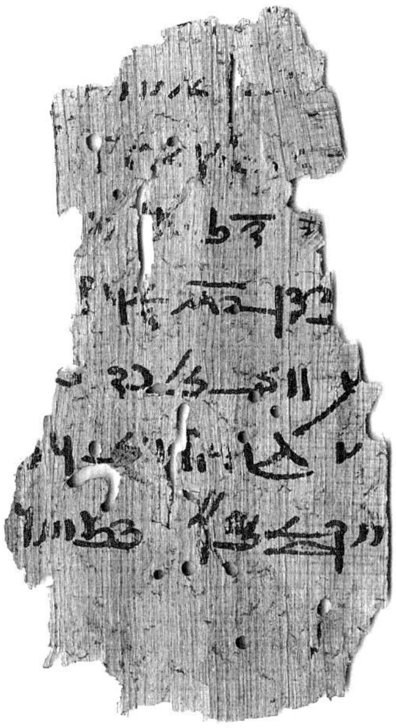

Fragment 41

Fragment 42

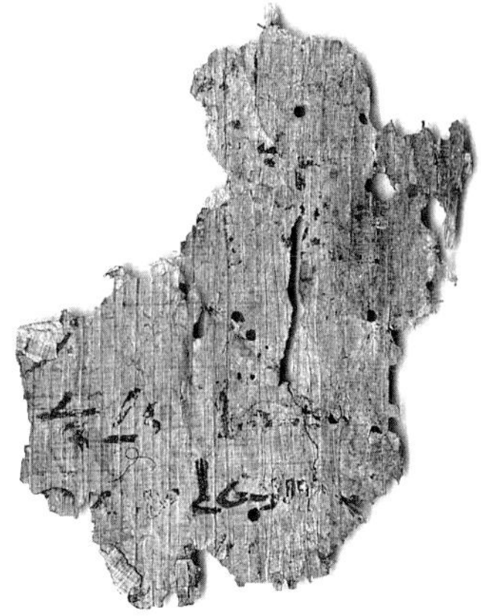

Fragment 43

Fragment 44

Fragment 45

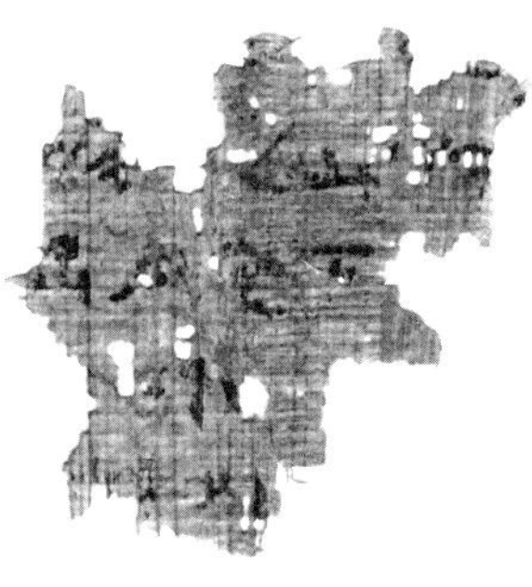

Fragment 46

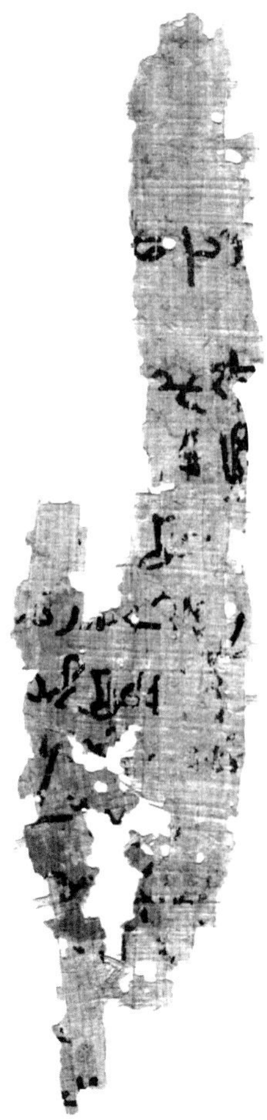

Fragment 47

Fragment 48